The Nature of Difference

History and lithic use-wear at two Upper Palaeolithic sites in Central Europe

Silvia Tomášková

BAR International Series 880
2000

Published in 2016 by
BAR Publishing, Oxford

BAR International Series 880

The Nature of Difference

ISBN 9781841710815 paperback
ISBN 9781407352213 e-format
DOI https://doi.org/10.30861/9781841710815
A catalogue record for this book is available from the British Library

BAR Publishing is the trading name of British Archaeological Reports (Oxford) Ltd. British Archaeological Reports was first incorporated in 1974 to publish the BAR Series, International and British. In 1992 Hadrian Books Ltd became part of the BAR group. This volume was originally published by Archaeopress in conjunction with British Archaeological Reports (Oxford) Ltd / Hadrian Books Ltd, the Series principal publisher, in 2000. This present volume is published by BAR Publishing, 2016.

BAR

PUBLISHING

BAR titles are available from:

BAR Publishing
122 Banbury Rd, Oxford, OX2 7BP, UK
EMAIL info@barpublishing.com
PHONE +44 (0)1865 310431
FAX +44 (0)1865 316916
www.barpublishing.com

Table of Contents

List of Maps

List of Tables

List of Graphs

Acknowledgments

This monograph stems from a Ph.D. dissertation thesis in the Department of Anthropology, University of California at Berkeley, submitted to my committee in 1995. I am most grateful to my advisors Meg Conkey and Ruth Tringham for many years of generous support. I am indebted to all the archeology faculty in the anthropology department, and to my fellow graduate students, especially Heather Price, Mira Stevanović, and Dana Lepofsky, for an environment that encouraged and challenged me. Barbara Voytek and Kent Lightfoot have been particularly kind and questioning of my work, forcing me to always strive for higher standards.

The archaeologists at McGill University, especially Mike Bisson and Bruce Trigger, started my interest in prehistory and history, and have remained staunch supporters over many years.

During my fieldwork in Czechoslovakia and Austria, the late Dr. Bohuslav Klíma, Dr. Jiří Svoboda, Dr. Walpurga Antl, and Dr. Gerhard Trnka provided invaluable assistance, and this project would have never been accomplished without their help. I am grateful to Gerhard for providing me with further materials and data related to Willendorf in the years following my original research.

A postdoctoral fellowship in the Anthropology department at Harvard University allowed me to look at my work through a different lens. I am most grateful for this experience to Ofer and Danny Bar-Yosef who offered a kind welcome and much support.

Parts of this work were presented in the anthropology departments at the University of Washington, Yale University, University of Toronto, and the University of Texas at Austin. I thank all participants, and especially Sam Wilson, for valuable questions and comments.

Financial support for background research, two years of fieldwork, dissertation writing and postdoctoral fellowship came from the Social Sciences and Humanities Research Council of Canada Doctoral and Postdoctoral Grants, a Dissertation Research Grant from the Wenner-Gren Foundation, an Andrew Mellon Fellowship in East European Studies, a UC Berkeley Fellowship, a Chancellor's Dissertation Writing Fellowship, Robert H. Lowie Graduate Scholarships, and a Stahl Endowment Fund grant. A UT-Austin Summer Research Fellowship in 1998 provided me with funds needed to finish the writing of this manuscript.

Finally, I wish to acknowledge the contribution of Peter Redfield, whose help extended over many years while this monograph has been in the making. I will always remain grateful to him for saving me from all the maps, drawings, illustrations, and graphs that he has created and recreated over many days and nights, overcoming the mysteries of vanishing files and powerful computer bugs.

Zoë Sofia Tomášková Redfield, born 11.11.1996, was instrumental in delaying this production for at least two years, while giving me a richer perspective on time, work, and being.

This monograph is dedicated to the memory of my father who left too soon, and to my mother who has managed to carry on with a remarkable strength.

Foreword

"I am away from home and must always write home, even if any home of mine has long since floated away into eternity. All this writing is nothing but Robinson Crusoe's flag hoisted at the highest point of the island."
— Kafka to his closest friend[1]

On August 7, 1981 at 11:30 in the morning, I sat on a train, waiting for the armed soldiers to finish a border control between Czechoslovakia and West Germany. This border was quite visible, made palpable by electrified barbed wire, a sand field that showed foot prints of anyone who tried to cross, and a contingent of German shepherd dogs trained to catch anyone who tried to run. Throughout my childhood I believed that all borders in the world looked like this; it was the "natural border" for me. On this day the boundary was particularly terrifying, as I was crossing it with a one-way ticket, knowing that I did not plan on returning into the space that had until then been a natural place for me — my motherland and fatherland.

Since that day I have acquired a new identity, and new labels have been attached to me at every subsequent stage of my life. I became stateless for a number of years, as my "former" native land stripped me of citizenship, disqualifying me from a membership in the category then called "Czechoslovak citizen". I was at first a political refugee, then a landed immigrant in Canada, before those labels were dropped, and I acquired a new identity, becoming a Canadian citizen. In 1990 Czechoslovak authorities, due to the fall of the Communist regime in 1989, again deemed me one of their citizens, awarding me their passport as a proof of belonging, only to request in 1992 that I identify myself by a narrower category, as either a Czech or a Slovak. A new border between these two nations was raised, and my father, without leaving his house, became a foreigner, a Czech living in Slovakia. My siblings and I were not members of either group, being products of a mixed marriage. We could no longer belong to both categories — Czech and Slovak — as the new boundary between the two groups did not recognize exceptions to the rule of exclusivity that defines each class in any classificatory system. Thus my brother became a Czech citizen and my sister a Slovak on the basis of residence, while I found myself again alienated, a Canadian studying in the United States.

By then I was working on the problem of modern day borders in archeological practice, regularly crossing borders between Austria, the Czech Republic and Slovakia. It appeared to me, looking at the issue from the perspective of North American academia, that the distance between Dolní Věstonice and Willendorf was so small that it could not have been a case of two different archeological traditions, as it was then claimed, but most likely just one. After years of research, I wish I could state that the problem was solved, and the sites indeed were so similar that they should be considered of the same cultural group. However, as my own personal history indicates, simple categories may apply to some phenomena but rarely to the enigmatic domain of "culture", unless we reduce it to a few basic (and hence banal) measurable dimensions.

I have designed this project to discuss not only the classification of prehistoric material remains, and their role in understanding the human past, but also ways in which the process of classification itself limits our vision, particularly when the boundaries between individual categories are viewed as natural, permanent, and transcendentally meaningful. I travel through the present, the recent past, and the distant past, suggesting that the divisions between them may not be as simple as we may sometimes wish to think. Objective scientists as we may strive to be— and I would by no means disregard attempts for neutral objectivity — we are all products of a particular social, cultural, and historical background. Moreover, we are always working through layers of history and culture, the deeper stratigraphy of each and every archeological site[2].

Notes:

[1]Quoted on p 109 in Damrosch, Leopold Jr. 1988. "Myth and Fiction in Robinson Crusoe," in Harold Bloom, ed. *Daniel Defoe's Robinson Crusoe.* New York: Chelsea House Publishers, pp. 81-110.

[2] In revising this work, I have dropped a chapter of general literature review of East European archeology in order to focus more directly on the comparative case involved. I have, however, included my field journal entries at the start of each chapter, in order to provide a sense of place in which I conducted this research.

Chapter 1: The Past

"Do you believe in reincarnation?" the old man asked. He was leaning against a scrawny, young apricot tree. I was weeding a zucchini plot. The plants were small and tough; the weeds were tall and abundant. But this was the only vegetable that I had.

"It's an interesting issue," my American self said, avoiding an answer. The sun was hot, piercing, and I did not look up.

"It would be good to come back again," the man said, as I worked and he watched. "Let's say if I were to come back as your little boy." I stood up and straightened my back. "But you would not know it was me," he added quickly.

"It is interesting to imagine another life in another time, coming back to the same place," the old man said. He was not from this place, this village. Neither was I. Nobody here was "from here". The old man's parents moved here after the war when the German inhabitants had to pack their possessions into two suitcases each and leave in forty-eight hours. I had come here to study another life in another time. My parents had never lived here either. Some of their ancestors were from villages upriver, on two sides of a border

(May 1993, Dolní Věstonice, Czech Republic).

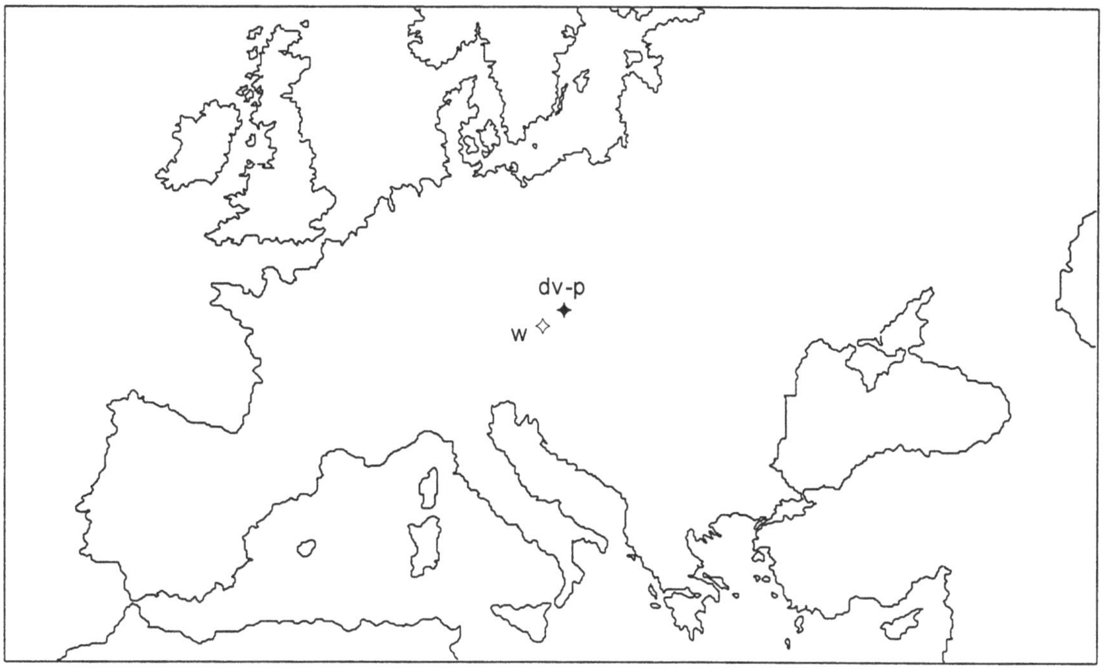

Map 1.1. Dolní Věstonice/Pavlov and Willendorf in Central Europe.

Introduction

On the most general level, this text addresses two central themes: the relation of attitudes towards the past to distinct cultural and historical contexts, and the practice of categorization and classifying phenomena, be they material objects, natural features, archaeological sites, cultures, countries, or people. In examining both topics as they are intertwined in the particular cases of Austria and Czechoslovakia, I present the frame of my research as well as its results, suggesting paths around some of the obstacles, offering a few direct answers.

Generally speaking, classification is one of the basic tools that we all use in our interactions with everyday experiences, in "making sense" of events, encounters, material items, as well as ideas. It would be therefore rather naive to suggest that classification is bad, and we should try to stay away from it. On the other hand, as I discuss later on, categorization not only helps us in organizing reality and reveals recognizable patterns, but also channels our attention in a certain direction. Recognizing that classification is as much a social process as any other activity that enables us to interact with reality, opens the possibility of studying its social contingency, rather than accepting it as a given outside of social and historical influences. Removing classification from a pedestal of objectivity, allows us to question the reasons why categories are constructed in a particular way, investigate their internal consistency, and to wonder how many "pieces of reality" actually meet the criteria set up by definitions of any class. Above all, it allows us to scrutinize the socio-historical contexts in which these categories were constructed.

In this text, I am interested in examining the points or seams that are identified as the locus of difference, on the basis of which objects can be distinguished from each other through the process of classification. Following that, I look at the effect that the positioning of "clumps of reality" in different discursive fields has on subsequent discussions of them. It is my intention to show that the process of categorization, particularly when dealing with cultural reconstruction as the past, is complicated by a number of factors that do not separate smoothly, and whose influence, even when identified, may not be so effortlessly corrected.

This study looks at two opposite ends of the spectrum that every archaeologist has to deal with - the present historical and social context in which the archaeological knowledge is produced, and the prehistoric past that is brought to light with a great deal of guesswork, estimates, and suggested meanings, all based on limited evidence. Hopefully the results of this study encourage self-conscious reflection on modern contexts of archaeological work, and suggest optimism for our ideas about the very distant past.

The focus of the study is a specific case of a selected group of lithic artifacts from two central European Upper Palaeolithic sites - Pavlov and Willendorf II. By examining tool edges for traces of use, I suggest some reasons for differences between the lithic collections from the two sites. The approach that I propose rests on the interpretation of possible past use, and calls for other lines of evidence to strengthen the claims. The main promise of such a focus is the suggested way out of the discussion of the meaning of archaeological typology - the eternal question whether a tool type is a reflection of cultural preferences or functional uses.

The missing link

The starting point of my investigation was the striking absence in the archaeological literature of any detailed comparison between two significant Central European sites - Willendorf II and Dolní Věstonice/Pavlov - despite their proximity in both geographical (less than 100 km) and temporal (roughly 30,000 to 25,000 years before present) terms. This silence was even more startling considering the fame that these two localities have enjoyed over the century since they were excavated - Willendorf II primarily in 1908, Dolní Věstonice starting in 1924 and continuing into the present.

As a recent study has shown, the Willendorf II female figurine has been the single most commonly used Palaeolithic image in textbooks and popular discussions of prehistoric art, only closely followed by the second most common representation - the female figurine from Dolní Věstonice (Conkey pers. comm.). Considering the shared context in which at least some of the material from these sites has been presented, the lack of scientific comparison between the sites puzzled me. Only in general discussions of Upper Palaeolithic sites in Europe did Willendorf II and Dolní Věstonice appeared together, and such work placed them in the general European context without necessarily pointing out their proximity (see general works on the European Palaeolithic, e.g. Gamble 1986, Otte 1981; or a regional survey in Montet-White 1994).

My initial reaction was that this must be a simple case of a present boundary projected into the past, the border between Austria and Czechoslovakia (and simultaneously the post-W.W.II border between the West and the East), overshadowing the study of prehistory. It appeared that all that needed to be done was to examine the archaeological materials from both sites, and look for similarities. Thus I selected a sequence from Willendorf II (layers 5-9) that was temporally in the same range as the complex of sites in Dolní Věstonice and Pavlov (30,000 - 25,000 BP), with the intention of comparing the lithic artifacts from both localities.

Initially, I planned to examine them solely for traces of use-wear, and compare the acquisition and treatment of raw materials, as the basis for evaluation of differences and/or similarities between the sites. However, this seemingly clear case became much more complicated once I ventured into the historical background of the excavations. Although I intended to map the history of the research only to position my study, I soon found myself amidst historical details and disruptive circumstances that led me to investigate the recent past more thoroughly. I also began to realize that my simple comparison of the two sites was not going to unfold as smoothly as first thought and that I would have to face problems associated with the comparative method itself.

The difficult role of comparison in anthropology is not a new issue, as Franz Boas would attest (Boas 1948). His call for the use of historical rather than comparative method made little headway beyond cultural anthropology, and direct comparison remained a basic tool of archaeological method (Eggan 1953; Hammel 1983). There the sporadic archaeological record for a particular temporal sequence seemed as sufficient justification for comparisons over vast geographical distances (e.g. Gamble 1982; Otte 1981; Soffer 1985). However, upon closer look, a reader may notice that the sites, localities or artifacts chosen for comparison were not selected at random. A shortage of sufficient data in one area would be filled by data from other area, close geographically, temporally, or ecologically.

Modern day concepts of geography, and especially political geography, quietly move into the analyses and comparative studies of the past, focusing on geographical areas in their present day configurations (for an excellent discussion of modern day borders regulating human evolution models in the Middle East see S. Rosen in Clark 1992). The comparison of the prehistoric record in Central Europe then appears to be as much a construction of that record as a comparison. Thus my initial question - why these two sites were not compared - dissolves into questions about the manner in which comparisons are conducted.

Before considering interpretations that are offered for any archaeological pattern, we need to step back and examine what is being compared, and what were the options for the study. The comparative method is a highly selective tool that involves choices on the part of the researcher, and can always serve more than one purpose, whether these are conscious or not[1]. Thus the presence of facts in any one area does not automatically lead to their comparison, since we view that area through the shifting lens of the present, and the topography before us depends as much on the way of seeing, as on the forms of the evidence itself.

The problem of history

Having outlined the problem with the comparative method and its current uses, I chose to follow Franz Boas' (1948) suggestion and examine the evidence from a historical perspective, trying to understand why a particular classificatory system was accepted in first place.[2] The first steps of my research led me to the Czech and German literature from the 1920's until the 1990's. Uncovering the numerous layers of historical contexts, I began to realize that one of the reasons why the two sites were not compared stems from the different social and historical contexts in which they have been placed from the beginning, before they were even excavated. Their location in completely separate discursive fields - natural history and cultural history - precluded any discussion or comparison. This was only accentuated by different national approaches towards the past, the respective histories of the people living in Austria and Czechoslovakia, and their respective positions in European history.

The historical chapters (Chapter 2, 3) presented for Dolní Věstonice and Willendorf II are given extensive space here not only to provide the necessary background to my own work, but also to raise central theoretical issues. In the Czech case, I focus on the political context of the tumultuous twentieth century, from the early excavations in the 1920's, across the German occupation and the Communist regime to the present day. The goal is to delineate the manner in which not only the interpretive frame, but also the selection of techniques shifts between periods, all under the name of rigorous science. In the Austrian case, I focus on the architecture of Vienna, and particularly the Museum of Natural History, so as to point out that the place and space of science play a significant role not only in the presentation of the final results but throughout the whole process. The discussion of the museum touches upon one of the central themes of this thesis: the classification of materials. Drawing upon Michel Foucault's The Order of Things (1973), I place archaeology in the middle of broader historical systems of classification and representation, in order show the effects it has on our understanding and presentation of the past.

Culture in archaeology

From the historical framing of the claims about the prehistoric past, I question the notion that the Pavlovian of Moravia is truly a unique cultural phenomenon in Upper Palaeolithic Europe. A number of studies (e.g. Parkington 1986; Shennan 1978, 1989) have suggested that a reappraisal of the concept of "culture" as it is used in Palaeolithic archaeology, has been long overdue, being itself largely an artifact of an era of research when no other appropriate dating methods were available. Despite such occasional pronouncements, the culture concept in archaeology remains alive and well, and continues to lurk just below more technical designations based purely on typological variation, e.g. the "Aurignacian", "Gravettian", and "Magdalenian" styles of Palaeolithic stone tools.

The popularity of the label "Pavlovian" thus comes as no surprise. Rather than simply adopting it, I try to investigate what actually lies behind this name. The two data chapters on Pavlov and Willendorf II present the results of my fieldwork, examining the lithic artifacts from the two sites. In both cases I had initially noted the typological category to which each tool was assigned by its excavators, then examined the artifact in terms of the traditional Bordesian Palaeolithic typology, measured it, and examined whether the same standards were observed within each assemblage.

Once I had embarked on such a skeptical review, it did not come as a surprise to find that typological criteria were applied quite generously. A close examination of the lithic artifacts made me much more aware that many of the tools appear in fact to have been either unfinished, broken, or reused. The final results of the detailed artifact investigation suggest that the typological approach is ultimately not very informative about the prehistoric behavior behind the tools, and furthermore, by applying standardized labels, it frequently fails to accurately capture even the appearance of the range of recovered lithics in descriptive terms. Classification of archaeological materials is necessarily used to organize recovered artifacts in the initial stages of any archaeological project. Yet it frequently ends up being also used in the interpretive stages as the basis for explanations of prehistoric behavior. It is this process of translation by which tentative assortments of fact become assumed data that I most want to call into question.

I have oriented my study in the direction of prehistoric behavior, trying to examine how contexts of consumption may inform us about the social contexts of the past. The microscopic examination of the lithics from the two locations allows me to propose seasonal activities, and possible seasonal occupations of the sites over time. By looking at the artifact from the opposite end of production, its use and consumption, I suggest the possibility of outlining a sequence of tasks, actions that could tell us more about sharing, individual participation in processing and working acquired materials.

The final comparison of the interpretations of seasonal activities during the occupation of the individual layers in the Willendorf II sequence and Pavlov highlights the variation in the evidence, and the possibility of numerous events leading to the same final effect in the archaeological record. Throughout the discussion of the results of my observations, I made an effort to propose several alternative scenarios that might account for the pattern, and convey the sense how much more work needs to be done before any particular interpretation can be strengthened. The study presented here is only an initial step in this direction. Raw material sourcing, palaeoenvironmental reconstruction, and a more detailed archaeozoological analysis would contribute to a richer case that would strengthen the claims made here, and shed light on some aspects of the lived and experienced prehistoric past of central Europe.

An outline of the text

The chapters that follow trace first the most recent history of both sites - Dolní Věstonice/Pavlov in Chapter 2: A Site in History: Dolní Věstonice / Unterwisternitz through Time, and Willendorf in Chapter 3: The Museum: Architecture, Representation and Culture in Vienna. Setting the collections in their historical contexts, I proceed to outline methodological steps that I followed in this study in Chapter 4: Methods: Stone Tools and Archaeological Approaches. On one hand, this chapter provides a guide to the typological and use-wear analysis applied. On the other hand, keeping with the contextual approach, I also situate these methodologies, even if only briefly, in a wider theoretical and historical debate, in order to suggest the tentative nature of the interpretations that follow. Chapter 5: Pavlov: The Site and the Collection, gives a detailed account of the typological and use-wear analyses of the Pavlov collection. It is the most data-dense chapter in the book, providing tables and graphs in order to present the results in a more user-friendly fashion. Chapter 6: Going to Austria; Willendorf II: The Site and the Collection, deals with the Willendorf II collection in the same fashion, presenting the results in a graphic form as well. The results of both chapters are then united in Chapter 7: Seasons of Difference: Interpretations of Pavlov and Willendorf II. Here I consider the results of the analyses of both collections from the perspective of seasonality. The explanation of each collection is based on the results of the use-wear analysis, presenting an alternative to traditional functional interpretations, while recognizing that the results would be strengthened by additional palaeo-environmental evidence. Chapter 8: Conclusion: Meanings and Differences, considers the meaning of the suggested differences between the lithic collections from Pavlov and the layers from Willendorf II., and suggests further work that would enhance the present study.

The chapters are introduced by entries from a diary that I kept during my fieldwork. I conducted the fieldwork during two consecutive summers and winters in 1992 and 1993. My experiences at both sites and with both collections differed quite dramatically. I worked at Dolní Věstonice alone for long stretches of time, while for a time I commuted to Vienna from Czechoslovakia then lived in the city for several months and walked to the Natural History Museum. The diary reflects to some degree the cultural and political contexts in which I worked, and partly I hope it sheds some light on my position as a native/foreign archaeologist working in Central/Eastern Europe.

Notes:

[1]I am by no means suggesting any conspiracy theory, or crude ideological masking of reality, but rather a more mundane and subconscious operation of "common" sense and cultural knowledge.

[2] I am here adapting Boas' approach to my needs, since he was advocating the study of cultural histories of particular societies as a reaction against cultural evolution, not the historical context of the research itself. Boas stressed the idea that similarities in traits were often the result of historical connections. He therefore emphasized the need to investigate customs in relation to the cultures of neighboring groups in order to assess the impact of historical connections upon culture (Boas 1948). Here I consider archaeology as a peculiar form of culture, and examine its history in two neighboring political contexts, Austria and Czechoslovakia.

Chapter 2: A Site in History: Dolní Věstonice through Time[1]

The sleeping dog

I woke up cold, with my bones aching. The heater ran out of steam in the middle of the night, just as it had every night for the past month. I got up and put more clothes on. It was dawn, everything was peaceful, quiet, the cold air behind the windows stood still. I sat on the windowsill for a while, watched the backyards of the houses around, and listened for the first morning sounds. It was a little after six. I had to go to the store to get bread, delivered every morning. They might even have yogurt, I thought, just this one time.

In keeping with communist architecture, the general store looked like every village store in the whole country, or even the entire Eastern bloc for that matter - a box with a flat roof, its big windows stuffed with odds and ends that fell behind unwashed curtains.

A sand colored dog was sleeping on the ground in front of the store. Everyone walked by without paying any attention to it. It must be really cold, lying on the frozen ground, I thought, wondering whose it could be. I walked around it carefully, just in case it woke up and leapt at me; dogs here were not pets but guards to deter strangers. I got the bread, warming my hands by holding it. No yogurt. I exchanged the usual few words with the cashier, who was again disappointed with my uninteresting purchase. The whole village knew what I bought every morning. I exited and noted that the dog was still there, ignored by everyone, fast asleep. Later that morning when I walked to the post office to get the newspaper, I saw it still in the same position, on the same spot.

The old man knocked on my door: "Were you cold last night?" I was permanently cold in this house. "Whose dog is it, sleeping in front of the store?" I asked instead of answering. He did not know, so he went out to look at it. It started to snow, quietly, covering all the dirt in the backyards, the main street, and the dog in front of the store.

At noon the man came by again. "Nobody knows whose dog it is," he said. "But it's not sleeping, it's dead."

I sat on the windowsill, looking outside as I ate my lunch. The whole village was quiet, white with the fresh snow. It looked like it was sleeping, I thought, exactly like the dog...

(December 1992, Dolní Věstonice, Czechoslovakia).

> Whatever the journalist's definition of eastern Europe might be, let us state, as introduction to this Special Section, that Czechoslovakia is a country in Central, not Eastern, Europe. It is somewhat controversial to speak about its "return to Europe", as some politicians would have it, as it has been there all the time. (Venclova 1991: 306)

Despite archaeology's focus on space and time, it often ignores the exact location in space and time of its own research. This chapter seeks to examine varying grounds on which archaeological knowledge-claims have been made over time in relation to the well known Palaeolithic site, Dolní Věstonice, in the former Czechoslovakia. While it is necessary to consider the physical context of the artifacts that validate a particular hypothesis, at the same time it is also essential to unravel the social, historical, political, and personal contexts within which the site, the materials, and the investigators themselves are submerged.[2]

Palaeolithic archaeology, rooted in geology and positioned directly on the fault line demarcating the furthest edge of the humanities, combines the methodological aspects of natural sciences, and the interpretive aims of social sciences. Its heavy reliance on stone tools, soil samples, and geological profiles gives Palaeolithic archaeology the image of "hard", or by extension, "objective" science. The appearance of agriculture, architecture, and cities in subsequent time periods substantially increases the distance between natural sciences and archaeology, framing other archaeologies with a more humanistic mind-set. Thus Palaeolithic archaeology offers a pivotal case study on the border between the historical development of a natural science, and a humanistic pursuit of interpretation of factual evidence.

Contexts

Archaeological research operates within two contexts: one archaeological, the physical circumstances of the finds, and one culture-historical, the socio-political climate of the country in which the site is located, together with that of the researcher's country. These two contexts are inseparably bound. Giving meticulous attention to the archaeological context is itself no guarantee of a value-free interpretation, for all interpretations claiming to be factual based, and scientifically tested, are constructed in relation to both material evidence and social reality (Gero et. al. 1983; Pinsky & Wylie 1989; for a different approach to similar questions see Patrick 1985). What constitutes a location worthy of a survey, a site in a need of excavation, materials worth collecting, identifying, and analyzing varies over time, and is culturally determined, constructed, and reconstructed.

A good example is the claim that the 1853 drought in Switzerland exposed the "Lake Dwellings", resulting in a large scale excavation and the discovery of the Bronze Age settlements (Trigger 1989:83). The drought certainly occurred, and aided in revealing the wealth of finds, yet one still has to question why the Swiss scientists were searching for ancient settlements in such a meticulous way, in that particular geographical location, at that point in time.

It does matter after all, as the historian of science L. Pyenson asks: "why astronomy rose to such prominence in the Netherlands, despite a climate entirely unpropitious for star-gazing, while no such tradition emerged in a topographically well suited country of comparable size and temperament, Switzerland" (Pyenson 1989: 379).

Most discoveries in general cannot be attributed to a phenomenon that has not been previously noted but rather, circumstances have risen which encouraged and supported an investigation of a particular kind, questions probing certain answers. The fact that the lakes themselves were a focus of attention at the time, is as significant as the persistent search for ancient remains in the caves of France or Germany.

Romanticism at the end of the 18th century was combined with the growing interest in the origins of diversity in the animal and plant worlds, and the popularity of the evolutionary theory during the 19th century. It had a significant influence upon the social imagination in terms of the extent to which humans came to be viewed as a part of the natural world, and the places where the relics of the past could be found. Switzerland, particularly the Alps, had by then become a popular area visited by upper middle class European tourists, and the country constructed its national identity primarily in terms of the landscape. The cultural mapping of the geographical location would have certainly encouraged identifying features within the landscape in terms of antiquity, according to the, then burgeoning, idea of equation of nation and land (Lowenthal 1985).

Archaeological excavations and research, it is often written, should be problem-oriented, addressing specific questions formulated to deal with issues pertaining to the past (Renfrew and Bahn 1991; Thomas 1989). Yet, one has to pause and consider the meaning of such a bold, basic, introductory-textbook statement. What constitutes, at a specific point in history, in a specific geographic location, a question that needs to be dealt with? Presumably, a question is posed so as to obtain an answer, and an answer that is seen as vital and interesting. The question itself is a lead toward an answer that acquires interest and meaning in a particular context. The specific historical and social context gives meaning to the social construction called 'the past', be it local, regional, national, or universal human past.

History of European prehistory

Palaeolithic archaeology operates within a time-bracket labeled as *prehistory*, the period before written records (Renfrew & Bahn 1991: 17). The term prehistory, as opposed to history, appears in England and continental Europe during the second half of the 19th century, signifying a separate, conscious aspect of the study of human existence (Chippindale 1988; Clermont & Smith 1990). At the time, prehistory was called *préhistoire* in France, *preistoria* in Italy, *Vorgeschichte* in Germany (Daniel 1963:14). All these terms suggest a similar optimism: there was a tangible past before the occurrence of written records, knowable and worth naming.

In Germany, with the concurrent rise of the nation state, the ancient past was viewed not only as something tangible but also something significant to the national identity, and prehistory "changed its name" in 1886 from *Vorgeschichte* to *Urgeschichte* - indicating that it was the source (the prefix *ur-*) of history, rather than its antecedent (Sklenár 1983). Neither England, nor France followed suit.

Czech intellectuals, on the other hand, in the midst of battles for language rights in the Austro-Hungarian empire, avoided using the existing loan word *historie* (history) that would have suggested links to Western traditions, and instead adopted the term *pravek* with its strong Slavic ring. The created word evoked the image of ancient ancestral time (the prefix *pra-* indicates antiquity, old age along direct kinship lines), a Slavic past which the present population descended from (*vek* is the Slavic root for age). These three different words - *prehistory*, *Urgeschichte*, and *pravek* - may serve as the background against which the materials from Dolní Věstonice were interpreted in the 20th century.

Setting

The site of Dolní Věstonice is located near a village of the same name, in the southern part of the central section of former Czechoslovakia known as Moravia. The country itself was created in 1918, as one of several small nation-states that rose from the disintegrated Austro-Hungarian Empire after the First World War. In many respects its making was more a result of a political failure in the sphere of multiculturalism, particularly mishandling the language question, than a mature decision of all parties involved. Political boundaries were created that did not correspond with ethnic territories, resulting in the incorporation of long-settled linguistic minorities into the new national entity.

Czechoslovakia was simultaneously part of Central Europe and the western-most Slavic country, possessing at the time of its creation large German and Hungarian populations in addition to Czechs and Slovaks[3] Historically the country has never been comfortable with its geographically expansive and threatening, but also inescapably influential and appealing, German-speaking neighbors to the west, south, and north, an unease that is counteracted by an equally ambivalent relationship with its eastern neighbor - the former Soviet Union, heir to the Russian empire. This historically shaped and ideologically charged geographic location, with perceived ethnic attachments and emotional value, played a major role in the collective national identities of the Czechs, the Moravians, and the Slovaks.

Dolní Věstonice lies in the southern part of the Moravian Karst, traditionally a German region, later part of the contested Sudetenland. The village was known under its German name Unterwisternitz, and the overwhelming majority of the population was ethnically German (Frolec 1985). Their identity within the newly formed state became unclear, since they had traditionally identified themselves as Austrian Germans under the Empire. As of 1918, they were no longer Austrian, and yet could not claim to be German either.

The system that had allowed the ambiguity of a figure like the writer Franz Kafka - who lived in Prague, but was Jewish in ethnic tradition and religion, and German in the literary language - had collapsed. Where the old imperial German language had distinguished between *tschechisch* (Czech) and *böhmisch* (Bohemian), the new national Czech language recognized only one state identity - Czechoslovak. This only contributed to the German sense of alienation from the Czechoslovak state's identity, an issue that developed into a serious problem in the 1930s, and an irreparable crisis by the beginning of the Second World War, with generous assistance from German and Czech political parties.

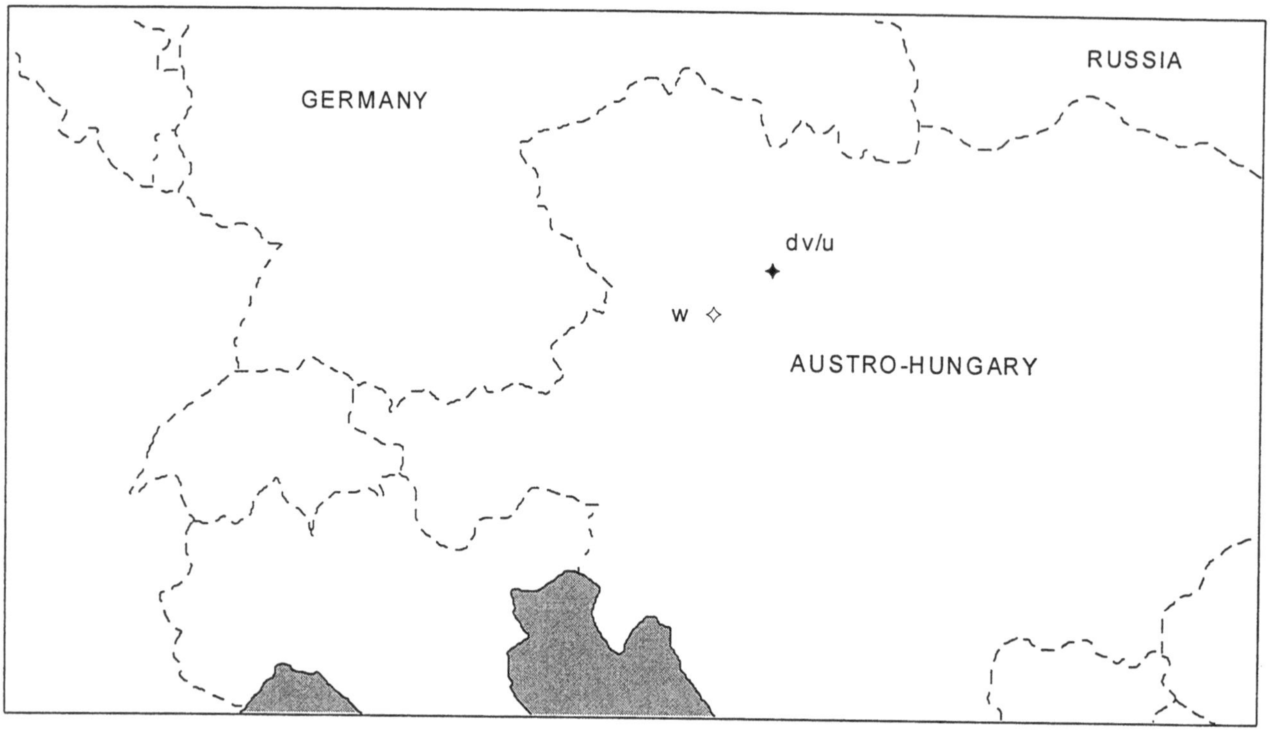

Map 2.1. Dolní Věstonice/Unterwisternitz in situ prior to 1918 (Willendorf in white).

The site located in the nineteenth century

The hilly region of Southern Moravia was of historical interest for centuries, mainly due to its claim of having been the location of the first Slavic state - the Great Moravian Empire (9-10 c. AD). The search for the exact site of this once powerful community captured the imagination of all local patriots from the 18th century, when land and nation became inseparable concepts in the popular imagination (Gojda 1991). The first mention of prehistoric finds is dated to the 17th century when a doctor at the emperor's court recorded mammoth bones recovered in the vicinity of the village (Absolon 1938, Klíma 1954, Schwabedissen 1943, Svoboda 1991). Throughout the following centuries local doctors, teachers, botanists, and naturalists made numerous records of extremely large bones, later identified as mammoth. The hunt for ancient fauna was replaced, in the second half of the 19th century, by a search for ancient human inhabitants, partly due to increased awareness of the antiquity of humanity:

The swelling tide of Darwinism in the '60s and '70s even exceeded the impact of Hegelianism in the '20s and '30s. Psychology, anthropology, economics, geography, philology, ethical philosophy, and even epistemology were remodeled on Darwinian lines. The idea of natural evolution came to dominate the thought of the later nineteenth century in a way comparable to the fascination exerted by classical mechanics in the seventeenth century (Ermarth 1978:68).

At the same time, optimism associated with the possibilities of science in answering questions previously in the realm of myth and fiction ran particularly high in the increasingly powerful German state. Bohemia and Moravia, still a part of Austro-Hungary at the time, were strongly influenced by these intellectual movements, in part due to their desire to gain independent recognition outside the empire and, in part, due to the traditionally strong ties with the German universities.

The turn of the century was also marked by the growing nationalism and ethnic strife for recognition in many European countries. Consequently the intelligentsia in Central Europe, the local doctors, botanists, naturalists, geologists, as the practitioners of science, came to play the role of national revivalists, by providing scientific evidence for the importance of a local place. The Museum Society in Brno (the urban center of Moravian intellectual life) was established in 1888, with a twelve member board of governors, including ten amateur archaeologists, and two historians (Frolec 1988).

The acceptance of evolutionary theory as a research framework, and as a new philosophical outlook on the questions of human antiquity (Grayson 1983), led to a "Naturalization of Spirit" (Ermarth 1978). The search for the most ancient remains in caves was not coincidental (see Rupke 1983: 391-395, for a discussion of fascination with caves since the mid 18th century in France and Germany). Humanity no longer, or at least for the time being, stood in opposition to Nature but rather had its origins in the natural world. Caves were an ideal setting that the new scientific geology could study, and at the same time presented an ideal metaphor for the birth of the "modern" humans, the opening in the earth from which people came.

The Czech lands of the Austro-Hungarian Empire were no exception to such an interest. The first Palaeolithic finds in Moravia came from the excavations at Bycí Skála in 1867, and later other cave sites were discovered and recorded: Kulna, Pekárna, and in 1880 the Sipka cave (Sklenár 1983:119; Svoboda 1991). In the process of this meticulous search, Karel Maska, the foremost Moravian prehistorian at the time, discovered the Předmostí open-air site with twenty skeletons in 1894 (Sklenár 1983; Klíma 1951; Absolon and Klíma 1977).

This spectacular and unique find diverted the attention from the caves that have yet to render anything comparable to paintings of Altamira found in 1878. Yet the French Palaeolithic has been considered the ultimate measure of achievement in prehistoric times, and most archaeologists in Central Europe focused on later periods, the Neolithic and the Bronze ages, that appeared to have been much more impressive in this area. The majority of the Neolithic sites excavated around the turn of the century contained artifacts that seemed suitable for interpretations related to the ethnic identities of their makers. Pottery, art objects, weapons, and house remains were viewed as possessing style unique to ethnic groups, present and past.

The difference between the mid nineteenth century, when the first sites were located, and the beginning of the twentieth century, when they were excavated, marks a crucial divide in the social climate, and in the sense of crisis in the political, cultural, and historical expectations of most European states (Crawford 1992; Jalevich 1987). The initial fascination with, and unequivocal faith in, positivism at the end of the nineteenth century seemed to have faded, and the stage for the new century was set amidst an "anarchy of world views" (Ermarth 1978:80-1):

Increasingly after 1880 intellectuals in many fields began to question the ultimate validity and grounds of scientific conceptions and came to challenge the effects, which such conceptions had upon the conduct of life. In contrast to the earlier positivism, one is confronted with the singular spectacle of science brought to witness against itself. In some circles confidence in science began to wane and belief in the objective truth of science was questioned... These decades witnessed various intellectual movements of revival and innovation which went under the various names of neo-idealism, neo-romanticism, vitalism, voluntarism, the "new irrationalism", and the "new metaphysics".

The original archaeological research at Dolní Věstonice, carried out by Karel Absolon, took place in a very specific historical and social climate of which it was a product, yet played an immensely powerful role in its construction. Understanding this relationship is essential, in order to unravel some of the reasons that stood behind the interpretations of the particular site and its materials later in the twentieth century.

The site Dolní Věstonice / Unterwisternitz excavated

A large scale excavation began at the location in Southern Moravia in the summer of 1924. Dr. Karel Absolon, a professor at the University of Prague and a curator of the Moravian Museum at Brno, led the project, and for the next fifteen years was its director, its successful publicity manager, its chief fund-raiser, and the writer of all the research reports (Absolon 1925, 1927, 1938).

Dolní Věstonice turned out to be an immensely rich find with large numbers of artifacts; the first evidence of portable art, a head of a bear, was located in the initial weeks of excavation in 1924. The foremost prehistorians of Europe, gathered for the International Anthropology Congress in Prague, in September 1924, were immediately invited to inspect the site, and participate in the experience. A black-and-white photograph shows a large group of bearded men in three-piece suits lying on the ground, crowded in a narrow gully but with solemn expressions, pursuing science. Count Bégouën, we are told, was lucky and found a piece of a mammoth bone, and a micro-blade (Absolon 1938).

Considering the local politics in the 1920s, it is noteworthy that both the Czech and the German press in Moravia covered the excavation extensively at the time, without mentioning any national issue at all. The same can be said about Absolon's 1924 report, which is remarkable in its complete absence of any nationalism, though full of patriotism and pride in the region. He includes both the Czech and the German place-names, and from his published diary it is clear that, but for one, all the workers at the excavation were local Germans (Absolon 1938). The report written in 1938 in Czech for the local professional audience does not offer any racial or ethnic theories, and stands in a sharp contrast to the articles that Absolon published at the same time in French and German anthropological journals.

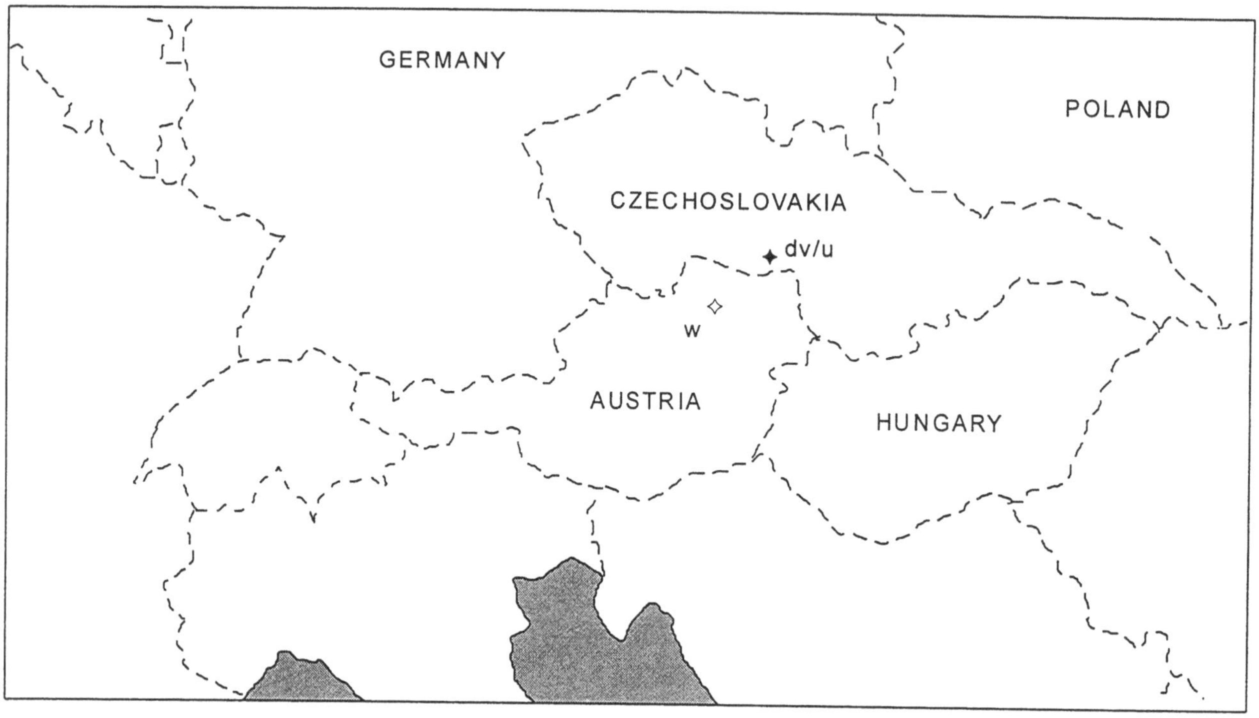

Map 2.2. Dolní Věstonice/Unterwisternitz in situ, 1918-1938 (Willendorf in white)

The site and the noblemen

The first brief note in foreign press appeared in *The Illustrated London News* on 31 October, 1925, under the most expressive title: "A discovery as wonderful as that of Tutenkhamen's Tomb. Moravia over 20, 000 years ago" (Absolon 1925). The impressive finds of the Egyptian tomb were replicated in Moravia, placing them on the same cultural level, while the British archaeologists were as successful as the Czechs in uncovering such treasures. This magazine later carried extensive coverage of the project, serving as the main source of information about it for the general public in the English speaking world, and for Western Europe in general.

The second extensive report appeared a year later in German in the Acta Musei Moraviae, a local periodical put out by the museum of which Absolon was a curator (Absolon 1926). It may seem unusual that the report was not in Czech, the established official language of the newly independent country. Yet there were several reasons for this rather typical Central European phenomenon. Despite its newly acquired state of independence, Czechoslovakia continued to be dependent on Austria and Germany in many aspects, and scholarly language, education, and research remained a powerful legacy of three hundred years of cultural dependence. Most university educated researchers, particularly those in physical and natural sciences, were trained in German. At the same time, even though Czech was then the accepted scholarly language in Czechoslovakia, such a spectacular site as Dolní Věstonice had to be made known to the very community from which the young archaeological establishment was trying to make itself independent.

This peculiar situation persisted throughout the century, with different foreign languages playing the role of the master language. It continues to this day, and is not unique to Czechoslovakia, or even to Eastern Europe, but rather is expressive of the scientific capital of various countries, and the neo-colonial dependency of those that do not possess it.

Shortly afterwards, Absolon published articles for the professional audience in the French *Revue Anthropologique* and the German *Tagungsberichte der deutschen Anthropologischen Gesellschaft*. In both reports, he stressed the importance of the site itself as a new Aurignacian station in Moravia, and one of the largest in the world (Absolon 1927, 1928). A detailed description of the now-famous clay figurines appeared here for the first time, with the animal statuettes discussed as much as the (at the time) only female figurine, and the torso whose sex Absolon did not suggest.

The description considered the figurine "*fortement stéatométrique*" [strongly steatopygic], a view that later changed, depending on the researcher (Absolon 1927:86). None of these reports gave the statuettes any religious significance, and even though they were mentioned and described, they were not considered the highlight of the finds; thirty-one mammoths, after all, outweighed a 10 cm tall female figurine. The primary focus was the impressive number of mammoth bones recovered, the "kitchen midden", with at least 31 complete individuals (Absolon 1927). The slaughter at the location was compared to the then-current extermination of elephants in Africa, where Absolon calls for a protection law so as to avoid the fate that the mammoth met in Moravia (Absolon 1927:88). Analogy with Africa and Australia was a frequent explanatory device in Absolon's writing, its connotations varying immensely over time, as did perceptions of race, and the primitive.

The most extensive popular coverage appeared in a series of lengthy articles in 1929 in *The Illustrated London News*. The articles contained most of the photographic evidence cited and widely used during the following decades by researchers discussing the site. The whole series carried an impressive title, "An Amazing Palaeolithic 'Pompeii' in Moravia", in a far from subtle comparison to the famous Roman site:

> As we learned from Pompeii how the Romans lived, this 'prehistoric Pompeii' will reveal how primitive Europeans lived under the glacial conditions; the readers are introduced to the hearths and homes of the mammoth-hunters of Moravia. (Absolon 1929, 1936:499)

This introduction carried several messages addressed to varied audiences, first and foremost in Europe. When Absolon points out later in the article, "great light has been thrown on the antiquity of humanity", it is clear that it is only a certain part of humanity on whose antiquity light has been shed - the Europeans (Absolon 1929:875). The "homes and the hearths" make it a familiar sight, particularly to the British readers that are introduced to it. Further, the links created between England, where the article appeared, Pompeii, where the grand Roman history was suspended under the ashes of the volcano, and the "prehistoric Pompeii" in Moravia covered with sand, so as to evoke the image of one grandiose, uninterrupted European past just waiting to be uncovered. At the same time, Absolon makes sure that Moravia is placed on the map of Europe, and of the world:

> Moravia has thereby become one of the most important countries for the study of the origin of man and human culture, as here under loess, we have the largest Palaeolithic stations in the world ... the mammoth hunters are, in fact, the outstanding feature of Ancient Moravia...their habitations are of world-wide importance, because at the time of their occupation, Moravia was the scene where man passed through important phases in his development. (Absolon 1929:875)

Patriotism on one hand, and regionalism on the other, permeated all the writing. The newly established Czechoslovak republic was promoted vigorously, at the same time as it was subjected to internal battles between central Prague and provincial Moravia. This first article in the series is in particular a strong appeal for acceptance of Czechoslovakia, and Moravia, as European countries with the same glorious past, and for recognition of the local scholars as sophisticated archaeologists who know their field.

> I am fully aware of my great responsibility to future generations...and have adopted a firm scientific basis for research...the location has been explored methodically and in accordance with modern views," wrote Karel Absolon, proceeding to describe his excavation methods in detail, while leaving it to the specialist reader to decide who the future generations would be who would consider a Moravian archaeologist their forefather (Absolon 1929:875). A similar appeal to the authority of science is also to be found in the 1927 French article, but none of the local ones, indicating the desire for Czechoslovakia to be included in the scientific circles of modern Europe. His concluding remarks echo this wish for recognition: "I should be happy if some larger specimen of these Moravian prehistoric records could become part of the collections of the British Museum of Natural History, as a proof Czechoslovakia's respect for the great English nation (Absolon 1929:875).

The site and the primitive

The entire series of the articles presented Dolní Věstonice to the European audience, both lay and specialist, as a camp of mammoth-hunters, who caught the animals by cunning, appreciated their brains as a delicacy, and created large numbers of decorative objects (Absolon 1929, 1936, 1937). Despite the occasional stress on sound scientific methods, Absolon constructed the whole interpretation as a suspenseful discovery narrative. He incorporated into the interpretations most of the current anthropological theories that suggested migration from Asia to the west, and later with the retreat of the ice-sheets northward. Absolon suggested that the skeletal material from Moravia, in combination with the artifacts, constituted sound proof that the Eskimos were descendants of fossil men of Magdalenian Europe, supporting his argument by citing a study in *The Annals of Eugenics* from 1926 (Absolon 1929).

The issue of race would develop over time, and one can notice a radical difference between the writing from the twenties, and that from the late thirties. As mentioned earlier, from the beginning Absolon considered the occupants of Dolní Věstonice to have been "primitive Europeans", yet drew his analogies from African and Australian cultures. The female figurine had four small holes on its head which, Absolon suggested, had feathers stuck in them as in African fetishes (Absolon 1929). The recovered pigment was associated with various stone slabs and round stones as toiletries, and interpreted as evidence of body painting, similar to that of the Australian aborigines. The reasons for making such analogies were never stated, much less discussed or substantiated.

In 1935, Dolní Věstonice first appeared under the German name, Unterwisternitz, in a short report in the Zeitschrift für Rassenkunde about human teeth recovered that were perforated in the same way as the animal teeth that presumably formed a necklace. This brief note itself is not in any way an indication of Absolon's adherence to racist views, yet the fact that he did publish the piece, short as it may be, in a journal that at the time already focused intensely on studies of racial differences and their implications for social policies is indicative of a greater historical milieu that he did not escape. It is important to remember that it was not only under the extreme and catastrophic developments of Nazi regime in Germany that race became an issue.

The then-common belief in human differences rooted in biology surfaced the following year, when another report in *The Illustrated London News* discussed the latest clay figurines. Absolon expressed his hope that the most recent scientific methods would enable the study of finger-prints, so as to determine "whether they differ from those of living men today" (Absolon 1936:547).

As much as African and Australian cultures were considered comparable to the fossil record in the 1920s and early 1930s, by the mid 1930s they were seen as different. Previously, "European civilization" evolved where that of Africans and Australians stopped. In the later years, a serious attempt was made to prove that the "Europeans" had completely separate ancestry. When a small ivory head was found at Dolní Věstonice in 1937, the news was announced under the heading "By a Prehistoric 'Leonardo da Vinci' of 30 thousand years ago: the Earliest Known Portrait of a Human Being - an Ivory Head" (Absolon 1937:549). The importance of the find was expressed in unambiguous terms by Sir Arthur Keith, who wrote the introduction to the article:

It is of particular interest to us Europeans - for it is a portrait of one of the white or Caucasian pioneers who began to colonize Europe in the later phases of the Ice Age. It is not merely the early history of Moravia that is opening up, but the history of our European forefathers - and foremothers![4] - soon after their first arrival in Central Europe. (Keith in Absolon 1937:549)

With race becoming a major issue in the European context, Moravia had been recognized as a part of the "civilized" world, a bearer of European history. The East Europeans did not have to ask for recognition when their material was useful in promoting a particular ideology, that of Caucasian Europeans. Local scholars did not mince words in making racial comparisons either: "How crude are the faces of present-day Australian aborigines, or the pygmies of Central Africa, compared with this 'classical' portrait from old oriental civilizations, or even a modern drawing such as the head of Christ by the Dutch painter Jan Toroop in 'Night in the Cathedral'," wrote Karel Absolon in 1937, and he expressed belief that the necessary exact scientific somatological analysis, and anatomical construction would show measurable racial definitions (Absolon 1937:552).

The narrative to promote motherland and the native region was replaced in his writing during the 1930's as Czechoslovakia, and Moravia, were already accepted as a part of Europe. A peculiar combination of a pan-European historical narrative and scientific discourse emerged, aiming to provide a story of European racial purity, especially that of Caucasians of Christian faith. As mentioned earlier, the different treatment of the material before different readerships was most striking. The local audience was offered a slightly romantic, suspenseful account of six weeks of labor, discoveries, and a vision of future (archaeological and by extension also national) that reinforced the pride of place (Absolon 1938). The international readers were given proofs of all the current racial anthropological theories (Absolon 1927, 1936).

Less than two years later, Czechoslovakia was invaded and broken up, and Bohemia and Moravia, as Böhmen-Mähren, became a protectorate of Germany. Karel Absolon ceased to be the director of the excavations at Dolní Věstonice, and new racial questions came into focus. These questions were equally rigorously explored with the latest scientific methods, but Slavic populations were no longer included in the sample of the Caucasian race.

During the first 15 years of excavations and research, Absolon and his team mainly aimed to uncover as large a space as possible and find out the size of the settlement. Originally, the physical space, its size and the mammoth-bone accumulations composed the context. The site was one large camp of mammoth-hunters. For these reasons, stratigraphy was observed only marginally, and relationships between artifacts were noted only toward the end of Absolon's research at Dolní Věstonice. The main goal was to put this Moravian site on the map, and prove the coexistence of mammoths and humans. Once that had become an accepted possibility in the mid 1930s, the clay, bone, and ivory objects acquired a new meaning. They were no longer the mere decorations of the mammoth-hunters whose sheer existence had been in doubt. Rather, these objects themselves grew meaningful, as they could now reveal the identity of their makers. Relationships between the artifacts became the essential interpretive clues, and the association of the art objects with the mammoth bones was reduced to that of supporting context.

The need, and desire, to prove the contemporaneity of the ancient fauna and humans led to an inclusion of specialists on the team. Geologists, pedologists, and palaeontologists participated in the analysis, mainly to assess the period of occupation, and their contribution consisted mainly in establishing credibility for the narrative that surrounded the prehistoric settlement. Their analyses were not included in the international publications, remaining only a part of the Czech technical reports.

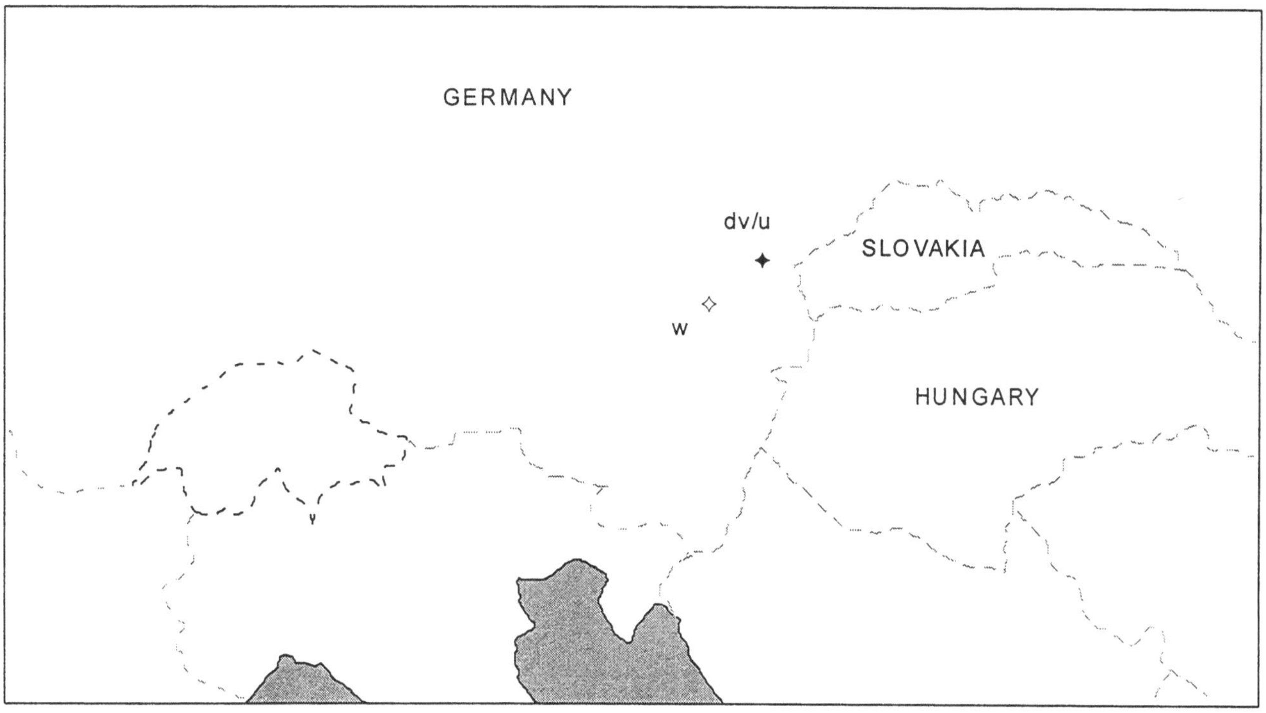

Map 2.3. Dolní Věstonice/Unterwisternitz in situ, 1939-45 (Willendorf in white)

The site Unterwisternitz excavated

In 1939, following the Munich agreement of the previous year, Czechoslovakia ceased to exist, and Bohemia and Moravia became a German protectorate, a situation that would last for the next six years. Charles University in Prague began teaching in German; and prehistory, labeled as "the national science", became one of the most popular subjects. The head of the department Prof. Lothar Zotz published a number of articles on the prehistory of the region, arguing for the German ancestral ties to the land (Zotz 1940, 1944). His best-known student Gisela Freund, the only published woman archaeologist to have worked in Moravian Palaeolithic research, undertook a comparison of the Unterwisternitz/ Dolní Věstonice material with other Palaeolithic materials from the wider central European region, including the Willendorf collection for her dissertation research (Freund 1944). This project never materialized, possibly due to a struggle between the University and the Forschungsstätte für Urgeschichte des Ahnenerbe (Research Center for the Prehistory of Ancestral Heritage), a new research organization established by Himmler (Freund 1944: 3-10).

The research center achieved a complete control over the site, the excavations, and the retrieved materials from previous years, considering it too significant for its own goals to share it even with other German archaeologists. During the three following years, Dr. Assien Bohmers attempted to "correct the neglect of scientific techniques over the past fifteen years" (Bohmers 1941a:45). Consequently, the foremost question that needed to be answered was that of stratigraphy, and the specific geological age of the culture located at the site (Bohmers 1941b). The German team shifted the focus from the historically oriented prehistory, to a natural science centered investigation that renewed the close ties with geology. In collaboration with specialists on pollen, mollusks, and geologists, Bohmers announced that for the first time, the Würm Interstadial had been discovered and scientifically proven at this location (Bohmers 1941a:45). The same site then was again a source of national pride, only of a different nation - German, and this time as a place of an overtly scientific breakthrough.

Yet this concern for science was not isolated from human issues. Bohmers ventured into the area of the cultural materials as well and offered an interpretation, equally linked to the issue of race. According to him, the art objects, rather than the mammoth bones, were the most important find, as from these one could get a glimpse of the world-view (*Weltanschauung*) of their makers (Bohmers 1941a:47). Bohmers further explained that the characteristics of these artifacts, combined with the geological assessment of the deposits, confirmed the hypothesis that the people who inhabited the site were of the Cro-Magnon, Aurignacian or Brno race (Bohmers 1941a:47). This, according to him was significant as these people appeared after the "annihilation" of the Neanderthal on the European, Asian, and the African continents, which was one of the greatest conquests of all times. These people were of a "europide" shape and from them later came the "Indogermans" (Bohmers 1941a:47). Coming in the midst of a world war, an interpretation justified in such explicit racial terms needs little commentary.

Bohmers, just as Absolon, remarked that the ivory head reflected the racial features of the population of Unterwisternitz/ Dolní Věstonice and it "was not for example Mongoloid or Negroid but rather a fully Indo-European racial group" (Bohmers 1941a: 51). Mainly due to its fairly distinct facial features, this piece attracted far more attention than other figurine fragments that may have had bodies but no faces.

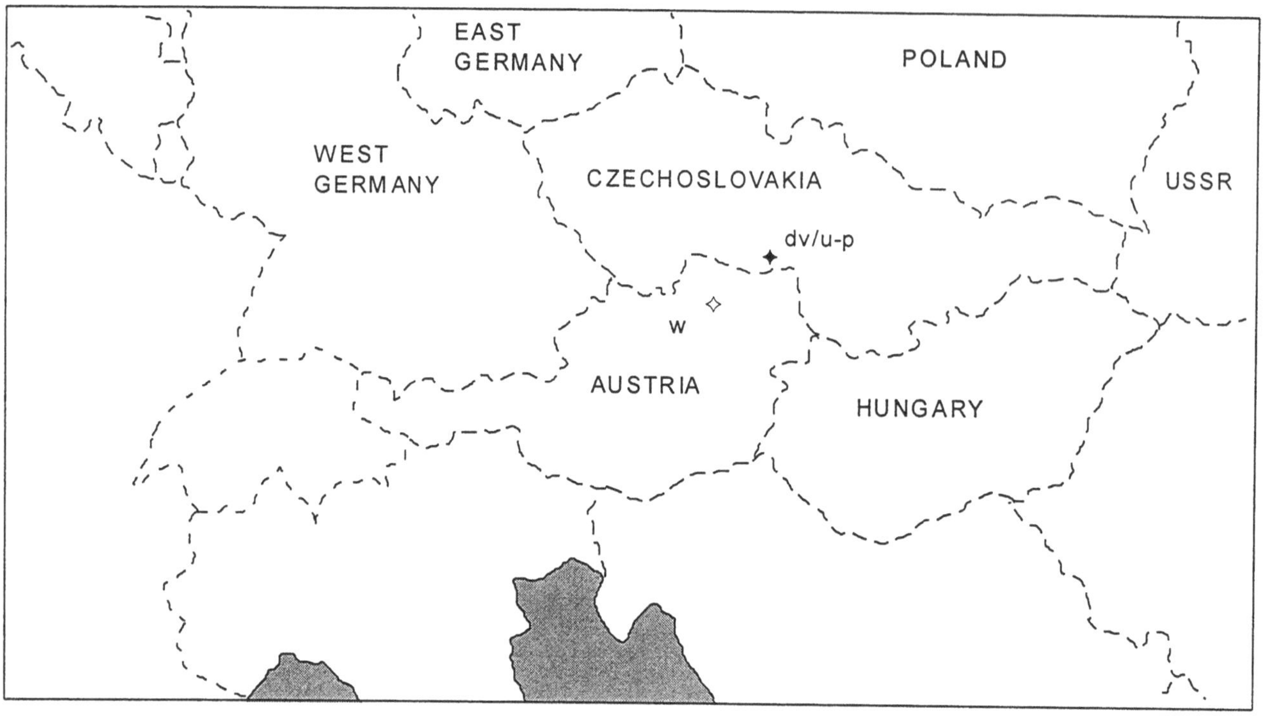

Map 2.4. Dolní Věstonice/Unterwisternitz in situ, 1945-92 (Willendorf in white)

Bodies were assumed to be universally undifferentiated; racially, ethnically indistinguishable, they only varied in the levels of fatness. The Vestonice figurine seemed steatopic to Absolon, but fairly slim to Bohmers. Yet they both agreed that faces were the markers of difference.

In Bohmers' view, the physical space was the crucial context, but this time, the vertical aspect mattered more than the horizontal. Where Absolon wished to impress with breadth, Bohmers tried to secure the depth. The mammoth bones were no longer a major attraction; rather, racial definition and its dating dominated. Hence, the art objects in context - carefully positioned in a scientifically verified stratigraphic sequence - became the focus of research. The ancestors of the present population were "Indogerman"; after all this was the German village Unterwisternitz.

The site Dolní Věstonice excavated again

Seemingly uninterrupted by international turmoil, research returned to Dolní Věstonice shortly after the end of the Second World War. However, the village lost its German connotations as the entire population was expelled in 1945-6, and re-settled in Austria and Germany, while new Czech inhabitants moved in. Unterwisternitz became the Czech Dolní Věstonice, lending the illusion of Slavic permanence, while purging all memories of the German settlement in an effort to forget the horrors of the war. Needless to say, along with the village, control of the site passed back into Czech hands.

In 1945, continuing what the German team started, an interim Czech team carried out a very detailed geological and pedological study. During the war the entire area, as well as many others all over Europe, had been carefully mapped and surveyed for military purposes. A large number of scientific instruments that the army had for these purposes became available to geologists who carried out survey for mining companies, and to construction companies that were re-building the roads in the post-war period. As a result, most research by archaeologists in Czechoslovakia at the time was closely tied with these enterprises, due to a lack of other equipment and funds.

The original reports from 1945-7 are not available, probably due to the transitional character of the post-war régime. In 1948 Czechoslovakia fell firmly under the domain of the Soviet Union, and anything written in the post-war period was either left unpublished or disappeared, only to reappear in the 1950's in an altered form. This gap is so striking that the proliferation of works that came out afterwards, dealing solely with stratigraphy is even more astonishing. All of prehistory became one complex stratified deposit that needed to be investigated. The group of researchers that worked at Dolní Věstonice after the war, Knor et. al., summarized the situation very succinctly in the foreword to their 1953 volume:

> The entire problem of the Upper Palaeolithic and its typology, as well as the questions of the geological periodisation of the last glaciation, has come to be addressed, together with the methodical issues of open-air sites, so

successfully by the Soviet archaeology. It became clear to us that further research at Dolní Věstonice must look for new approaches to work. (Böhm in Knor 1953:5)

The last sentence sounds especially striking, since Böhm does not explain why it became clear, or what the new approaches were, yet it is obvious from the brief foreword and the reports themselves that "further research" depended on this premise, or at least on its inclusion in the foreword. Following this Czech foreword we find a Russian translation and, as the manuscript continues, a Russian translation for each contribution as well.

We should remember that the years 1949-56 were politically severe, with show trials, labor camps, and arrests of innocent people whose crime was their supposed "betrayal of the republic". Making one false step may have had severe consequences, particularly in fields so highly ideological, and at the same time public as archaeology.

The adoption of the new approach was also followed by a removal of Dr. Absolon from his research position, and his forced professional insignificance until his death in 1960, when his son, by then living in the United States, was not even allowed to attend the funeral. Any bibliography of Karel Absolon's work is a clear testimony to his fate, the last date in it being 1947. The monograph on Dolní Věstonice that he was planning to publish never came out, and his monograph on Předmostí was posthumously edited in German by B. Klíma (Absolon & Klíma 1977).

In this era, complete immersion in the soils and their sequences was the safest approach to take. The scientific writing that previously supported and justified the dominant ideology was now used to protect the scholars from the ideology of their own society. Geology, pedology, palaeontology of micro-vertebrates became the safe haven for all those who supposedly studied past human societies. The justification for this complete avoidance of interpretation, and the exclusion of people from archaeological consideration, was found in adopting an extreme ecological perspective:

> Today, we are no longer interested in collecting objects, but rather in scientific observations, which can be done only through a systematic survey of all aspects of the Palaeolithic sites.... we were therefore not interested in flint artifacts, bone objects, figurines, skeletal remains of humans and animals, but rather in their depositional environment and how they related to it... (Knor et. al. 1953:8)

In order to be able to say anything about past societies, went the interpretative logic, one needs to know intimately the natural environment in which they lived. In the process of such detailed research, no one will talk about the people, their relations, or their social system, so went the underlying political logic. Consequently, the stratigraphic sequence became the context that needed to be analyzed before anything else, a process that could - conveniently - last *ad infinitum*.

The situation in Eastern Europe stabilized after the mid 1950's, and the Cold War became less of an immediate and obvious presence. This loosening of political tensions opened the door for the prehistoric people to return to Dolní Věstonice, and the new work re-discovered humans in the past. However, these were not from "*pravek*" (ancestral past) any more but rather from the international Palaeolithic. Science had arrived to stay. In contrast with the earlier wartime writing, at least the prehistoric populations were not placed in the wide and general Würm-Interstadial, but in the more personal Palaeolithic. They were defined not in terms of the weather but the means of production, in the form of stone tools.

Archaeology under these new conditions accepted the existence of a rigid Marxist interpretation as the inevitable framework into which everything had to fit. For that reason, a perfect fit was somewhere in the middle, no elaboration or imagination involved; as long as a detailed description of the material remains is accomplished, the framework will provide the necessary explanation. Materialism and means of production were combined, and resulted in the major focus of Czechoslovak archaeology until present: typological study of tools. Most of Eastern European archaeology paid lip service to Marxist theory without making any attempt either to understand it, to consider its relevance for social theory, or to contribute to it (Laszlowsky & Sikoldi 1991 provide an excellent discussion of the absence of theory in Eastern European archaeology see; see also Kobylinski 1991, and compare with Neustupny 1991 in the same volume). Palaeolithic archaeology solved the dilemma of balancing methodological and theoretical needs by opting for the former, and was in effect reduced to a technique of recovery, assuming a scientific appearance.

Any attempt at interpreting the social context was bracketed with Soviet sites where similar situations occurred, or which Soviet archaeologists considered significant. Evoking this master reference and others within the eastern alliance became a necessity, as most archaeologists had very little access to foreign materials, and theoretical developments in the West were unknown. All theoretical discussions of this particular time period cited the same work, Engels' *The Origin of Family, Private Property and State*, accepting his scenario of matriarchy, equality, and no private property. For example, Klíma's interpretation of the social context reads:

> The immense bone accumulation is not merely a sign of the abundance of animals in the area, but more importantly, it is the sign that numerous members of a matrilineal clan lived here for a long period of time, in prosperity, and sometimes even an overabundance of food, and that they did not feel excruciating, long-lasting hunger. Economic well-being was secured, besides gathering, by collective hunting, whose success depended on good organization, hunting strategies, and effective weapons. Even though these were in personal, individual possession, the fact that hunting was a collaborative effort, meant that the kill was communal property. The large size of the settlement and the large number of the accumulated bones fully prove the

> correctness of interpretation based on historical materialism. (Klíma 1963:97)

Stabilization of political boundaries in Europe since the second half of the 1950's resulted in a renewed cooperation with some western countries. In scientific research this led to a new attention to international publications. A longer overview of Moravian Palaeolithic research was published in 1957 in the German Quartär where Lothar Zotz was the editor, and it was in this journal that Moravian research was regularly presented throughout the 1960s, a tradition that continued when Gisela Freund succeeded Zotz as editor in 1968 (Klíma 1963, 1965, 1968, 1991; Valoch 1959). Excavations at Dolní Věstonice in the 1950's and 1960's recovered remains of a kiln with fired ceramic pieces (possibly covered by a hut-like structure), a burial of a woman, several pieces of human skeletons, and more mammoth bones (Svoboda 1994). All these attracted sufficient interest in Czechoslovakia, as well as abroad, and served as a vehicle to strengthen contacts with the scientific community in other European countries.

The most striking find came in 1986, as a part of a salvage operation during the construction of a controversial mega-dam project that flooded a vast area in the vicinity of Dolní Věstonice. A burial with three perfectly preserved skeletons drew instant attention, not only in Czechoslovakia but also in the wider archaeological community in Europe and North America. The three individuals were presented as a replacement for the skeletons from Předmostí that perished at the end of the war in 1945, allowing again for interpretations of the prehistoric people's lives: their difficult survival, their "irrational religious beliefs", their "strange burial practices" (Klíma 1987: 252). Not unlike the significant finds at Dolní Věstonice in the 1930's, the three skeletons were presented as unique objects in Palaeolithic research, providing direct information about the "prehistoric inhabitants of the European steppes" (Klíma 1987: 252). The re-establishment of Dolní Věstonice as one of the more significant Palaeolithic sites in Europe on this new empirical basis created a new interpretive space for the researchers, as well as popular writers (Auel 1990: 492-647). Yet in scientific terms the space was largely to remain empty; the human finds generated much interest, but most technical work continued to focus on stone tool typologies.

The dramatic political events of 1989 in Eastern Europe brought about remarkably little change in archaeology, particularly Palaeolithic archaeology. The site of Dolní Věstonice is now frequently visited by Western researchers, but the theoretical approach remains relatively unaffected. Typology combined with scientific techniques dominates the research, and a basic evolutionary paradigm is the motivating force. Ideology or history, whether current, past, or even prehistoric, is considered irrelevant. The recent stasis of Palaeolithic archaeology in Czechoslovakia, and Eastern Europe in general, finds support in the continuing preoccupation with stone tool typology in Western European archaeology. It also lends credence to Communist-period research, and allows the possibility of claiming scientific neutrality all along.

In 1992 a remarkable short article entitled "Who was Lothar Zotz?", by K. Sklenár, appeared in a small newsletter of the speleological society of Czechoslovakia. The author discusses the contributions of L. Zotz to "Czechoslovak archaeology" (the entity which, if archaeology were to operate without a social context, would exist as a unified, local branch of independent science), arguing that his membership in the Nazi party was only formal, and any extreme pro-German rhetoric in his articles should be ignored for the sake of the scientific contributions made:

> It is clear that he [Zotz] could not have reached his position without a membership in the ruling Nazi party; however, we all know from our own experience, how many capable people consider joining the party as a condition and a price for the opportunity to have a professional career, without internally identifying with the principles or even practices of the party. (Sklenár 1992: 16)

This article closes the circle between the practice of wartime science under National Socialism, and the post-war science under Communism, while retaining the perfect illusion of a neutral, rational scientific tradition.

Traumas born of the Second World War paralyzed most social sciences and humanities in Central Europe, a condition only augmented in Eastern Europe by the experience of the Communist experiment. The consequence has been a theoretical silence, in need of a historicizing, self-reflexive discussion. Only such debate would enable social scientists to move from disciplines defined only by techniques, to more theoretically engaged studies that directly contribute to our knowledge of the human condition, past and present.[5]

Conclusion

Of all the branches of archaeology, work on the Palaeolithic has been positioned most closely to the natural sciences, due to a combination of the material available for study and the techniques involved in its investigation. This close connection to neighboring realms such as geology has lent Palaeolithic archaeology the aura of objective science, even while the ultimate subject in question remains human and social. Within the Palaeolithic tradition this scientific connection has frequently blurred into scientism, and a belief that interpretations represent the stumbling block of proper neutrality. The stories that social scientists spin lead them astray from neutral facts and the material evidence; narrative obscures what would be otherwise clear data.

Critics of the concept of neutral science have directed most of their attention towards textual deconstruction of interpretations, analysis of the structure, and the underlying logic of the argument, presuppositions, and hypotheses. Sociological and historical studies of developments in archaeology have shown the extent to which these are imbedded in a socio-cultural context. However, this type of analysis leaves the choice of research techniques, material evidence, collaboration with other sciences, and what counts as data largely intact. A major insight of recent social study of science (e.g. Latour 1987; Shapin & Schaeffer 1985) is precisely that techniques are not neutral, and that facts, however factual, are also social constructs, and have social and political lives.

The case of Dolní Věstonice shows that Palaeolithic archaeology is just as susceptible to social influences as any other branch of archaeology. In addition, the prospect of deciphering the enigma of human origins, the solution of the time and space question of when "we" became what we are today, adds to the possibility of strengthening particular origin myths, territorial claims, and justification of modern-day social phenomena. The interpretations of the site have clearly been used to promote different agendas over time. Research in the 1920's tried to promote the newly created state of Czechoslovakia, place it on the map of Europe, and secure its position in the origins of civilization.

Nationalism played an indirect role when the site was officially located in Moravia, Czechoslovakia; its location in an ethnically German part of the country was never mentioned. The overriding sentiment was less anti-German than a desire to emphasize that the Czech lands belonged to a pan-European civilized world. This claim of European unity turned towards a Caucasian racial affinity in the 1930s, and then, after the German occupation, the key racial category was narrowed down to the exclusion of Czechs. German research focused on the origins myth, as the question of "Indogermans" came into prominence, and the site was in a sense rediscovered, and found to have been located in a German part of Europe all along. After the war, the historical and political context of communism framed a concentrated study of the means of production, affirming the correctness of the crudest form of Marxist theories. Internationalism and economic production were re-told against the background of the earliest human history.

However, as I have shown in this historical study of Dolní Věstonice, it is important to note that not only the interpretations of the site varied over time, but that scientific techniques, methods of excavation, and the choice of data shifted as well. As meticulously as Absolon uncovered a large horizontal space with numerous faunal remains in the 1920s and 30s, the German team carefully focused on the vertical space, the geological sequence, and its dating. Both methods were scientifically solid and justified, as was the post-war preoccupation with stone tools and typology. Moreover, this consistent tradition of sound methodology gave the illusion of a single, neutral, apolitical scientific enterprise evolving all along, even as researchers and local people come and go. Yet technology is no more neutral than interpretations, and facts are as social as are the words that talk about them; all can be alternately emphasized or ignored. Alison Wylie (1989) has noted that there are material constraints to interpretations, and that evidence is not infinitely plastic. In agreeing, I would add, however, that what counts as evidence seems to be more so, and that different contexts allow for the collection of different data. As the case of Dolní Věstonice/Unterwisternitz shows, even while keeping to the facts, and nothing but the facts, we are always working through history.

Notes:

[1] An earlier version of this chapter appeared in Antiquity (vol. 69/263)

[2]In following the case of Dolní Vestonice over time, I will take a historicist position in assessing the value of its interpretations, suggesting the problematic nature of the notion of "good" and "bad" science. It is important to note that such a stance does not imply the non-existence of bad or unsuitable techniques, procedures, or methods, only the absence of transcendental certainty beyond the reach of context or time. For a sampling of work in historical and sociological studies of science that informs this essay, see Bowler (1983, 1989), Clark (1991), Crawford (1992), Fortescue (1987), Foucault (1970), Haraway (1989), Latour (1987), Longino (1991), Lynch and Woolgar (1990), Rudwick (1985), Shapin and Schaffer (1985), Weiner (1989) and, of course, Kuhn (1970).

[3]In 1930 Czechs officially accounted for 51.1% of the population and Slovaks another 15.8%, while Germans (22.3%), Hungarians (4.8%) and Ruthenians (3.8%) made up the most substantial minorities, in addition to Jews (1.3%), Poles (0.6%), and, most likely underreported, Gypsies (0.2%) (Magocsi 1993: 132-3).

[4]The remark refers to an argument between Absolon, who insisted that the head depicted a man, and Keith, who was convinced that it was a portrait of a woman. Neither provided an explicit rationale for his respective attribution of gender to the object.

[5]Addendum: President Václav Havel offered in 1992 an official apology to those Germans who were expelled at the end of the war, calling it an unjust punishment based on a faulty presumption of collective guilt. His speech was not welcomed on either side: the Czechs fear retribution and possible loss of property and land, the Sudeten Germans desired far more, beyond the act of recognition of injustice.

Chapter 3: The Museum: Architecture, Representation and Culture in Vienna

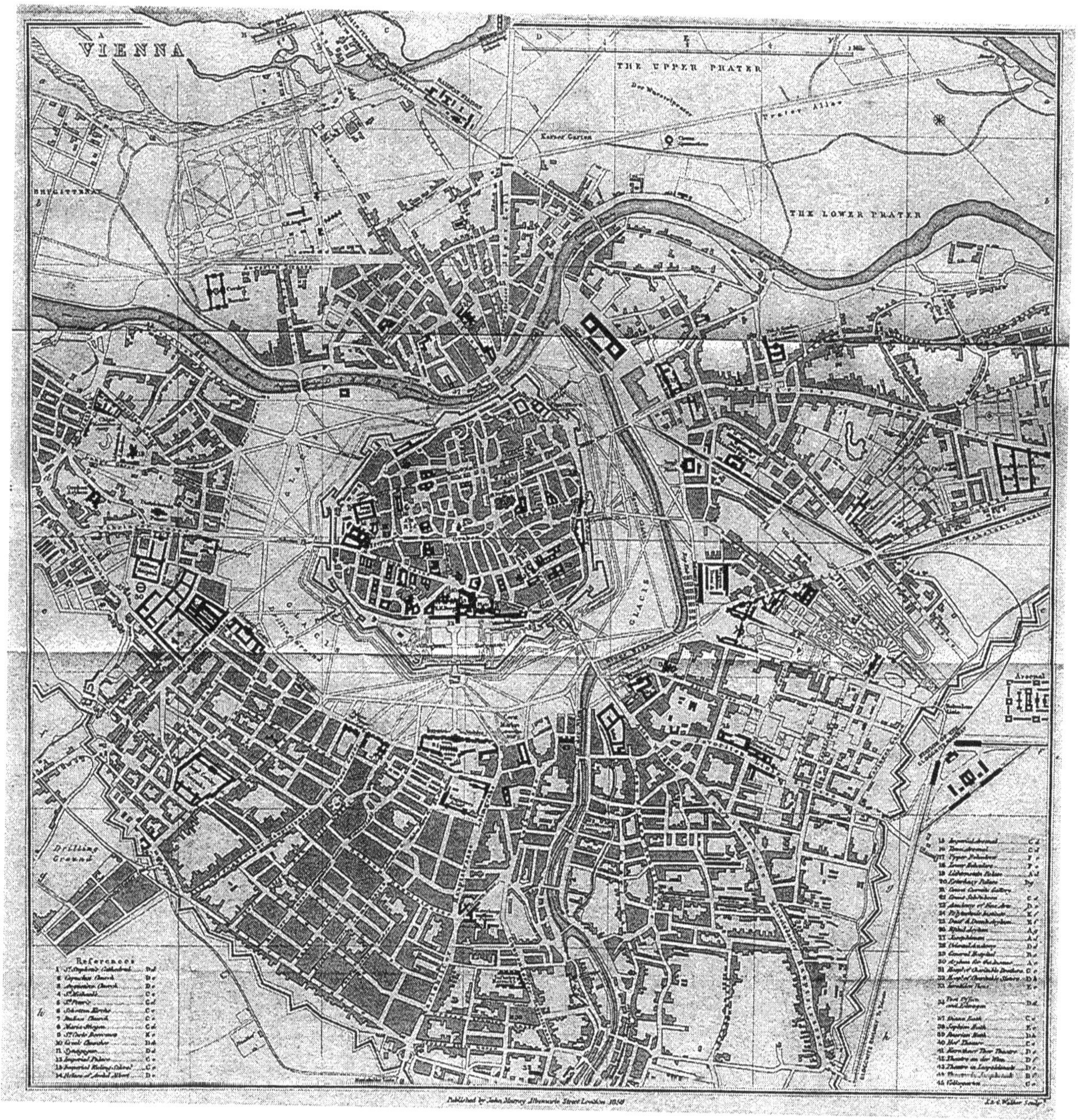

Map 3.1 Vienna in 1858, showing city walls. (source: University of Texas at Austin Library Map Collection)

The architecture of culture

On Christmas Eve 1857, Emperor Franz Joseph of the Austro-Hungarian empire issued an imperial decree, ordering the demolition of the defensive walls surrounding the city of Vienna (Springer 1979). Centuries of struggle against the Turks were now officially over, and Vienna was to become a European city, self-confident, unthreatened, no longer an eastern outpost, a bastion against the Oriental invasion, but a part of the grand classical tradition of Europe. The Austro-Hungarian Empire had numerous reasons for wanting to assert its importance and identity. Anxious over its own geographical position, tilting on the eastern edge of Europe, it had indeed been threatened historically as well as metaphorically by Turkish invasions and the Orient.

In the absence of any clear geographical boundary separating Europe from the East, the countries occupying the edge of that constructed continent, Europe, live the conflicting historical reality of being the closest "other" to Western Europe still within "Europe". Uncertainly civilized, not a part of the same social space, yet part of the same geographical space, a mix of physical allure and sophistication that changes in proportion the more east and south one travels, the loss of Christianity that increases oriental features in each nation placed on the geographic margins.[1] The Austro-Hungarian Empire established its character on a certain level of otherness and exoticism compared with other European states at the time, playing the role of the multiethnic, racially and religiously mixed oriental land.[2] Representations of southern, non-Christian peoples of the Austro-Hungarian Empire had been dominated by themes of passion overcoming reason, nature ruling over culture, sexual undercurrents constantly threatening civilized stability.

Vienna was the ultimate border, permeated from both sides. Thus the decision to destroy the overt sign of this constant threat, embodied by the defense walls against the Turks, was a major historical event. The replacement of the walls by a wide boulevard with the most notable cultural institutions: an opera house, a theater, an archive, a library, an art history museum, a natural history museum, a royal palace, and a parliament was a conscious public statement, a permanent architectural expression of the victory of Culture over Nature. The oriental threat was overcome by cultural institutions that created a boundary between Europe and the East. Looking east-ward, the Austro-Hungarian Empire was certain of its European aspirations, and made all the necessary gestures to assure its recognition among the civilized nations of the West. Yet it would take the rest of the 19th century for this major step to actually take place, and the buildings to be built, for the complete shift in social aspirations towards Europe to come to life in a new self-perception of the empire and its place in history (Springer 1979; Jelavich 1987).

Besides the larger picture of the position of the Austro-Hungarian Empire in Europe, the new city planning simultaneously addressed internal issues of class relations. Long after all other European capitals razed their medieval fortifications, Vienna had maintained them. The third largest city in Europe after London and Paris, it had grown far beyond its limits, surrounding the old town with developments, and habitation quarters. The layout stressed the preservation of aristocracy concentrated in the fort of the City, surrounded by a wall and a broad glacis, a wide stretch of open land and a deep ditch that insulated the inner town not only from the invading Turks, but more recently from the growing populace moving in from the countryside. The threat that this represented to the established order seemed similar in its consequences to that of the invasion by foreign troops from the east (Olsen 1985).

The development of the Ringstrasse, a wide boulevard with cultural institutions built to replace the old city-walls, was an architectural expression of the Austrian ambivalence towards any major change in social stratification, in the complicated relationships among the traditional aristocracy, the new bourgeoisie, the rising class of intellectuals, and the large number of lower class citizens. The historian Carl Schorske argues in *Fin-de-Siècle Vienna* (1980) that the construction of the new area around the old town was a marker of a new era, a break from the imperial Hapsburg Austria to a liberal constitutional monarchy, the accession of the middle class to power. While agreeing that the Ringstrasse project was a marker of a new era, I would suggest that the urban reconstruction, although a major shift in Viennese social history, was one still dominated by the monarchy, firmly tied to the past, and celebrating itself, more than witnessing its own decline, ambivalent about the modernity project. Vienna did not solve its identity crisis but made a bold statement about its position in Europe, and strengthened the tie between the new wealth and the old nobility by accumulating the symbolic representation of this class in the space surrounding the imperial palace, and separating them from the suburbs.

Yet the resulting accumulation was clearly not a unification, or cancellation of social boundaries. The individual monumental buildings stood in clear separation from the housing projects in the outlying areas, each designed by a different architect in a different historical style including neo-gothic, neo-romantic, and neo-classicist. Beyond Schorske's claim of the victory of middle class values, it appears to me that the Ringstrasse project made a clear statement about the connections to the classical and royal traditions of the past, even while giving Vienna a new modern look at the same time.

The new architecture of culture embodied the idea of the state as supported by the Austrian politicians, in opposition to the rising nationalism within the empire. All the significant cultural institutions in the capital Vienna, and in the empire, were crowded onto one boulevard, reproducing classical European architectural styles in the face of the growing interest in particular national styles propagated especially in Prague. The most serious cultural competitor to Vienna, Prague was increasingly becoming the center of Czech nationalism, and architecture in Vienna provided the ultimate public and permanent medium for displaying sentiments towards the past, the present, and the future of the state and the entire Austro-Hungarian empire. Vienna needed to assert its role as the capital, and to make a statement on the crucial questions of statehood. Thus the Ringstrasse project took on multiple layers of messages and meanings, even while requiring fifty years, dozens of famous architects, personal involvement of the emperor, and extraordinary financial and labor cost to reach fruition.

Besides the political edifice, the parliament, and the cultural stages, the opera, and the theater, the two largest and most important museums in Vienna, and in the Empire for that matter, were a part of the Ringstrasse project: the Museum of Art (History), Kunsthistorisches Museum, and the Museum of Natural History, Naturhistorisches Museum. They were both an important part of the project and played multiple roles on the national, international, as well as professional stages, simultaneously addressing numerous audiences. The first major enterprise carried out under the auspices of the Museum of Natural History was the excavation at Willendorf that started haltingly, without any major publicity in 1890. The archaeological site was the location of a new scientific study of geological and faunal developments through prehistory, with human antiquity only as a sidebar. Keeping this in mind, let us look at a museum from the inside, as a locus of knowledge production and representation.

The museum

Since archaeology as a scientific enterprise does not operate in a vacuum, the architectural setting as well as the display role that a museum performs are crucial not only in the final, post-excavation stages but in much earlier stages of a project, framing the questions that drive any recovery of prehistoric materials, especially those meant for museum collections. A number of critical studies have examined museums as social institutions as well as architectural sites and places of national narratives (e.g. Karp and Lavine eds. 1991; Sherman and Rogoff eds. 1994; Vergo ed. 1989). Museum practice is a rich field for lengthy debates, and thus I will limit myself here only to exhibits of prehistoric artifacts, as displayed in natural history museums. The seam between Nature and Culture is most tenuous here, challenging the ease with which such a distinction has been accepted:

> The very nature of museums as repositories for knowledge and objects of value and visual interest makes them key institutions in the production of social ideas in many nations. Museum collections and activities are intimately tied to ideas about art, science, taste, and heritage. Hence they are bound up with assertions about what is central or peripheral, valued or useless, known or to be discovered, essential to identity or marginal. (Karp et. al. 1992:7)

Only two buildings from the entire Ringstrasse project, the court museums, the museum of art and the museum of natural history, were originally designed to be linked to the imperial residence with a massive arch. The symbolism of the permanent and firm control of culture and nature by the imperial line is clearly far from subtle. All the royal art and nature collections were to be stored there, and for the first time made available for public viewing, while making it clear that they were still a part of the imperial house. The collections were divided into two distinct sets, culture and nature. All artistic achievements of Europe were deemed as Culture, and were moved to the museum of art history. The stuffed animals, geological specimens, herbaria, and objects from non-European peoples were all classified as Nature, and moved to the museum of natural history (Springer 1979). This 19th century division was not unique to Vienna, and was repeated in museums all over Europe and North America. The separation of the imperial collections into these two museums was not only a necessity created by a large and growing number of objects, but also a conscious decision that these objects belonged into separate categories. Culture was the expression of spirit that was to be aesthetically appreciated, nature was the embodiment of matter to be scientifically studied.

The two museum buildings face each other and are mirror images, their sameness unites them, makes them interchangeable. Yet at a close inspection, the decorative details, sculptures, and building materials are completely different, revealing the difference between culture and nature in the executed minute details, not the overall impression, but obvious to those who can identify the gods and goddesses of the mythical past that decorate the buildings. The art history museum is guarded by classical Greek gods in warrior like defensive poses, and goddesses embodying graceful beauty and serenity. The natural history museum is flanked by danger personified: teeth baring animals, and threatening and savage looking natives of other lands. The empress Maria Theresa, seated on a pedestal in a park between the museums, is not looking at either, facing the royal residence, the arch, the gate to the old city, the neutral ruler in between, separating nature and culture with imperial power.

The viewer and seeing

The display of artifacts reveals the role of science as a spectacle, presented in a museum for an audience that comes for the purpose of viewing. Yet the difference in such a form of spectacle, as distinct from theater, opera or film, should not be ignored. The "dialogue" between the objects and the viewers is not of the same kind as it is in a setting where both sides communicate with the same means of expression, mainly because of the passive role the materials play, but also due to the underlying system that all museums impose on artifacts. In order to bridge the temporal and spatial gap between the original context of the object, and the contemporary context of the viewer, an organizing system is imposed that erases the original context, and presents the artifacts in a rectangular table that the viewer can grasp without being distracted by extraneous cultural "noise" (Bohrer in Sherman and Rogoff 1994: 197 222 provides a discussion of Assyrian artifacts in the British Museum from a similar viewpoint). The objects by themselves, devoid of any cultural or historical connotations, appear familiar to the viewer, mainly from a functional point of view (their assumed use), and simultaneously distant, evoking a clear sense of progress that the contemporary Western world has achieved. Thus classification of artifacts on display is crucial in making them intelligible to the visitors.

Classification, then, constitutes one of the major functions of any museum, the creation of order, and the presentation of this order as somehow inherent in the objects themselves. The first basic division of items into cultural and natural domains, and their subsequent separation (spatial as well as into distinct discursive fields), sets the stage for further classification into groups that are set up in a hierarchical relationship, which is then presented as the "natural context"

(Jordanova in Vergo ed. 1989). This process not only invests the objects with authenticity but also presents that authenticity as the natural state of their existence. Moreover, one of the achieved effects of such a museum display is the creation of the viewer's context, a sense of community evoked by distancing and exoticizing, or in the opposite case, creating a sense of familiarity and intimacy with the displayed objects. In the context of the natural history museum where ethnographic objects from non-Western societies and prehistoric materials are on display, the setting evokes a feeling of inseparable community between the natural world, prehistoric societies, and non-Western cultures that remained "on the same level" as prehistoric peoples, and a sense of community that separates Western countries from the rest of the world. The distance between the nature exhibits and the prehistoric and non-western materials is only a few steps, from one display room to the next, while "real" culture is housed in a completely different building across the street.

Most archaeological exhibits of Palaeolithic artifacts are presented in the nature category, as objects from the time when culture was non-existent, when people were not only driven purely by the forces of the environment they inhabited but also they themselves were incapable of any "artistic", read purely cultural, production. Stone tools, the ultimate Palaeolithic objects, are seen to connect humans to nature, their modified forms taken only to represent a struggle for survival. The classification of these artifacts as suited for a museum of natural history preempts any discussion of their role in human cultural production, and endows them with meaning in the context of natural history.

The more subtle message simultaneously transmitted is the equation of culture with art, specifically art of the Western tradition. This move denies the existence of culture not only in the prehistoric past but also in present-day countries that have not produced the equivalent of Western paintings or sculpture (Price 1989). A similar process operates on an inter-societal level, tying cultural production to an elevated class position. The subtlety of such a meaning strengthens the power relations in modern day societies, even while it presents a very convincing image of the past.

Natural history, classification and order

In recognizing that a museum is simultaneously about science as a knowledge producing practice, and a spectacle where this practice is performed in public view (see e.g. Karp and Lavine eds. 1991; Sherman and Rogoff eds. 1994; Vergo ed. 1989), we need to avoid the all too easy generalization that would homogenize all spectacles as performance with actors and audiences involved to the same degree, or in the same type of relationship, same kind of representation. As discussed previously, the variation depends on the type of objects on display, and the contexts into which these are placed. Furthermore, the discursive field in which these objects are positioned, privileges certain aspects of perception, an outcome of a fairly recent history of classification and representation.

As Michel Foucault points out in *The Order of Things*, Western systems of classification have shifted dramatically over the last five centuries:

> To the Renaissance, the strangeness of animals was a spectacle: it was featured in fairs, in tournaments, in fictitious or real combats, in reconstitutions of legends in which the bestiary displayed its ageless fables. The natural history room and the garden, as created in the Classical period, replace the circular procession of the 'show' with the arrangement of things in a 'table'. What came surreptitiously into being between the age of the theater and that of the catalogue was not the desire for knowledge, but a new way of connecting things both to the eye and to discourse. A new way of making history. (Foucault 1970: 131)

Prior to the 17th century, the role of the historian was to re-tell all the resemblances that were hidden under a name, to evoke all the fables, legends and stories with which it had been involved. History was the entire semantic field that connected this name to the whole world without separating that which we see, what others may have observed, and what others may believe or imagine, such divisions did not exist (Foucault 1970: 129). The following century transformed entirely the way nature, and then objects of the past, were represented and displayed. Zoological and botanical gardens, herbaria, and museums arranged objects for viewing in such a fashion as to provide only rhetorical spaces for naming, so that the items themselves would speak in comparison with adjacent items. Not only was the formerly extensive narrative of verbal context forced into the background, privileging sight, but so were other senses, smell or taste, also deemed too variable. Thus classification and the rise of the natural history museum as a separate institution accomplished the ultimate visual victory. Foucault convincingly argues that in the quest for knowledge, nature was the most obvious object for "purification" and separation from other senses that were judged as distracting, and could not be so easily manipulated. The world of animals and plants became the focus of intense scrutiny not because of a new interest in exotic species, that had been there for a long time, but rather because this focus permitted a creation of a new type of controlled environment, the zoological and botanical garden, where the original context was erased, and each specimen could be studied in close proximity with other specimens in isolation (Foucault 1970; O'Brien 1989).

The main theme of Foucault's work that changes the way history may be considered, is that power is not only as a controlling process, but also a productive process that leads to "truth" (Foucault 1970: 10, 1980: 142-144). Thus the creation of the natural history museum, and its principles of organization and classification not only reflect the action of power to control nature and society, but also the creation of a new discourse about these same subjects, and created new discursive fields into which the objects of study were placed. This perspective allows us to think of science not as "representing" but rather to think of "representation" as a part of the intervention (Keller 1992:4). All forms of representation, linguistic or visual, may be perceived then as tools of intervention that reveal as much as they control.

A natural history museum is a prime example of such a site of scientific practice and a form of representation created specifically for public viewing. One step further removed from the already decontextualized zoological, botanical, or mineralogical garden, the museum displays nature in a lifeless form, stressing only the visual aspect, since the stuffed animals do not move, eat, or roar, and the desiccated plants do not bloom or smell. The ultimate objectification, removing life from the displayed nature, allows not only its study without resistance, and a display of subordination to power, but also a much easier juxtaposition with other lifeless objects and materials from human societies. The context in which they are placed is devoid of culture in the sense of live, propagating, continuing form (for discussions about connections between nature, life, women and science see e.g. Haraway 1991; Harding 1986; Keller 1992). Thus, as Evelyn Fox Keller stresses, we need to consider not only the fact that science works, classifications and organizations of objects in particular ways *do* reveal patterns not noticed before, but we also have to make the next step and ask what does it work for. It is not only the magnitude of the action that deserves attention but also the directionality (Keller 1992: 75). We need to take the next step after observing the patterns created by classification, and ask in which direction is the interpretation channeled. Placing material culture of non-Western and prehistoric societies in the same space as natural phenomena may evoke some similarities observable in all living creatures but it also tightens the link between the animal and non-Western worlds, while distancing cultural production of Western societies.

The construction of the Museum of Natural History in Vienna at the end of the 19th century illustrates the background, and the social and historical stage on which Austrian archaeology took place. The excavation of Willendorf was the first major project that the Museum undertook after its construction (Felgenhauer 1959). This did not happen independently of the reconstruction of the city of Vienna and the whole Austrian society, asserting its position in the European classical tradition.

The research at Willendorf focused on the study of ancient geology and fauna, and was thus not seen as directly related to the past of the country. The prehistoric artifacts were as much a part of nature as were the recovered animal bones. They occupied the same discursive field, and were placed not only in the same museum but also in the same context, depicting the past that lacked any connotations of culture. Culture, in the German sense of the word (*Kultur*), was restricted to an expression of refinement that not all societies possessed, tracked through art history, and ranked above superficial forms of civilization (see Elias 1939 for a detailed discussion of the development of the German concept of *Kultur* as distinct from the French concept of *civilisation*).

The actual excavation started as a salvage operation during a construction of railway tracks linking the provinces to the capital Vienna. With specific "nature oriented" scientific questions in mind, the research continued very slowly without much publicity, until 1908 when the now famous figurine of the "Venus of Willendorf" was found, and confused the research slightly, as it did not quite fit with the nature/culture division. Thus photographs were taken, and the figurine was safely stashed away in the Natural History Museum, only to be sold in reproductions as early as 1910 and a larger than life statue was placed at the location

Once the excavations were finished, a stratigraphic sequence of layers was reproduced as evidence of soil development as influenced by prehistoric human activities. Research at the site ended then, as scientific evidence was collected, and the site was not seen as significant in any way to the Great Civilization of Europe — defined then in terms of the classical tradition of Greece and Rome.

Consequently, when several decades later, Palaeolithic sites in Moravia were uncovered and discussed in terms of unique cultural phenomena, there could not have been any response from the Austrian side. The prehistoric objects were not considered products of Culture, and, unlike in Czechoslovakia, they were not in any way the pride of national cultural heritage. Inadvertently, the Palaeolithic finds from the Austrian territory were not even assigned to the same category as the same materials just across the border, preventing any comparison even on the most basic discursive level. The Austrian past of archaeological interest started only much later with the Iron age site of Hallstatt (700, 500 BC) that was clearly tied to the Mediterranean world. This social context of archaeology was set at the end of the 19th century; its legacy would be felt throughout the following decades.

Notes:

[1] Although written in a much later time-period — post World War I — an elegant example of such glaring stereotyping of this loss of civility and an increased natural and passionate savagery among southern Slavs can be found in Rebecca West's 1940 work, *Black Lamb and Gray Falcon.*

[2] Race and ethnicity were frequently interchangeable in most European discussions of European multiculturalism in the 19th and early 20th century, depending on the scope of comparison. Within Europe ethnic differences then played the discriminating role that race did in other contexts (Anderson 1983; Gellner 1983).

“What I like about archaeology is the instruments that it uses. All those magnetometers and such that can tell you what's under the ground." The old man stood in the doorway, leaning against the frame, just as he did every day when I had to work. I had the microscope, lights, computer, all set up but instead I listened to him. He had no one to talk to all winter, and neither did I. "Could a magnetometer tell what's behind a wall?" he asked, clearly thinking of a specific wall. "It depends what you want to know", I answered vaguely. "I have a wall in my wine cellar that was built in much later, " he continued. "I don't know whether there is anything behind it."

I nodded, watching him and pondering the immense distance between us. I was the opposite of anything he could have ever met. I had read about people like him, but had never known anyone personally who had been such a life-time believer in communism. Whenever he did not know how to respond to my criticism of the former regime, he cited disconnected sentences from Marx. Neither of us understood the other.

"Maybe there is a treasure", he continued, half-seriously, wondering if I could be trusted. "Just take the wall down and look", I suggested in an absurd fit of rationality. "Why wouldn't you just look?"

I was being mean, since I knew the answer. But he had me caught yet again in another web of this village's past. His house had been built in the last century by a German family. It was not large, only a spacious room with two windows, a small kitchen and a water pump in the back-yard. But the wine-cellar beneath was enormous, much larger than the whole house. Cobble stones paved its ground, leading past the huge muddy pile of winter turnips for the rabbits, on into the darkness where the wine barrels sat, and where the wall stood.

"What if it's designed to collapse on whoever tries to take it down, and get the treasure?" He was serious. Revenge reaching from the past into the present seemed natural to him. He had come to the house as a little boy, with his parents and a brother in 1946, when the Germans left the village. Empty, furnished houses were sold for low prices, and were quickly occupied. Some talked about suicides by those who could not part with their homes, and said that spirits of the former inhabitants lingered.

"What if there is someone buried with all the treasure, someone who died when they were leaving?" It was the first time he had mentioned those who lived there before. "I could never take gold jewelry from a skeleton. Well, not one in my house". The people who used to live in that one room house could not have been particularly wealthy; perhaps they had some family jewelry, but perhaps they had none.

"Archaeologists find skeletons with gold on them all the time. I saw it on television. So they could help me find out if I have one in my cellar. You have all those techniques to see what's behind walls, under the ground, where the treasures are. So what do you think? Whom should I ask?"

I wondered suddenly if there was a skeleton in his cellar, or in many of the cellars that lined the road. The old man wanted the treasure, but not the body.

"Maybe a geologist, they have the technology to find precious metals," I suggested feebly, seeking to disassociate archaeology from selective memory

(November 1992, Dolní Věstonice, Czechoslovakia).

Timeframes

Major debates in Old World Palaeolithic research concentrate on milestones that mark the origins of "modern humans". Thus the early hominids, the Neanderthals, the transition from Middle to Upper Palaeolithic all are tied to the search for the beginnings of humanity, however loosely defined. The evolutionary paradigm outlines the narrative trajectory of these debates, either explicitly or implicitly. Although the center of the earliest finds is unstable and shifts temporally as well as geographically depending on the level of humanness in question, Europe and the Levant still occupy the lead roles on the Palaeolithic stage.

The period between 35 - 30,000 years BP stands as the generally agreed upon time boundary in this area that allows for modern humans to firmly cross the line between the Middle and the Upper Palaeolithic, and permanently establish their presence. The Upper Palaeolithic opens a time slot of about 20,000 years when people move into an imaginary — as well as real - open wild space, colonizing uninhabited regions, successfully, (or occasionally less so) overcoming forces of nature.

With major steps in physical evolution accomplished and taken for granted, humans of this time period are under scrutiny for cultural achievements, progress, and general success in adapting to the new surroundings. The issue of how "new" these surroundings may have been for the Upper Palaeolithic groups is a complex one, as it is at the core of the debate of migrations or indigenous developments of the anatomically modern humans in Europe (see e.g. Mellars 1990, 1996; Mellars and Stringer 1989).

However, once established in the region, the former inhabitants of central Europe have been credited with a very few first time accomplishments, and have received outside attention only recently (Soffer and Vandiver 1994, Vandiver et. al. 1989). On the other hand, they were given autonomy by their excavators, and labeled as distinctive, different from the other groups as not simply Gravettian, but rather Pavlovian in nature (Klíma 1963; Svoboda 1990, 1991). The question then is how very different the Pavlovians were, judging from what we know about them.

Methodology and theoretical background

Prehistorians in the Old World have traditionally identified and described Palaeolithic cultures in terms of stone tool types associated with specific dated sites (e.g. Bordes 1961, 1973; Klejn 1982; Kozlowski 1986; Svoboda 1991). Analysis of the morphology of intentionally modified lithics recovered in excavations has formed the basis for the creation of discrete tool types. Cultural traditions came to be characterized by the frequency of those tool types - blades, cores, scrapers, burins, or leaf points - found at different archaeological sites (e.g. Kozlowski 1986; Otte 1981). This typology-oriented approach has dominated European Palaeolithic research since its inception (e.g. Barton 1988, 1990; Clark 1991; Sackett 1981, 1985, 1991), and has provided an initial foundation for the analysis and interpretation of the stone-tool rich archaeological record of the Palaeolithic.

At least two reasons can be suggested for the prevalence and endurance of this approach. First, on a conceptual level, prior to the development of techniques of absolute dating, a logic of typology - especially as derived from palaeontology and geology - directed efforts to place sites into relative chronologies for the purposes of comparison and identification of a space-time grid (Daniel 1963; Sackett 1981, 1991; Trigger 1989). With the increasing use of sophisticated dating techniques, typology is assumed to be just the initial step in organizing recovered materials in the field. Frequently though, final interpretations are driven by these initial categories, and the transitional nature of a stone tool type turns into an inherent characteristic, which in turn is taken to embody the maker's intention.

Yet typology poses at the same time an interesting challenge as a comparative base for dates achieved in laboratories. Frequently the stone tool types fit within a defined time frame, and correspond with the laboratory dates, confirming archaeologist's sense as to where to place the site chronologically, as well as regionally. On the other hand, all archaeologists are familiar with at least some site or collection where the dates do not correspond with typological expectations (e.g. Trnka's recovery of typologically Aurignacian material from the northern Austrian site Alberndorf, dated to 20,500 BP Neugebauer-Maresch 1993, Trnka pers. comm.)

On a socio-political level, regionalism in scientific practice in Europe led to the frequent passive assumption (e.g. see individual entries in Soffer & Gamble eds. 1990) or active insistence (e.g. Felgenhauer 1959; Dolukhanov 1989; Kozlowski 1986) on separate ethnic heritages and discrete Palaeolithic groups defined by their technology (Sklenár 1981, 1985; Trigger 1989). Since Kossina's use of the "cultural region [*Kulturgebiete*]" (Meinander 1981; Trigger 1978; Veit 1989), the territorial inscription of culture has remained in use either explicitly or implicitly in the use of the concept of archaeological culture (Klejn 1982).

Archaeology and the "cultural tradition"

Acceptance and development of the concept of the archaeological culture took different routes in American and European prehistoric studies (Meinander 1981; Shennan 1989b; Trigger 1978; Veit 1989). My research, informed by professional training in North America, primarily focuses on the European aspect of the question. Traditionally, archaeologists addressing the problem of culture focused on the material remains, and their spatial distribution was interpreted as an "archaeological culture". Where typology provided the temporal frame and ascertained the antiquity of the people being studied, the concept of archaeological culture added the spatial component that was provided by the geographic distribution of the identified artifacts (Miller 1983).

G. Kossinna's proposition in 1911 of a "settlement archaeological method", based on typological method, was met with wide acceptance in Germany and Kossinna introduced the notion that "sharply defined archaeological culture areas correspond unquestionably with the areas of particular peoples or tribes" (Veit 1989: 37). Most archaeological finds in Germany at the time were post-Palaeolithic and the recovered pottery, ornaments, and house forms found recurring together were interpreted as displaying distinct ethnic styles that were traceable geographically (Arnold 1990; Eggert 1978; Girtler 1982; Meinander 1981; Veit 1989).

"Archaeological culture and settlement" was subsequently adopted by European and American archaeologists and has influenced most interpretations of recovered assemblages (e.g. Childe 1935, 1956; Willey 1956). The interest and preoccupation with ethnic differences remained the underlying premise in European archaeology, and although Palaeolithic research has been largely spared of the direct historical approach, it accepted the methodology that is rooted in it. Boundaries between different groups are drawn on the basis of changing proportions of tool types and ethnic labels are replaced by the names of the major sites in the region. All the classic types of "industrial cultures" have eponymous sites; for example, the Gravettian is named after the French site La Gravette, hence the Pavlovian (named after the site Pavlov) to which Dolní Věstonice at the Pavlov mountains is assigned, and the Willendorfian, after Willendorf II (Broglio & Laplace 1966b; Felgenhauer 1959; Klíma 1963, 1976; Kozlowski 1986; Svoboda 1991; Valoch 1986). An opinion that is held among many archaeologists is that there exists a direct relationship between archaeological cultures, on the one hand, and linguistic and ethnic entities, on the other (Dolukhanov 1989:265).

The typological and archaeological culture tradition approaches that have served methodological purposes of recovery and analysis have channeled interpretations into a delimited range of possibilities (Sackett 1985; Shennan 1989b; Wobst 1990 a, c). They tend to treat archaeological sites in isolation and have mainly served to confirm the constructed discrete categories, leaving very little room for models of social behavior of the prehistoric peoples who were after all, more often than not, active participants in their interactions with the natural and social environments. My research employs the same empirical evidence used in previous research but it examines the acquisition, production, use and discard of lithic material as a social activity in a specific environmental context (see e.g. Montet-White and Holen 1991). The case of comparison of Pavlov and Willendorf II offers an excellent opportunity for such a study, since differences in geography and time are minimal, while the amount of material recovered is quite large.

Typology as a theoretical subject and as a method.

Despite archaeology's heavy reliance on classification, it has had an ambivalent relationship with the subject from a theoretical standpoint (Dunnell 1986; Ford 1954; Hegmon 1992; Hill & Evans 1972; Spaulding 1953). Only a modest amount of the literature addresses the topic, generally treating it as an unexciting, if necessary evil (Binford 1965; Watson et al. 1984), although some have argued for the critical importance of classification for the field (Dunnell 1971; Whallon and Brown 1982). Generally, typology is considered to be a subject within the domain of methodology with little attention given to the theoretical issues that underlie the choice of a particular kind of a classificatory system (Adams & Adams 1991; Dunnell 1971, 1986; Gardin 1980; Klejn 1982).

Types, constructed as discrete units, which archaeologists devise in order to sort artifacts into manageable groups, are based on the notion of boundedness. Empirical entities - artifacts - are categorized on the basis of some criteria of recognition that involve a clear cut separation from other entities. Ever since the Linnaean classificatory system, there has been a tendency to equate boundedness with natural types; the sharper the boundary between categories the more natural and thus, the more real the boundary appeared (Adams & Adams 1991; Dunnell 1971, 1986).

Archaeological typology accepted this premise as a methodological tool but implicitly incorporated it into explanatory models of prehistoric social systems, structures and behavior, despite the fact that cultures do not display the sharp boundaries suggested by the biological analogy (e.g. Binford 1973; Chang 1967). Dunnell has argued that the problem lies in the confusion of type as empirical entity and type as a tool of measurement, designed to make the empirical experience meaningful (Dunnell 1986:152). Thus hypothetical models that are built to account for the patterning of the archaeological record are constrained and bracketed by the categories that were set up in the process of recovery and analysis. The boundaries are written into the materials only to be subsequently discovered and explained. When a group boundary is the research objective, then classification methods are employed to locate the precise point of disjuncture (Conkey 1982; Dunnell 1986; Green and Perlman 1985).

In the late 1960's, typology and classification came under criticism from practitioners of the "New Archaeology" (see discussion in Conkey and Hastorf 1990; Trigger 1989). The classificatory approach used previously for chronological reasons to design a temporal sequence of cultural groups, was now seen as inadequate for dealing with issues of human behavior and cultural processes in the past. In an attempt to strengthen the scientific and behavioral value of archaeological methods, several leading American scholars suggested that the different categories and groups of tools reflected different functions rather than ethnically linked styles (Binford 1973; Leone 1972). The categories themselves were not questioned, rather their meaning as cultural markers was replaced by a functional interpretation. At the same time, statistical approaches to variability validated not only the notion that the boundaries between categories were real but also that they were measurable which only further strengthened their perceived existence.

During the 1980's the stylistic-functional debate has been focused in a new, promising direction that stresses the production, use and maintenance of lithic artifacts rather than formal tool types (see e.g. Barton 1988, 1990; Dibble 1984, 1987, 1988; Geneste 1990; Jelinek 1988, 1991; Knutsson 1988a, b; Kuhn 1991, 1992; Rolland 1981, 1988; Stiner 1992)[1]. These studies, dealing primarily with materials from Middle Palaeolithic sites, have shown convincingly that behavioral inferences can be made if the narrow categories of "tools" are reconsidered (e.g. Barton 1990; Dibble 1988; Neeley and Barton 1994, Rolland and Dibble 1990). Although this research has been concentrated on the

Middle/Upper Palaeolithic transition, theoretical issues addressed in these studies have a much wider applicability, and I draw upon them in my work. By means of an empirical study of lithic materials from Pavlov and Willendorf, this research tests the validity of stone tool typology as an analytical basis for definitions of archaeological culture. In effecting a comparison between the lithic remains of two site complexes relatively close in temporal and spatial location, the study contributes to the issues involved in Palaeolithic style and/or function debates.

Use-wear analysis

Over the last 100 years, numerous studies have attempted to determine the purposes for which stone artifacts were made or used (see reviews in, e.g. Olausson 1980; Sieveking & Newcomer 1987; Vaughan 1985). Methodological approaches to prehistoric stone tools have undergone a remarkable expansion since the 1960's, when Semenov's Prehistoric Technology (1964) appeared in English translation. Despite the fact that a number of researchers in the former Soviet Union were already working on the problem of use-wear traces, particularly G. Korobkova in then Leningrad[2], it was Semenov's work that became the most influential and is cited as the classic text in use-wear studies.

Historical materialism, the only theoretical approach conceivable in the former Soviet Union in the late 1950's, channeled all studies of prehistoric technology into investigation of means of production, as a way of revealing the classical stages of societal development as described by Morgan and adopted by Marx and Engels. Sterile as this approach may have been from a theoretical viewpoint, it resulted in innovative studies of prehistoric tools, particularly the examination of the traces created in the process of tool use. Microwear studies have gained momentum and contributed significantly to the potential of lithic analysis.

Microscopic observation of technological, natural and use-wear traces on experimental and archaeological artifacts have provided insight into stone tool manufacture as well as into the manner of use and materials that were worked (e.g. Amick & Mauldin 1989; Crabtree 1972; Hayden 1979; Keeley 1980; Moss 1983; Odell 1977, 1982; Swanson 1975; Tringham 1971; Tringham et al 1974). The initial enthusiasm for the method as an omnipotent answer to all questions related to lithic materials has abated in recent years due to some inconclusive and problematic studies that have been a result of an uncritical application of the technique (for criticism see e.g. Cross 1983; Gero 1983,1989; Hayden 1979; Odell 1982). Nevertheless, increasingly complex behavioral questions may benefit from the insight of use-wear analysis.

Primarily, wear traces such as edge damage, polishes and striations that result from a contact between the tool and the worked material, and their characteristics provide a solid ground for behavioral inferences that are conducive to interpretations of the social context of technology (e.g. Odell 1980, 1988; Gero 1988; Perles 1987; Wright 1973). Microchipping and striations, though not use-specific, serve as a good general indicator of the mode and intensity of edge utilization (e.g. Dumont 1982; Mansur 1982; Odell 1982; Tringham et al. 1974). Microwear polishes tend to correspond with the type of worked material, providing valuable information regarding the nature of subsistence and materials with which the prehistoric people came into contact (Keeley 1980, 1987).

Keeping in mind the complicating factors of post-depositional damage and the technical problems inherent in low and high-power microscopy (see e.g. Juel Jensen 1988; Keeley 1980; Odell 1979, 1982), I am confident that a cautious application of medium range magnification (90x and 300x) using an incident light microscope provides us with the best compromise between time invested and the acquisition of desired results.

Study collections and lithic assemblages

The lithic collection from Pavlov is located at the Archaeological Institute in Dolní Věstonice, the Czech Republic. I was based there for the duration of 10 months in 1992, during the first part of this project. The Pavlovian assemblage that I selected for study is a part of the material recovered by Bohuslav Klíma during the 1952-53 excavation seasons. The entire recovered collection amounts to 52,883 lithic pieces (Klíma 1954). Klíma selected, typologically identified, and entered 5379 pieces into the site catalogue. This catalogue inventory constituted the study sample that I analyzed in its entirety. In addition I selected a sample of 400 lithic pieces from the remaining material so as to reduce the bias involved in the original selection, since this was made solely on the basis of stone tool types. My project was a part of an on-going overall site analysis directed by Jiri Svoboda at the Archaeological Institute (Svoboda 1994).

The material from Willendorf II is permanently stored at the Naturhistorisches Museum in Vienna where Dr. Bart, the head of the Prehistoric Division, and Dr. Walpurga Antl, who is a resident Palaeolithic specialist, were of great assistance. During the original excavation in 1908 and subsequent studies, a clear stratigraphic sequence was recorded only at Willendorf II where Layers 1-9 were identified (Felgenhauer 1959: 91-4). Layers 5-9, which I have selected as a focus of my study because of their contemporaneity with corresponding layers at several Pavlovian sites, were identified as containing a "Gravettian culture of western as well as eastern types with specific Willendorf culture in Layer 9" (Felgenhauer 1959: 211; see also Broglio & Laplace 1966; Kozlowski 1986; Montet-White 1988a). The entire collection from this portion of the sequence contains 7000 lithic pieces (formal tools and débitage), out of which 2712 pieces have been typologically identified and recorded (Broglio and Laplace 1966). In 1992 and 1993 I analyzed the entire recorded collection that (with the help of Dr. Antl) I was able to locate in the storage of the Museum of Natural History in Vienna. This totaled 2631 lithic artifacts.

Methodological steps

The analysis of the lithic collection from each site, Pavlov and Willendorf II proceeded in a number of linked steps. After the initial raw material identification, I recorded and examined all the lithics, including both formal tools and débitage. Since the aim of this research was to compare the two collections, I assessed and recorded specific attributes in each assemblage. The data recording procedure followed that designed by G. Odell (1979), constructed parsimoniously while capturing all the necessary information for the questions addressed by this research.

I. The following formal attributes were noted:

a) size — all artifacts were measured to the nearest 0.1 mm using sliding calipers, recording length, width, and thickness

b) retouch — recording location as well as faces of retouch (uni- and bifacial) (discussion of reasons see e.g. Barton 1988, 1990)

c) angle of the working edge — acute and obtuse edges as these are subject to different patterns of use-wear (discussion e.g. Vaughan 1985)

d) fragmentation — noting the location on the flake blank (proximal, medial, and distal)

e) hafting — lateral or proximal as an indicator of possible functions (discussion e.g. Keeley 1987)

f) rejuvenation — recording all traces of curation as an evidence of intensity of use (discussion of reasons Barton 1990; Binford 1979; Dibble 1988; Hayden 1975)

The use-wear analysis proceeded in two stages. First I spent ten months at Dolní Věstonice, due to the size of the Pavlov collection, which was followed by a six month stay in Vienna at the Museum of Natural History, working on the Willendorf II collection, examining lithics from layers 5 — 9. I used Meopta microscope with an external light source with two magnifications 75x and 200x for the use-wear examination of the Pavlov collection. The analysis of the Willendorf II materials was carried out with a WILD M32 Kombistereo microscope that had a working distance of 11 mm. Four external light sources provided sufficient illumination for the observations needed. As with the previous collection, I used two magnifications — 60x (field diameter 3.3 mm) and 160x (field diameter 1.3 mm).

II. The following microwear attributes were recorded (following e.g. Dumont 1988; Moss 1983; Perles 1987; Vaughan 1985):

a) discrete utilized portions per utilized piece — an indicator of use intensity

b) use-wear traces — distinguishing four broad categories: microchipping, striations, micropolish, and edge rounding that correspond to general categories of worked materials

c) location of the use-wear traces — lateral, distal, ventral, and dorsal as indicators of possible functions

d) characteristics of the use-wear traces — relative size, distribution, and orientation of the microchipping, striations, micropolish, and edge rounding as indicators of possible functions, worked materials, and use intensity (Dumont 1988; Vaughan 1985)

e) intensity of wear traces — only general categories of weak, medium, and strong were recorded as this attribute served only as a control variable, due to the high frequency of non–use causes (Amick and Mauldin 1989; Moss 1983; Keeley 1980)

Use-wear traces that resulted from the suggested prehistoric activities could be best described in the following terms:

a) polish — gradual loss of surface material and smoothing of the surface, resulting in a reflection of light. Varying intensity, appearance, and extent, most common in woodworking and plant working.

b) rounding — consistent loss of a sharp edge, caused by the contact material and use motion. Location (ventral or dorsal surface) and degrees of rounding were noted, as they contribute to information particularly in transverse actions. Varying appearance was noted depending on the worked material.

c) striation — appearance of linear scratches and grooves in the direction of use motion, noting width, depth, and quantity. Depending on the amount of abrasive particles, it is pronounced with dry and frozen materials.

d) microflaking — various forms of crushing, nibbling, step and hinge fractures, that leave scars along an edge, common in bone work, distinct in frozen materials. In addition to worked material and use motion, the main variables affecting microflaking are edge angle, pressure, contact angle. Consistency and localized scars were considered as major criteria, so as to eliminate random microchipping caused by natural forces and trampling.

e) residue — organic remains adhering to lithic artifacts, observed in the Pavlov collection in the form of pollen and hafting resin.

Experimental study

Prior to the examination of the archaeological lithic artifacts, I carried out an experimental project that was further supported by evidence from similar work by other researchers (e.g. Keeley 1980; Knutsson 1988; Tringham et. al. 1974; Vaughan 1985). Blades made from cobbles of flint recovered at Pavlov were used in the experimental work, so as to achieve results most closely resembling those observed in the archaeological materials. Motions parallel to the edge included cutting, sawing, and slicing. Motions perpendicular to the used edge included scraping, planing and chopping. In each case the type of activity was recorded, and carried out for a predetermined duration, and measured in terms of minutes on a cumulative basis (10, 20, and 30 minutes). The initial observation point was prior to the use of the implement, and was followed by regular observations, in order to determine developmental patterns. In addition to the worked materials, I noted all the variability involved in the different activities, recording edge angle, utilized edge, contact angle, contact surface, and the mode of action.

Worked materials were chosen so as to include a wide range of potential prehistorically used substances. Soft wood (pine) was selected on the basis of the palaeoenvironmental evidence for mainly evergreen trees present in the area. Non-woody plant materials included wild grasses (Gramineae), acorns, dandelion leaves, and the pith of elder. Butchering was performed on fresh meat products, long cow bones, and deer antler. The antler was soaked for five days prior to the experiment. Cow hide was scraped and cut in a fresh state. Frozen materials were also subjected to the same type of activities as the fresh materials, and those in the experiment included hide, bone, and meat. All materials, prior to use, were exposed to temperatures below -15°C for five consecutive days.

The soft plant material had to be worked for at least 20 minutes in order to show even minimal microscopic effect (under medium magnification) on the edge used, underlining the necessity of approaching the category of "unused" lithics cautiously. When interpreting the spatial distribution of activities that left traces on stone tools, it is thus crucial to consider "empty spots", areas with unused lithics, as possibly areas of sites with activities of a short duration. The differential formation of edge damage may also influence the interpretation of the types of activities that were carried out at the site, giving undue importance to processing of harder. materials, e.g. bones, antler etc. (for discussion see e.g. Bamforth et. al. 1990, Levi-Sala 1986).

A total of 50 pieces were incorporated into the experimental study, five each for each category: bone, antler, wood, reed, plant matter, meat, hide, frozen hide, frozen meat and frozen bone. The results enabled a broad separation of the worked materials into five groups:

1. coarse, hard/frozen material — e.g. shell, frozen meat, bones, and hides

2. soft resistant material — e.g. wood, tough roots

3. soft material — e.g. plant, fat

4. fresh tough material — e.g. meat, fresh hide

5. fresh hard material — e.g. bone, antler

Data presentation

Measurements and observations were entered into a data base program that provided all the basic and necessary tools for sorting and grouping of the data. A spread sheet program was used for calculations and graphics. Standard deviations were calculated for all measurements and are included in the discussion of size and use-wear correlation. Since the research focused on the theoretical issues underlying the typological debates rather than a search for normative patterns, I did not enter into any statistical manipulation of the data base beyond necessary pattern identification and analysis.

Interpreting use-wear in space and time

There are several ways how one could approach the interpretation of results of a use-wear analysis. Most common is the functional interpretation that describes the possible uses to which lithics at a site could have been used, either as individual pieces, or as composite tools (a number of excellent studies are available see e.g. Hurcombe, 1992; Knecht 1988; Plisson 1985; Shea 1993, 1998). More technically focused research contributes to our understanding of the trace formation on the lithics (the number of experimental studies include e.g. Keeley 1987; Knutsson 1988).

I have examined the Pavlov and Willendorf II assemblages from the perspective of the worked materials, so as to investigate the possibility of interpreting site activities not as much in terms of site function, i.e. hunting or residential camps, but rather in terms of seasonality. I would argue that such an approach may be more productive in terms of prehistoric behavior, and would at the same time circumvent the typological quagmire. The interpretation of the use-wear in the context of seasonal activities offers a new perspective on some of the noted differences among the sites, and the occupational layers. A much more detailed study with numerous additional lines of evidence, including faunal, macrobotanical, pollen analyses, would be necessary to provide a solid case. Yet I feel optimistic about such a possibility, and consider the study described in the following chapters as a contribution to such a larger project.

In the case of the Willendorf II research, I was working with museum collections, and I could only rely on information available to me from the records of the excavations, without being able to either have any impact on the way the information was obtained, nor being able to collect additional data. Therefore I was not able to place the Willendorf II collections into any further context, as the excavations focused on the fauna and geology from a comparative perspective in the broadest sense, comparing the prehistoric species with those from the present, without keeping any detailed record for individual layers (Felgenhauer 1959). Furthermore the recovered lithic artifacts were only identified as to the layer from which they came but not in any detail in terms of their spatial location within the occupation itself. The detailed spatial study presented in this text includes then only the Pavlov assemblage recovered in the 1953 excavation season when the site was mapped in 1 meter squares, and all the related features were recorded as well.

Notes:

[1] However, it has to be noted that post-Pleistocene research addressed these issues much earlier (Tringham 1974, 1978), due to the fact that social factors in the Neolithic time period were considered more accessible and worthy of study.

[2] I thank Ruth Tringham for this information.

Chapter 5: Pavlov: The Site and the Collection

The wine country

Turning off the main road that links Vienna and Brno, we pass through vast expanse of vineyards. Brownish, thin vine branches stretch as far as the eye could see, a barren slope all winter. Vienna, Mikulov, Horní Vestonice, Dolní Věstonice, Pavlov, Mílovice...

Mikulov, a border crossing to Austria nearer Vienna, is now mainly used to get across the border without a long line of cars that most other checkpoints have. The head rabbi for the entire Moravia settled here at the beginning of the 18th century, a member of a large Jewish community that lived in town since the 14th century.

I knocked loudly on a heavy wooden gate. No one answered. I stood outside for a while, when a young woman leaned out of the window "What do you need?" I asked for the key, and she sent me several streets further on, to another house. My mother did not think they would give us the key anyway, so she settled under a tree by the cemetery wall. I knocked on the gate and an older woman came out. She looked at me without a trace of friendliness, but went back to get the key. I signed my name and my mother's address into the slightly worn out notebook, and received the key. I could have written any made up name, I thought to myself. Strange mistrust combined with a total trust in the written word.

I opened the lock on the metal gate, and we went in, locking again behind ourselves. "This grass is in a bad need of a scythe," my mother said, as we walked across the cemetery, uncertain where the path once might have been. We looked at the tombstones, most fallen, hard to decipher some from the 1930's, but many centuries old. The whole place was overgrown by bushes, trees, and wildflowers, very quiet, hardly ever visited. "The second largest Jewish cemetery after Prague, the burial place of numerous Talmudic scholars from the local Yeshiva school that functioned here for centuries..." my mother, unsure what to look at, was reading from a guidebook. A few pebbles lined one burial, memories revisited. The Catholic cemetery was on the other side of a crumbling stone wall, paths well kept, flowers and trees watered in the summer heat. It was still alive in local memory, as there was a large community to keep it.

Afterwards I knocked on the gate again but nobody answered. I waited for a while, then dropped the key into the mailbox. An old woman watched me from behind the curtains as I walked by her windows.

Continuing along an empty concrete road, I watched rabbits running criss-cross through the fields, pheasants blending with the surroundings, hawks slowly circling in the sky, before throwing themselves head-down to catch mice. In summer the slopes turn thickly green, rabbits confidently sitting up, deer hiding in the rows, and tractors going up and down spraying vines with nasty smelling pesticides. The road, lined by a few old stone or metal crosses, and wine dungeons locked with rusting padlocks, follows the river just below the slopes of the Pavlov hills.

(Summer 1993, Mikulov, Horní Vestonice, Dolní Věstonice, Pavlov — Czech Republic)

Pavlov: An overview

The excavation started in 1952, revealing a promising site with a large number of recovered artifacts and features. As of 1953 systematic research began, uncovering a large settlement in 1x1m squares. Within the first season this resulted in an extraordinary 39,142 pieces of lithics, and by the end of the excavation (after 21 seasons), rendered over 60,000 catalogued and numbered items, and a total of close to a 1,000,000 artifacts (Klíma 1994). As a site, Pavlov was much larger than any of the Dolní Věstonice finds (or most Palaeolithic finds, as a matter of fact). Unlike the earlier excavations at Vestonice, it was excavated and spatially recorded in a systematic fashion that even forty years later it could be reconstructed and analyzed. Up until 1994, however, all published statements about the nature of the collection, its relationship to other Moravian, as well as other central European sites, were based on preliminary results obtained in the process of cataloging. Due to the large size of the recovered material, the actual analysis of Pavlov began only in 1992, and the thesis proposed here is based on the material from the 1953 season.

The term "Pavlovian" was first defined by B. Klíma in 1959 as a specific local variety, or a cultural phase of what was then known as the "central European", "eastern", or "lower Austrian - Moravian" Gravettian (Felgenhauer 1952, 1959, Klíma 1959). During the 1950's, the main issue in European Palaeolithic archaeology was to refine the chronology of the Aurignacian techno-cultural definition that spanned across the Middle and the Upper Palaeolithic, and distinguish it from the subsequent Gravettian sequence.

Despite Felgenhauer's call to accomplish this task in terms of settlement patterns, economic, and social life of the prehistoric populations, as well as typology the differentiation continued to be based on typological grounds (Felgenhauer 1952). In addition, observed local variation of the Aurignacian — leaf shaped points that did not occur in Western Europe during that time period — gave grounds for proposing independent cultural evolution in central Europe[1]. The "Szeletian", named after the Hungarian cave Szeleta where a large number of the leaf points was found, was coined in 1953 by Prosek, providing a convenient stepping stone for a claim of separate development in east central Europe (Prosek 1953).

Once Felgenhauer outlined the concept of the eastern Gravettian (1952), and Prosek suggested a central European variant even in the previous Aurignacian period (1953), it seemed almost obvious that local Gravettian groups should be accounted for as well. Klíma proposed that the eastern Gravettian should be understood as a large, territorially bound, unified cultural complex that consisted of smaller local units, groups, and types (Klíma 1959:41).

The only remaining task was to choose the eponymous site that would name the local groups. Despite the fact that the term "Aggsbachian", proposed by the Austrian archaeologist J. Bayer in 1928, was considered outdated, the new proposed cultural group generally comprised the same territory, but now carried the name "Pavlovian". It was suggested that Moravia was the center of the area, and Pavlov, together with Dolní Věstonice, had the longest continuous sequence, and in addition to being systematically excavated and well recorded. Pittioni and Delporte were credited with proposing the term (Klíma 1959:43). Besides Dolní Věstonice, Pavlov and Mílovice in southern Moravia, Předmostí (one of the largest, and possibly most complex sites)[2], and Petrkovice in northern Moravia came to be considered as settlements of the Pavlovian culture.

Archaeological layers at Moravian sites under discussion were all formed during a long time span (29-20,000 BP), with a minimal deposit of loess (Smolíková 1991). All layers are composed of matter that is mainly a result of anthropogenic activities, making differentiation between separate occupations rather difficult (Smolíková 1991). Consequently undisputed contemporaneity, or its absence, is much harder to establish at the Moravian sites. This contrasts with the situation at Willendorf II, where the subsequent layers are separated by thick soil deposits, creating a very clear stratigraphic sequence. Horizontal movements within the layers at Pavlov are apparent but archaeological features were not affected by them to any extensive degree (Svoboda 1994:199).

Three samples for C-14 dating were taken in 1956, producing consistent results (Svoboda 1994:210):

GrN 1272: 26,620 ± 230 BP
GrN 1325: 25,020 ± 150 BP
GrN 4812: 26,730 ± 250 BP

Raw materials

The raw materials from which all the artifacts at Pavlov were made can be described in the most general terms as silicates. This group, the largest and most common among minerals, includes 40% of all known rocks, making them very hard to identify with a high degree of specificity. In terms of characteristics, silicates average higher in hardness (7 on a scale of 10), and lower in specific gravity than most minerals (Pough 1971:229, Sieveking and Hart 1986).

Quartz is the most common mineral in this group, found in every class of rocks, forming under numerous conditions, but always a result of a chemical process in sedimentation. Chert, flint, and jasper, all present in the Pavlov collection, contain microscopically grained quartz, a relatively high degree of impurities, and their fibers are not banded together in a linear pattern, which would result in translucency.

There does not seem to be a general agreement on the relationship between chert and flint. As Luedtke notes, some geologists and archaeologists have used the terms interchangeably, while others have argued that they are two different materials, although some have defined chert as a variety of flint, and others consider flint to be a variety of chert (Luedtke 1992:5). The major distinction in the use of the terms seems to be a result of different scientific traditions[3]. American geologists view chert as the more general category, with flint and jasper as its varieties. British researchers distinguish between dark, higher quality flint found in Cretaceous chalk formations, and the lighter colored, lesser quality chert nodules in limestones (Luedtke 1992; Sieveking and Hart 1986). The use of both terms is much more loose in archaeology where visible attributes such as color, translucency, or texture, are the main criteria for classification, as partly they are considered to have been more relevant to the prehistoric peoples, and partly the raw material alone is not the main subject of a study.

Jasper, another siliceous material that appears in Palaeolithic collections of central Europe, is a chert with a high iron content. However, even geologists do not agree on the percentage of iron to justify the label. In addition, archaeologists tend to use the term for cherts of mustard or reddish color, yet high iron content may result in green or black color (Luedtke 1992). These inconsistencies in terminology should be kept in mind when comparing collections described by different researchers, such as Willendorf II and Pavlov.

Sources of raw materials

Lithic source analysis, the determination of the geographical origin, has become one of the more worthwhile inquiries in Palaeolithic archaeology. It provides information that is potentially useful in approaches to such crucial topics as prehistoric cultural contacts, exchange, mobility, geographical sensibility, or access to resources. Sources of minerals can be determined through various analyses, including petrographic study of thin sections, X-ray diffraction, trace element spectroscopy of both the archaeological materials, and of samples from potential quarries, with the accepted level of uncertainty (Luedtke 1992:118). Since cherts are some of the most common minerals, a large number of source samples is always preferred.

Despite the fact that geology has enjoyed a favorable position in scientific research in central Europe, the situation is complicated by a lack of systematic tests of a large number of chert outcrops in a wider region. As Takács-Biró notes in her evaluation of work in Hungary, that is relevant to all of central and eastern Europe, little petro-archaeological study has been undertaken as yet, and this has led to a limited understanding both of the lithic raw material types used and of their sources (Takács-Biró in Sieveking and Hart 1986:120). More recent work conducted by Kaminská (Kaminská 1991) explored the importance of raw material sources in Eastern Slovakia for the wider Central European context. This work outlined an intensive long distance exchange/contact between Slovak, Polish, Hungarian, as well as Ukrainian groups that occupied this region during the Upper Palaeolithic. The detailed study of mineral resources in eastern Slovakia and their relationship to Upper Palaeolithic sites in the region suggested a far more central role of the area than has been previously known (Kaminská 1991).

Tests conducted in the Moravian context included 10 pieces from the Pavlov collection, and only 8 pieces from Polish and eastern Slovak sources. This low number does not warranty any plausible interpretation, as the results are too few and varied to be indicative of even tentative trends. Only the most general location of some long-distance chert quarries is known, and references are frequently made to the several broad areas in the same eastern and northern directions of Poland and Slovakia.

Although more geoarchaeological work has been carried out in Hungary, so far little effort has been made to collaborate on long-term problem-oriented projects. Apart from a systematic research of obsidian sources in the Carpathian mountains of eastern Slovakia and Hungary, carried out mainly by British and Hungarian geologists and archaeologists (Williams and Nandris 1977, Williams, Thorpe, Warren, and Nandris 1984), no Upper Palaeolithic archaeological materials from Moravia have been compared to Hungarian mineral quarries. This area to the southeast would have been geographically close, and the presence or absence of contacts should be investigated, rather than assumed. Very little is also known about sources in Austria, Italy, Bohemia, or even further in Germany. This still remains a question for future research, that demands attention, since the raw material used at Pavlov is not of local origin (Prichystal 1994), and the claim for long-distance travel or contact can be further strengthened by the presence of Mediterranean shell in the archaeological context (Svoboda 1991).

The typological approach

Analysis of the abundant Pavlov material has still to be completed, with data only from the 1952-3 excavation season completely processed. The results are the most recent addition to our knowledge of the Palaeolithic in eastern and central Europe, and have been summarized by J. Svoboda (1994). Therefore I will present his views as representative of the research carried out in the region, and will include some other recent opinions on the data from international scholars who incorporated his results into their work.

On the most general level, Svoboda combined typology of the Gravettian sites with environmental data, and divided the eastern Gravettian into three stages:

1. Early Pavlovian 30-27, 000 BP,
2. Evolved Pavlovian 27-24, 000 BP,
3. Willendorfian-Kostenkian 24-20, 000 BP

The first stage of this scheme is represented by Layer 5 at Willendorf II, together with lower parts of Dolní Věstonice I and II. Typologically, according to Svoboda, this phase is dominated by burins, backed implements, and endscrapers, with a few geometric microliths, and "Aurignacian-like" (meaning thick) endscrapers. Pollen analyses from Dolní Věstonice indicate temperate climate and steppe-like vegetation (Svobodová 1991).

The second stage is defined as layers 6-8 at Willendorf II, middle and upper parts at Dolní Věstonice I, and some parts of Dolní Věstonice II, Pavlov, and Předmostí. The one common feature at all these sites is a rich bone tool industry, but the lithic typology is further divided into two facies on the basis of retouch on blades and flakes. Geographically these two suggested divisions do not fall out evenly, as Willendorf II, Předmostí, and some parts of Dolní Věstonice are grouped together, and are contrasted with Pavlov and a few other parts of the localities at Dolní Věstonice. Svoboda maintains this division also in the perceived stylistic differences in "art", separating geometric decoration from curved linear design.

The third stage is represented by Layer 9 at Willendorf II, Petrkovice, sites in western Slovakia, Spadzista in Poland, Layer 7 at Molodova, layer 1 at Kostenki, and Avdeevo. The one typological feature that unites them all is the shouldered Kostenki point. Some sites have leaf-shaped points that used to be considered Szeletian, i.e. re-appearance of a pre-Gravettian type, but the new chronological division of the Pavlovian culture assumes them to be of a new, different kind than the Szeletian points.

Besides the Gravettian culture, the typological approach accounts for the existence (or survival) of the Aurignacian as late as 20, 000 BP at Stránská skála in Moravia, and Alberndorf and Langmannersdorf in Austria, explaining it as an evidence of two coexisting different cultures — the Pavlovians, and the Aurignacians (Svoboda 1994). This model strongly resembles the one presented in the Neanderthal — modern human debate, without strengthening its plausibility mainly because it does not define the concept of "culture". In addition, due to the description of archaeological materials as a substitute for interpretation, it remains unclear what the implications of the coexistence of the two cultural groups might be, either for the prehistoric occupants of the area, or for our understanding of the past. The source, or the meaning of the observed differences is never elaborated upon, combining thus the role of typology as a heuristic tool in organizing a vast amount of material, with an evolutionary explanation (the types change as a result of unidirectional evolution). Affirmation of "the type" prevents any interpretive move beyond its recognition in a collection, transforms the text of description into its explanation, and successfully avoids any "just-so-stories" that a narrative interpretation of the site remains might involve.

The application of the typological approach

The initial division of the recovered archaeological materials into distinct type categories is worth the attention, as this technique is the basis for any claim of significant differences between collections, and by extension cultural groups. Since one of the goals of this study was to evaluate the effectiveness, reliability, and foremost, information value of using stone tool types as a determining characteristics of any archaeological entity, I first focused on the internal consistency of categories, as empirically applied in organizing materials from the Pavlov site.

In order to evaluate the viability of typology, beyond the point of organizing and describing a large collection, I have followed type definitions as they have been used by most researchers working in European Palaeolithic archaeology. I recorded each of the 4081 stone pieces in terms of its spatial location in the site, which was divided into a 1 square meter segments, measured its length and width, noted its typological assignation in the index catalogue of excavated materials. I described forms of retouch of each piece, and after examining it under a microscope (for a detailed discussion see methodology section), I included the presence or absence of traces of hafting and use-wear.

The starting point in a classification of a lithic collection is visual inspection. Each stone piece is evaluated in terms of several observable features — shape, size, length and width ratio, presence or absence of retouch. At the same time, certain pieces are labeled with terms that are purely functional - assumed and not observed; yet these labels are applied in the same descriptive process, as the ones based on the observed features. Consequently, we end up with microliths of various shapes (e.g. trapezoid, half-moon), or backed implements, terms that are all based on their visible and measurable properties. On the same analytical level are scrapers, burins, points, or even blades, all relating to assumed function, and not any observable, measurable feature. In most traditional studies, this system was not only the basis for organizing the collections, but was then also turned around and used in the analysis of them, expressing statistically that which the organizing process achieved. After defining a category, say, *burins*, and identifying which pieces should fall into it, the site and culture associated with it can then be defined in terms of the burins they contain. The internal logic in such a study is self-referential, and cannot in any way strengthen, or invalidate the original proposal.

Whenever the need for systematic organization and categorization of a large number of archaeological materials arises, multiple criteria are evoked. The majority of materials at any Palaeolithic site consist of stone pieces; whether they are identifiably shaped pieces, debitage, or chunks of mineral from which the implements were presumably made. Even in the first basic description of finds, we are not only organizing the recovered material, but also channel the subsequent analysis in certain directions. A collection list that identifies, for example, formal tools, has accepted a number of assumptions about form and function that are built into the descriptive process. Any stone artifact labeled as a formal tool is assumed to have been a more-less finished object, and implicitly also the intended product. The presence of formal tools in a collection also implies that the rest of the associated artifacts are either production waste, pieces of raw material that could have been intended for future modification, and rejects not intended for use. In terms of use-wear, this assumption suggests that other lithics than the formal tools are unlikely to have any traces of wear.

This is not to say that classification of archaeological materials can be entirely avoided, should be eliminated, or that it is entirely faulty. Rather I would argue that the criteria used in description and categorization of archaeological matcrials should be made explicit, so as to recognize their imperfect and tenuous nature, allowing them to be better used by other researchers. The first step should be a clarification and specification of used categories, followed by an examination of their consistency, and uniform use throughout studies, so that diverse collections could be compared. This may appear as a very mundane point, yet everyone who has worked with museum collections, or materials excavated by other researches knows the frustrations of fuzzy categories, unspecified criteria, or a complete disagreement with someone else's system. All these observations, after all, are made on the basis of human perception, that is uniform only to a certain degree, and is subject to numerous influences.

Pavlovian: A type with limits[4]

Any lithic collection is first described in terms of shape, and divided into three basic groups — flakes, blades, and cores. This division was originally designed with the belief that predominance of flakes in a collection was a temporal indicator of older time period, than any material that was dominated by blades. The Middle Palaeolithic was considered flake based, while the Upper Palaeolithic was dominated by blades (Bordes 1984). This strict division has been shown to be erroneous, as new materials manifest presence of flake dominated sites even in the Upper Palaeolithic, and blade sites in the Middle Palaeolithic. Blades are generally defined as elongated flakes that are at least twice as long as wide, with sides that parallel each other (Bordes 1961:6, Tixier 1974:4-8). Applying the definition of blades strictly metrically, rather than by mere visual inspection, the distribution at Pavlov is as follows:

Table 5.1. Pavlov: blade/flake/core ratio

shape	blade	flake	core	total
count	3129	805	147	4081
%	76.7	19.7	3.6	100

In the whole collection, 4081 stones, 3129 (76.7%) met the criteria, while 805 (19.7%) should be considered flakes in terms of measurements, and 147 pieces (3.6%) were cores, or core residues. The Pavlovian complex, like most Upper Palaeolithic sites, has been defined as blade dominated, and the blade index at other related sites in Moravia ranged from 65% to 80% (Klíma 1959, 1961, Svoboda 1994). The range is statistically quite broad, expressing diversity among Palaeolithic sites, yet the particular meaning of the blade content would have to be explained at each site contextually. The proportion of blades in our sample (77%) seems to be in line with other contemporaneous sites in the area.

Two factors could contribute to the relatively high proportion of flakes in the sample (20%), which are not necessarily mutually exclusive. On one hand, most collections are largely classified by applying a general understanding of terms by visual inspection, rather than exact measurements, a practice, which may account for some of the differences between individual sites. The common understanding of what constitutes a blade is mediated through cultural norms of scientific precision, and the understanding of what an Upper Palaeolithic lithic assemblage ought to contain.

On the other hand, the occurrence of flaked pieces of stone at Pavlov is the result of human activity in a particular setting. A number of interpretations that would explain a particular composition of stone tool kits is possible. The most commonly evoked are the availability of raw materials in the area, the range of purposes for which the stone implements were used, as well as their treatment. These will be considered with the support of data from the use-wear study later in the text.

Once the general shape of pieces has been determined, closer scrutiny of retouch and possible function follow; so as to organize the artifacts into typological categories. The issue that comes to the forefront is that two different but not mutually exclusive criteria — function and form — are being used in the same classifying process. For example an endscraper is described in terms of its shape and retouch, "a blade or a flake with a continuous, non-abrupt retouch at one (or both in case of a double tool) of its more-less round, rarely square, extremities" (Bordes 1979:39), yet named after its assumed function. A backed implement is described in terms of a visible modification as a blade with abrupt retouch along one edge that is thicker than the presumed cutting edge. A microlith is any small retouched piece shorter than 5 cm, and narrower than 1.2 cm. A burin, the most ambiguous of them all, is named after a graving chisel, assuming a similar function in the past, but described in terms of a bevel from which a spall is removed (Bordes 1979: 40, Vaughan 1985:488). This mixed terminology provides a generous space for labeling variously shaped pieces as one category. On the other hand, the functional terms are far too suggestive, and have held a firm grip on our imagination when inferring tool use. Awareness and questioning of the issue has encouraged some alternative approaches to the accepted correlation between form and function, and has produced some interesting studies (e.g. Sliva and Keeley 1994, Odell 1994, Vaughan 1985).

Following is the formal typology of the analyzed sample of the Pavlov collection (Pavlov 1953), as described in the site report (Klíma 1952, 1953).

Table 5.2. Pavlov: Formal tool typology.

type category	count	%
blade	1620	39.7
burin	749	18.35
debitage	660	16.17
microlith	551	13.5
point	182	4.46
endscraper	177	4.34
scraper	63	1.54
backed	49	1.2
core	22	0.54
combination	8	0.2
total	4081	100.0

Individual typological categories:

Blades

The largest (1620 pieces) and the most general category, labeled as blades, was made predominantly on flint (1589 pcs., 98.08%), which is consistent with the distribution of raw materials in the collection as whole. The category included unretouched blades (887 pcs. or 54.75%), as well as blades with retouch (733 pcs. or 45.25%), that were not recorded in any other typological group.

Microliths, points, endscrapers, or backed implements were also made on blades, but had distinct features, that were seen as determining, and overriding any other criteria. What follows is an evaluation of the blades in terms of these features, to determine the consistency with which they were applied.

Table 5.3. Pavlov: blade modification.

modification	count	%
unretouched	887	54.75
retouched	733	45.25
count	1620	100.00

Out of the total 1620 typological blades, 356 (21.98%), strictly taken, did not satisfy the blade criteria, they were not twice as long as wide. Only 23 of these pieces were broken (a factor, which should be taken into account since the implicit assumption is that we are always dealing with whole and completed pieces). A large portion of the 356 pieces, 175 (49.16%), did not have any retouch, and were generally retouch flakes, or debitage, and should have been included in the debitage category. The use wear study confirmed this suggestion, since only 15 (8.57%) of the unretouched pieces; technically flakes, had any traces of use or haft.

If we follow the trail with the rules in hand, we find out that among the 1264 pieces that were technically blades, 712 (56.33%), had no traces of retouch, and thus could be labeled as blanks, presumably a reserve for future reshaping into tools, or blades resulting from a reduction sequence, a part of the production process. However, microscopic examination revealed that 105 of these pieces (14.75%) had traces of wear or hafting, 36 pieces had heavy traces of use, 69 pieces moderate use, and 42 pieces could not be evaluated, as they were severely weathered.

Retouched blades (552 pieces or 43.67% of the sample) had a much higher rate of use: 230 pieces (41.67%) had traces of hafting or wear.[5] The fact that such a high percentage of them display evidence of use calls into question their placement in the same category as the unused, unretouched blades. If the categorization is to be informative about functional differences in the lithic collection, a further breakdown by type of use is required. Upon closer inspection, it appears that the unretouched blades that had traces of use seems to have been used for different activities than the retouched blades (see Table 5.12 below).

The use-wear study highlights some similarities, and some interesting differences in activities for which the retouched and unretouched blades show evidence of use. In both cases moderate use dominated, with 139 (62.61%), and 69 (65.71%) pieces respectively, and a corresponding proportion had traces of heavy wear, 83 (37.39%) retouched blades, and 36 (34.29%) unretouched blades. The retouched blades were used on some of the same materials as the unretouched pieces, with the addition of fresh tough materials, such as fresh hide, or meat. However, the retouched blades were used for a wider variety of activities, including drilling, sawing, or shaving wood. This evidence contradicts the notion that an unretouched blade may have been a good for anything tool, or a "Palaeolithic pocketknife", that could have been used interchangeably for a wide range of activities. To the contrary, at this particular site it seems to have been suitable for cutting and scraping, but not much else. For any other activity, a more specified implement — one with some form of retouch — was used.

Unretouched blades appear to have been mostly cutting tools, and to some degree also scraping implements. The variety of activities for which they were used was limited, compared to other stone tools in the collection. In addition to cutting and scraping, only a small number were used for shaving wood (9 pcs.), plant working (7 pcs.), or chopping hard materials. The two materials most frequently processed with the help of unretouched blades were the extreme opposites of each other, soft vegetable matter and coarse tissue. On the basis of experimentation I would identify this latter group as frozen meat and/or bones. It would appear that a blank blade made a useful all-year cutting implement (see Table 5.13 below).

The used retouched blades were from a very similar size range across most activities: 20 - 100 mm in length (mean 46.8 mm), and 6 - 33 mm in width (mean 15.26 mm). Woodworking, such as shaving, was the only type of work where larger blades seemed to have been preferred, a pattern observed also with the unretouched blades. Size of the used pieces, particularly the standardized length, varied little throughout the collection. This suggestion is further confirmed by the minimal differences in standard deviation of used and unused blades.

Table 5.4: Pavlov: Size range of retouched blades used in different activities (mm)

activity	length	width
cutting	21 - 105	6 - 32
scraping	22 - 77	6 - 26
shaving	30 - 85	10 - 33

Table 5.5: Pavlov: Size range of unretouched blades used in different activities (mm)

activity	length	width
cutting	25 - 69	7 - 21
scraping	32 - 71	6 - 25
shaving	45 - 60	15 - 25

The variation between the blades in terms of size does not appear significant. Cutting implements ranged in length form 25 mm to 69 mm, and in width from 7 mm to 21 mm, while scraping tools were 32 to 71 mm long, and 6 to 25 mm wide. Larger blades were used in shaving wood, but even then only one blade was 100 mm long, while the rest was 45 to 60 mm long, and 15 to 25 mm wide, still within the general size distribution for unretouched blades.

Although these blades had no retouch, and could have been a byproduct of tool-production, they were of a remarkably standard size, with ∂ (st d) 10.7 mm in length and 4.42 mm in width. Only 17 (16%) of the 105 used unretouched blades had traces of hafting, while 138 (60%) retouched used blades were most likely attached to some handle. Thus it would appear that the hand held tools were more uniform in terms of length than those that were a part of an implement, although one should keep in mind that these differences are in the order of millimeters, and their importance should not be exaggerated.

The length of all blades ranged between 15 and 108 mm, where 45.2 mm was the mean length, and 10.95 was the ∂. The width ranged from 4 mm to 50 mm, with a mean 17.18 mm, and a ∂ (st d) 6.91. Since the lower ends of the spectrum in both length and width are far more like those of another category — microliths — it is worth examining the question of size in detail.

Table 5.6: Pavlov: Mean length, width, and standard deviation in all blades (mm)

	length	∂	width	∂
all blades	45.2	10.95	17.18	6.91
all blades: used	44.18	12.41	16.52	6.15
ret blades	47.64	11.83	15.55	5.09
ret blades: used	46.8	12.5	15.26	5.28
unret blades	45.73	10.08	14.28	5.04
unret blades: used	45.82	10.7	13.36	4.42

As a classificatory type a "microlith" is often loosely understood rather than specifically defined, and many researchers use different limits for what the term signifies. Pieces with low width but higher length are often classified with other lithic types. The usual length limit is between 30 and 50 mm (Tixier 1974). Microliths at Pavlov were between 5 and 50 mm long, and 2 and 17 mm wide, but the majority clustered in the 10 - 38 mm range in length, and 3 - 12 mm range in width. If we consider the category of blades from the perspective of length, 439 pieces (27%) out of the 1620 meet the criteria of a microblade, and 457 pieces (28.7%) would qualify if judged by width. Since length is usually the determining characteristic, we could consider 27% of the blades to be microliths, thereby increasing numerically that category. The result of such a shift would define the character of the Pavlov lithics as small in size, with microliths as the dominant category (990 pieces) among the formal tools.

Conclusion: blades as a lithic class

Numerous consequences and implications follow this step-by-step unpacking of a single common type. Several researchers have investigated the congruity of individual real-life stone artifacts with idealized categories, only to find out that frequently certain level of discrepancy was present (Adams and Adams 1991; Barton 1989, 1991; Dibble 1984, 1988, Kuhn 1991, 1994). A label such as "blade" by itself cannot deliver anything beyond the point of initial organization, and needs to be applied in a provisional manner, only to be removed once more particular information on the "life" of a blade is drawn out. It appears more productive to treat individual types in the most general sense possible, as temporary labels that are by no means fixed or determinant in behavioral interpretations. Moreover, this most general category is a good example of a need for flexible adherence to categorization, and an admission that any type or category is a heuristic device that serves an organizational purpose. The other option is to take scientific rigor as strictly as possible, apply typological criteria to all pieces as closely as possible, and reclassify a large portion of the collection. However, this move would only result in a new systematic organization of archaeological materials. Any explanations in social and behavioral terms would depend on additional lines of evidence drawn from outside observations, and not based internally on the typology itself.

Tying together several strands of information about one category of lithics creates a more nuanced picture, giving meaning to observed differences in a large group of blades. The point that underlies this exercise is a cautionary note: reality is more varied than a straight-laced definition allows. We need not just to acknowledge this fact, but also rethink the way material culture is categorized. The blade collection discussed here contained 1620 formal pieces, yet could be significantly reduced if all the criteria were enforced. Only 565, or 34.88% of the previous total number meet the criteria of blank blades. The large majority of the stones initially included in the blade category, 1055, or 65.12%, either satisfy the definitions of other classificatory categories, and could be transferred to those typological groups, or, being neither a blank blade nor any other typologically defined tool, could be separated out in the process of an analysis.

Formal tools

After the initial categorization of the lithic material from Pavlov, the largest initial category of blades (1620 pieces) was dropped, and attention was turned to a narrower group referred to as formal tools. These were basically all the pieces that had some form of standardized retouch that conformed to definitions of various Palaeolithic tools. The group included burins, microliths, points, endscrapers, backed implements, side scrapers, and a combination of any of the above. Pieces classified as debitage, a large segment of the collection (660 pcs.) were not evaluated in this group.

This restricted group of 1779 stone tools exhibited the following distribution:

Table 5.7: Pavlov: Formal typology

typology	count	%
burin	749	42.1
microlith	551	30.97
point	182	10.23
endscraper	177	9.95
scraper	63	3.54
backed	49	2.75
combination	8	0.45
total	1779	99.99

The new division highlighted features consistent with the described "typical character of the Pavlovian" (Klíma 1959, 1967, Svoboda 1994, Valoch 1981): a preponderance of burins (42.1% of the formal tool category), and a high percentage of microliths (30.97%). Let us turn our attention to these latter categories, and examine their typological consistency and properties from the perspective of use-wear.

Burins

Webster's dictionary provides us with the following definitions for the term *burin*:

1) a graver, an instrument for engraving on metal, made of hard steel, the cutting end being ground to a diamond shaped point;
2) the style of an engraver's work, as a soft burin; a brilliant burin;
3) a tool for engraving on stone (Webster's Dictionary, p. 243).

The prehistoric stone tool described by that term in archaeological discussions acquired its name on the basis of its assumed engraving function. Yet it remains an ambiguous implement not only in the mind of a lay person (whose imagination usually does not conjure up any image at all in relation to the term)[6] but even in the minds of many archaeologists. Burins, recorded as early as the lower Palaeolithic (e.g. some pieces at Olduvai Gorge, Phillipson 1985: 39), and commonly described in the middle Palaeolithic, constitute a class that is highly heterogeneous.

Although defined in terms of its function, it is generally recognized by the technique which produces it: the "burin-blow technique" which removes a long straight spall from an edge of a blade, or a flake (Vaughan 1985:488). Sackett, following Sonneville-Bordes, characterized the Standard Burin Group as consisting of three main types: dihedral, truncation, and break burins, distinguished by the manner in which the technique was applied to achieve the chisel shaped edge. The pointed dihedral burin results from the removal of one or more lamellar spalls from an opposing burinated facet that has been previously spalled in the same fashion; a truncation burin, by driving one or more such spalls from an abruptly retouched edge known as a "truncation"; and a break burin, by similarly spalling a transversally broken blank (Sackett 1989:52). Despite Sackett's insistence on the "intrinsic typological importance" of the three types, and the general acceptance of this statement, definitions cited above describe a wide variety of blades and flakes that were retouched in a certain fashion, without suggesting whether a particular burin is the intended end product, a step in a reduction sequence, a resharpening of a used edge, or a retouch of an unsuccessful edge. Any of these actions would have resulted in the same shape, creating a seemingly uniform category, and one which furthermore appears quite significant, due to its numerical importance in many collections.

The "Pavlovian," characterized as a separate subgroup of the Gravettian, is defined precisely in such terms: a high proportion of burins represented in the lithic collection, at some sites twice as high a frequency as the next category. If we are to accept the distinctive character of a collection on the basis of a type — a distinction that is meaningful in terms of activities that produced such a difference — we need to understand what the characteristics of that type are, how the object exhibiting them came into being, and, at least in some general sense, what it might have been used for. Yet the category of burin, typologically speaking, describes a vast, ambiguous assortment of pieces that are not always clearly described, or agreed upon. In the collection before us, then, this category plays a pivotal but problematic role.

Despite the heterogeneous character of the tool category, the burin has come to be considered one of the representative Palaeolithic types, frequently discussed in terms of numerous variations rather than questioning the category itself (e.g. Knecht 1988, Pradel 1971, Rigaud 1972, Sackett 1989). The notion of a well-defined type has been somewhat shaken since the introduction of new evidence from micro-wear analyses of burin edges (e.g. Moss 1983, Plisson 1985, Symens 1986, Vaughan 1985a). Use-wear studies have stressed the variability, as not only the tip of burins had traces of use, but also edges, and both proximal and distal ends. In addition, these studies have shown that engraving bone or antler were not the only activities for which burin could have been used. Keeley (1978) noted drilling function, and suggested that the burin edge could have been a result of resharpening. Vaughan (1985b) and Plisson (1985) went further, and actually questioned burin as a type.

Vaughan focused his inquiry on the technique by which a burin was created, pointing out that all functional studies of the tool, from the earliest Clark and Thompson (1953) case, to the more recent examples (Hayden 1977, Newcomer 1974, Rigaud 1972), accepted it as a finished product, and then only showed that which they presumed — that a burin was an efficient tool for engraving (Vaughan 1985b: 489). The technique that produced the tool received no attention, and burin spalls, even when bearing well retouched dorsal ridges, are counted as unretouched pieces (debitage), although the burin blow could have well been an eliminator of the used edge (Vaughan 1985b: 489). A use-wear analysis of numerous burins and burin spalls from Magdalenian sites in France (Cassegros), and Germany (Andernach, Zigeunerfels) has led him to conclude that in all three assemblages there are only rare examples of burin bevels (tips) that bear definite signs of having been used, while somewhat more frequent traces appear on the sides, and edges unaffected by the blow. A large portion of the examined assemblage did not show traces of wear after the removal of the edge. This Vaughan interpreted as evidence of rejuvenation, or elimination of the used edge, concluding "the burin blow was a secondary occurrence, a subsequent technique that could have been applied to any utilized or unused flint from the assemblage" (Vaughan 1985b: 492-494).

His conclusion was criticized on several grounds, mainly hypothetical (Juel Jensen 1988:73). The effectiveness of a cutting edge produced by a burin blow is difficult to estimate if the particular activity for which it was used is unclear, unless one goes back to the traditional hypothesis that engraving was the main purpose for which a burin was produced. Vaughan's attempt was to show the multiple utility of a burin, that was used not only in numerous ways, but also on various materials, such as bone, antler, wood, hide, meat, where cutting edge was not necessarily the primary focus (Vaughan 1985b: 492-494). A similar argument was made by Plisson (1985) in his analysis of the Pincevent material, where only a portion of burins had any traces of wear at all, and those significantly varied in terms of used material and performed activity. Plisson offered several possible interpretations for the form of burins, including recognizing them as unsuccessful production attempts, or as core residue.

As I will discuss later, a number of pieces from Pavlov seem to support this particular hypothesis.

In the years following these two brief inquiries into the nature of burins, nobody answered this challenge on either theoretical or empirical grounds, and the burin continues to be included in typological studies of Palaeolithic assemblages (Grigorev in Soffer and Praslov 1993, Knecht 1988, Rigaud and Simek, Mussi in Soffer and Gamble 1990, Sackett 1989, Svoboda 1991, 1994). Considering the strong focus on types and categories as temporal and space markers that dominates lithic studies, it may take a number of studies to inspire a serious reconsideration of the established methodological approach to classification in the discipline, where only interpretive frameworks are viewed as changing and developing.

The burin as the dominant feature of the Pavlovian

The specific character of the Pavlovian has been asserted by scholars working in the area on the basis of typological variation in the recovered assemblages (Klíma 1957, Valoch 1981, Svoboda 1994). The principal feature that is claimed to separate this particular cultural group from other Gravettian groups is the predominance of burins over any other category.

In the case of the analyzed Pavlov assemblage presented here, burins constituted 42.1% of the formal tool category (Svoboda 1994:149). In order to evaluate the category, I have examined all 749 burins for use-wear traces, as well as 74 burin spalls present in the collection. However, a large number of burin spalls were not included in the catalogued assemblage, as they are generally regarded as debitage. Results of the analysis provide a strong basis for questioning burin as a type indicative of cultural differences.

It appears that a number of steps involved in manufacture, as well as use, would result in a form that has features traditionally recognized in a "burin". Taking the category of burin apart, however, does not simplify the situation, rather it has the opposite effect, indicating that any one form may be a result of several actions, and frequently may not represent the final stage. Rather than simply explaining burins, this conclusion may serve as a starting point for examining the traditional concept of stone tools as discrete, predetermined, functionally specific forms, finished according to a single, accepted, culturally predetermined pattern.

Stone artifacts labeled as burins were made almost evenly on both blades (47.26%) and flakes (52.74%). Size was, as with the blades, fairly standardized, modal length 20 - 55 mm (mean 39.86 mm, ∂ (st d) 12.63), and 10 - 40 mm modal width (mean 21.54 mm, ∂ (st d) 6.65). Burins made on blades matched other blades in the collection in length, and mean width differed only by 1 mm. As mentioned earlier, microliths at Pavlov were in the 10 - 38 mm range in length, and 3 - 17 mm range in width. Since the mean burin length was 39.86 mm, it is obvious that a number of them must be in the microlith range, specifically 366 pieces (48.87%) were shorter than 38 mm. Consequently, I submit that almost a half of the burin category should be classified as "microburins". This move may appear as just another semantic shift, splitting a larger category burin, but, as we shall see, the implications are more substantive, revealing information about production sequences, as well as use of raw materials (and refuse) at this particular site.

Microburin technique, adopted more readily than in the case of a burin[7], is one of the suggested steps in production of bladelets, a method of segmenting a microblade which is then modified into its final shape. If this assumption is accepted, on the basis of ethnoarchaeological and experimental work (Barton 1994, Bordes 1957, Henry 1974; Tixier 1974), the large number of microburins at Pavlov may be linked to the high percentage of microliths present. Although a few microcores were recognized (16 pieces), the number of identified microliths (551) would call for some further suggestions as to their origin. It is plausible that tools were produced at another location than the site itself. However, the large number of recovered lithics (39,142 pieces), with a major share of retouch blades and blanks would indicate that at least some manufacture occurred at Pavlov. The microburins should then not be treated as a distinct tool type, but rather as a production byproduct, or one of the stages in a sequence.

Among the 366 microburins, I identified 22 as microblade core residues, a premise that was further confirmed in the use-wear analysis, as these pieces had no traces of any other use. In traditional studies of tool manufacture, it is usually taken for granted that a microburin is discarded without further modification, serving only as the medium to obtain the desired end product — a microblade. Yet when examined for traces of use, it became obvious that some of the microburins were themselves further modified and used for a range of purposes. Out of the total 366 pieces in this group, 151 (41.26%) had traces of use, 54 pieces (14.75%) were burnt or postdepositionally altered (i.e. their use could not be determined), and 161 (43.99%) exhibited no signs of wear observable under the magnification applied.

Evidence obtained in the microscopic examination underlines the observed lack of a specific character of microburins, and strengthens the proposition that they were primarily a residue in manufacture of other tools. A few of those pieces that exhibited traces of use had various forms of further retouch, either partial (possibly unfinished), or complete, not related to the burin edge. Only 5 pieces were bilaterally retouched, while 23 stones were modified unilaterally; 27 pieces were broken, either in use or postdepositionally, as they showed evidence of function. Three pieces were actually retouch or resharpening blades, removed from a used tool, without any subsequent utilization. In 11 cases the pointed end (a dihedral burin shape) that resulted from a removal of burin spalls, was the haft end of a tool, and the opposite end was used, mostly as a scraper. The burin bevel (tip) was used only in 11 cases for carving and drilling, in all other activities edges adjacent to it, or even opposite ends were employed. This evidence further undermines the notion that a burin blow was applied to create a specific tool used in engraving.

Compared to the previously discussed blades, possibly because of the wider edge that they offered, microburins appear to have been most commonly used for scraping (45% of activities), in addition to shaving wood (27%). Both these activities, it should be noted, are based on the same principle, placing similar demands on the shape of a tool, and involve similar motions. In terms of processed substances, this category of tools seems to have rather non-specific, as the marks of all types of materials are represented, although woodworking slightly more frequently (27.8% of all cases). Size of the implemented pieces was fairly constant, regardless of the activity for which the tool was needed. The observations support the proposition that microburins were primarily leftovers from blade production, either reduced cores, or flakes, and their further use was secondary, mostly nonspecific, cases of using a suitable piece of a stone for whatever need occurred, frequently without giving the tool a finalized shape. A number of other use-wear analyses have shown that many of the artifacts classified by archaeologists were probably used for numerous purposes, undercutting simple correlations between form and function (Moss 1983, Odell 1994a, Plisson 1985, Vaughan 1985). This conclusion would question the common assumption in lithic studies, that each type of stone tool served a specific function (or a related set of functions), indicating instead *a situated choice from a set of functionally equivalent morphologies.*

Burin use

Having eliminated 366 pieces out of 749 from the burin category, I now turn to the 383 artifacts that had some form of burin retouch — a spall removed from one or both edges with a single blow. Even this much smaller group consisted of a great number of reduced cores (50 pieces), with no traces of wear, and 133 various thick blade and flake debitage, most likely byproducts of manufacture, as only 4 of them had traces of having been used. The final group were 200 burin shaped stones, some with additional retouch, and 164 displaying clear signs of use.

The use of these stone artifacts resembled that of the microburins in the versatility of purpose, as well as materials processed. Again the largest portion appear to have been used for scraping hides, organic substances — not only meat or bones but a generous share of vegetal matter — as well as frozen, hard objects. In all of these activities the edges adjacent to the chisel end were used, frequently taking advantage of the straight, unmodified, but fairly wide edge created by the removal of the burin spall. Carving was one activity for which the burin bevel appears to have been used generously, more often than with the microburins. The pointed ends of several burins bore evidence of wood polish, but also traces of wear from bone, dry hide, and shells or a similar sharper material. A few pieces also displayed signs of heavy battering — either in chopping or another forceful action that left deep pits and scars. All eight such stones were former edges of a larger stone, removed either to resharpen it, or to create a new tool. This rejuvenation was achieved with a single blow, and a removal of a substantial thicker flake or a blade. Two other pieces apparently employed in working plant materials, presumably either scraping or mashing it, were actually core residues that had suitable surfaces used for subsequent work.

Table 5.8. Pavlov: Mean length, width, and standard deviation in burins (mm)

	length	∂	**width**	∂
burins	49.2	10.29	23.76	6.96
burins: used	52.12	12.64	22.05	6.17
burins: unused	47.23	7.51	25.07	7.23

The pieces used in different activities did not show any great degree of variation in size. Implements used in wood working tended to be slightly longer, a pattern already observed in the used blades; otherwise size did not seem to play an important role in the choice of stone for any particular activity. Overall burins were larger than other stone tools, but their size variation, in length as well as width, was approximately the same as in blades, keeping within the same range. This would correspond with the assumption that some blades and other further retouched pieces were struck off some of the burins.

As suggested for the microburin group, the formal category of burin seems in most cases to represent a stage in tool production, rather than an end in itself. At the same time, the technique creates an edge occasionally engaged for other activities as well. Depending on the activity, the materials processed, and most likely other factors — such as the time and context in which the production occurred — additional retouch refashioned some of the stones into objects different in appearance from basic debitage. However, the evidence presented here does not support the maintenance of a separate category of the burin as a formal tool. I would argue instead for the existence of a technique by which blades are removed from either cores, thicker flakes, or used tools. This blow leaves a remnant with a straight edge (or several in the case of multiple burins) which may be later used. With regard to the question whether a burin blow is a creator or an eliminator (Vaughan 1985), the answer would be that it is both, and more. A burin blow was applied in the sample of stones from Pavlov for several reasons: to segment blades and bladelets from cores (or other thicker flakes), to remove a used edge from a tool, and occasionally as one of the initial steps in a tool manufacture that was not completed.

Burin spalls

Upon closer examination of thc sample collection, a number of pieces (74), formally catalogued either as debitage (48), or blades (26), appear actually to have been burin spalls — blades removed by the burin blow from either a core, or a flake. Although such pieces are usually disregarded in typological studies, micro-wear analysis, combined with attention to manufacturing techniques, reveals information that allows for a better understanding of their use-lives, and helps explain the existence of burins in a collection.

One reason for the application of the burin blow technique was the decision to remove a used edge of a tool, presumably either because the edge has become functionally inadequate, or because the tool as whole was not needed anymore, and by removing the used part, it could be refashioned into a new implement. Traces of use were present on 21 of the burin spalls (28.38%), and in all cases the wear was on the edge opposite to the burin edge, confirming the suggestion that the technique was commonly applied to remove a used edge, rather than to create an edge that was then used. In most cases (9 pieces), it seems that the tool was heavily used on hard frozen materials, but the next most common group (5 pieces) was plant matter, and the additional pieces came from tools used in working fresh bone, hide, and/or meat. This evidence suggests that the tools were used on a wide variety of materials, and that this use was not restricted seasonally.

A majority of the burin spalls (50 pieces, 67.57%) had no traces of wear, and were either removed as a part of reshaping a core, or might have been later used for segmenting into bladelets. Thus the burin spalls provide further support to the hypothesis that the burin technique represents a way of shaping edges (used as well as unused), serving to eliminate worn out parts of a tool, and as a step in a production of blades. The group of tools traditionally labeled as burins appears highly variable in terms of its origin and stage in a production sequence. Frequently these lithic pieces may represent debitage rather than formalized tools, or a byproduct, which was then used secondarily. In either case their actual use is often independent of the burin edge which led to their classification in archaeological collections.

Conclusion: Burin as Illusory Type

Recall that the Pavlovian, as a particular eastern Gravettian group, was defined typologically on the basis of a high percentage of burins, and that this character was exemplified by the Pavlov site, where burins represented 42.1% of all tool types. However, now that we have subjected a sample of this collection to microscopic analysis, the burin category turned out to be not only highly varied, but also questionable as a typological unit, representing a multipurpose byproduct rather than a formal tool. Far from serving as a central functional or cultural marker, the burin as a typological tool alone tells us very little about what went on at Pavlov, and even less about its distinct character.

Microliths

According to the formal tool typology, there were 551 microliths (30.97%) in the Pavlov collection sample. These were all blades 5 to 50 mm long and 2 to 17 mm wide. However, the same criteria of maximum length and width applied to additional 700 pieces that were included in other typological categories. A small number of these (59 pieces) were broken blades, and should not be judged by the same standards as complete lithics.

The sample from the Pavlov collection then included 1192 lithic pieces (29.21%) that would satisfy the criteria of microliths. They were categorized by varying criteria, and labeled as combination (1 piece), scrapers (4 pcs.), endscrapers (13 pcs.), backed implements (29 pcs.), points (49 pcs.), burins (94 pcs.), debitage (95 pcs.), blade (356 pcs.), and microliths (551 pcs.). In most cases the presence of a particular form of retouch was viewed as the determining factor, although 222 blades (out of 356) were blanks, and had no traces of any modification.

Although the breakdown of all the pieces that met the size criteria of the microlithic group provides some insight into the possible use of some of the stones as types, in general this information does not have a high explanatory potential. The analysis lends itself to many more interpretive steps if the stone collection is judged from the perspective of its manufacture and use.

Table 5.9. Pavlov: Traces of use-wear on all microliths by typological groups

typology	count	use-wear	%
combination	1	1	100
scraper	4	3	75
endscraper	13	11	84.62
backed	29	22	75.86
point	49	30	61.25
burin	94	50	53.19
debitage	95	40	42.11
blade	356	103	28.93
microlith	551	392	71.14
total	1192	652	54.7

The proportion of use appears to be slightly higher in the general microlithic category (54.7%) than in the collection as a whole (42.22%). This is most likely a result of the relatively high use of microliths (71.14%), as well as backed implements (75.86%). I have noted a similar pattern in the Dolní Věstonice II (western slope) collection, where the microliths showed a strikingly higher rate of wear (57.3%) than the rest of the stone implements, particularly unretouched ones (1.18%) (Tomásková 1991: 100). I have argued that the raw material may not have been necessarily a focus of any "economic" behavior, rather a higher value appears to have been attached to the effort and time invested in the making of the finished minute tools, than the raw material itself. The results of the microwear study of the Pavlov sample further support this suggestion. The intensity of use appears much higher in the microlithic group, where 31.4% of the pieces had traces of heavy wear. This is much higher than the 12% of heavy wear traces observed in the rest of the Pavlov sample, analyzed where moderate use dominated.

Similar to blades, in terms of activity, a large portion of the microliths (51.83%) had traces indicating scraping and processing of several kinds of materials, most often plant matter (36.71%), or fresh hides, meat, or tougher organic substances (29.4%). The majority of the microblades that had traces of use showed signs of having been hafted (330 out of 392, i.e. 84.18%). This is not a very surprising result considering their size. However, a more interesting insight is given by the pattern of wear traces, that contradicts the common belief of microliths attached in line as a part of a spear, or a similar hunting implement.

The majority of the microblades in the sample from the Pavlov collection had perpendicular traces on both sides of one cutting edge, resulting most likely from a scraping brush-like motion (207 out of 330 hafted pieces, i.e. 62.73%). These implements seem to have been used mostly in processing (51.83%), and secondarily in gathering or cutting (39.87%), of plant and other organic materials. The activity was not restricted seasonally, as the third largest category (23.75%) was frozen and dry substance, suggesting a possibility of a consistent year-round pattern, proposed in the discussion of the previous groups of stone pieces.

Points

A much smaller group, 182 pieces (10.23% of formal tools, and 4.46% of the entire sample from Pavlov, were classified as points. Like most of the other typological categories in the collection, this one is not consistent in features that characterize tools included in it. The only modification that is present in all of them is a pointed distal end. However, this same feature appears on 185 additional blades that were labeled as blades rather than points. Presence of retouch is not a determining factor either, as 76 points out of the 182, were not modified at all. The remaining 106 points were retouched in a number of ways that included bilateral (38) and unilateral (15) retouch, backing (13), retouch at distal end (16), and retouch at proximal end (13), four pieces had a notch at one end, four blades had ventral retouch, and three blades were removed as a part of a retouch of an edge of a core or a flake. Size could not have been a discriminating factor, as mean length of points was 43.83 mm, (∂ (st d) 13.69), and width 16.04 mm (∂ (st d) 6.09), while the same measurements in the pointed blades were 44.88 mm (10.13) and 15 mm (5.9) respectively. We may then conclude that typologically there were 367 points in fact, a double of what was originally recorded. Out of this new total, 161 (43.87%) had discernible traces of use, and additional 8 had traces of hafting but use could not be determined due to postdepositional damage.

The points present in the sample from the Pavlov collection bore mainly traces of wear that could be interpreted as a result of cutting various materials (58.39%). This group of blades had the highest representation of a singular activity of all the categories examined. The observed traces appear mainly as produced by moderate use (108 pieces, 67.08%), while only 53 pieces (32.92%) were heavily worn out. This might be a consequence of the lower abrasion of fresh softer materials that could have been processed. I would suggest that the majority of the pointed blades were used for cutting up plant material (39.13%) and fresh meat and/or hide (29.19%).

It seems that the blades were either a part of a composite tool, several attached to a handle in some arrangement, or hafted singularly in a wooden or bone piece, as 104 points (64.6%) out of the 161 with traces of wear, had microscopic damage consistent with that left by hafting. In addition, these implements had traces consistent with processing of hard, frozen substances (15.53%), presumably during colder months. Pointed blades appear to have been used only marginally in woodwork (4.97%), where some points might have been used in carving.

Sizewise the points that had traces of wear tended to be smaller in length, than other retouched blades, regardless of the activity for which they might have been used. However, the mean length of the used points varied from other used blades only by 4 mm, and its standard variation was higher (13.79 as opposed to 10.7 and 12.5 for unretouched and retouched blades respectively), indicating higher variation in length rather than an overall smaller pieces used to produce points.

Table 5.10: Pavlov: Size of points (mm) used in different activities

activity	length	width
cutting	25-46	5-23
scraping	34-53	9-20
drill	32-53	7-27
carving	7-86	7-35

As with the previous typological categories, points were most likely used in activities related to processing a number of materials. Again the typological approach by itself does not appear to provide substantial new insights into the prehistoric behavior at Pavlov, nor does it support significantly claims of cultural exclusiveness and cohesiveness of a group that inhabited the region. These questions are better served by data gained from multiple sources, including use-wear analysis, and study of production and reduction sequences, together with the knowledge of other associated materials recovered at the site, and the knowledge of the environmental context.

Endscrapers

Endscraper, one of the least controversial categories, is generally made on a blade, but flakes do occur, where either of its extremities (or both in case of a double endscraper) is continually retouched in such a fashion that the end is rounded, occasionally rectangular (Bordes 1979: 39). Some earlier Lower or Middle Palaeolithic endscrapers have been identified but these rarely reached any significant numbers, and endscrapers are traditionally assumed to have been Upper Palaeolithic tools. They are usually divided into two categories — typical and atypical, depending on the shape, extent, and overall appearance of the retouch: '... each time when we think: this tool resembles an endscraper, but it's not very pretty, we should label it as an atypical endscraper.' (Bordes 1979: 39)

The sample collection at Pavlov included 177 pieces (9.95% of the formal tools, and 4.34% of the whole sample) that were identified as endscrapers. However, again a number of additional pieces from the more disputable categories, such as blades (12), burins (17), combinations (6), debitage (4), but even a microlith, a backed piece, and two scrapers satisfied all the requirements for endscrapers, some even passed the beauty test. Consequently, the group contained 220 pieces that could be labeled as endscrapers, evenly split between flakes and blades (110 each).

Endscrapers are not only a fairly easily identifiable group but in terms of use, they tend to be, besides microliths, most frequently identified as having traces of use-wear. The two characteristics: common understanding of what an endscraper might look like, and an identification of wear traces, may be connected, as one's perceptions of what might be a tool influence the hypothesis of its use, that is either confirmed or refuted by the microscopic observations.

In this group of endscrapers, 152 (69.1%) had traces of use and/or hafting, and 98 pieces out of this (44.55%) had traces of both. Blades were slightly more represented than flakes (81 and 71 respectively) but the difference seems rather marginal.

Only a very small portion of the endscrapers showed traces that could be interpreted as other activity than scraping, three pieces could have been used for carving and two pieces for cutting. Twice I observed traces of several potential activities for which the same stone was used. Once a burin had a retouched proximal end that was used for scraping while the burin retouched distal end had traces compatible with those left by carving, most likely bone. The other example was a flake that appeared to have been used as a side and an endscraper on the same type of material, probably a fresh hide.

All materials seem to have been processed by scraping almost in the same proportion, only fresh hide and/or meat, and frozen and/or dry matter were represented in slightly higher proportions. The size of the endscrapers was smaller, ranging from 14 to 107 mm in length, and 5 to 40 mm in width. The mean length of the stones that appeared to have traces of wear was 39.45 mm (∂ (st d) 15.17), and the mean width 21 mm (∂ (st d) 5.59). The closest other lithic category in terms of size was the burin (mean length 41.56 mm, mean width 20.21 mm) which was also mainly used for scraping various materials. On the other hand, the small size of the utilized pieces could be related to reduction sequences, of which the final artifact would be a result, rather than a purposefully produced tool. (see table 5. 18 below)

Sidescrapers

Sidescrapers are defined as stone tools made on flakes or blades by a continuous retouch along one or both sides, creating a straight, convex, or a concave usable edge (Bordes 1979:31). Four major classes were outlined in the traditional and commonly used Bordes' typology: single, double, convergent, and transverse sidescrapers (Bordes 1979). These have received heightened interest recently, mainly due to researchers working on sites dated to the Middle Palaeolithic, as sidescrapers have been considered the most characteristic tools of the Mousterian (Dibble 1984, 1987, 1988; Kuhn 1991, 1992). The attention has focused on the various forms and types of sidescrapers, suggesting that possibly a number of the forms might be considered a stage in the manufacture process, or a step in a reduction sequence (Dibble 1987). As Dibble remarked in his reassessment of scraper morphology:

> If correct, the reduction models ... would help to clarify the meaning of the scraper types defined by Bordes in terms of specific aspects of prehistoric behavior, i.e. the resharpening and continued utilization of lithic artifacts during their life cycles. This would not invalidate Bordes' typology. On the contrary, it would strengthen the use of the typology as an analytical tool in interpreting Palaeolithic assemblages. Accordingly, typological variability among the scrapers could be seen as reflecting variability in intensity of reduction (Dibble 1987:116).

Thus the focus has been to improve the typological approach, and to narrow down the variability in scraper morphology to fewer true types. The same attention has not been afforded to sidescrapers in Upper Palaeolithic studies, partly due to the fact that their representation is dramatically smaller compared to the former prehistoric time periods, and also perhaps due to a larger number of defined formal tool categories that could be studied.

As a category, sidescrapers were only marginally represented in the sample collection from Pavlov — 63 pieces (3.54% of the formal tool group, and only 1.54% of the whole sample). However, 17 of these had no retouch along either edge, having been misclassified, they appeared to be actually core residues. Only one of them had some discernible traces of use, and it seems to have been used despite the absence of any secondary modification. On the other hand, additional 9 pieces — 5 burins, 1 debitage, and 1 blade were functionally sidescrapers, and 2 endscrapers were used as both end- and side-scraper. Therefore, I considered the group of sidescrapers to consist of 72 pieces, out of which 32 (44.44%) had traces of wear that could be interpreted as a result of use.

The majority of the pieces were used for scraping, yet a small portion (9.38%) had traces that were parallel to the edge, and had the appearance of having resulted from cutting. Two pieces were heavily battered, and did not have any traces that would suggest scraping at all. All materials were processed almost to the same degree, except for wood that played a minor role in this group, keeping in mind the small number of used pieces as a whole.

The size of the sidescrapers was, similarly to that of the endscrapers, smaller than the size in other categories. The mean length of pieces with traces of use was 37.84 mm (∂ (st d) 9.83), and mean width 21.78 mm (∂ (st d) 6.73), suggesting that some scrapers were byproducts of reduction of larger pieces, particularly considering that the number of functionally specific sidescrapers in the collection was relatively small. (see table 5. 19 below)

One of the possible interpretations could be that a large number of lithics in several other categories was used for the same type of activity: scraping. Therefore, I would suggest that a style and a function of lithics were not necessarily always connected, and each may have been influenced by different factors. As the previous discussion has shown, frequently the same activity was carried out by stone tools chosen from a number of different morphologies. The relationship between an implement, its final shape, and its function appears to be more complex than a one to one equation — scraper means scraping — would allow.

Backed implements

Backing refers to a continuous, very steep retouch along a lateral edge, forming almost a ninety-degree angle with the dorsal surface of the artifact (Addington 1986:104). The result is a wider blunt edge that frequently shows traces of hafting, occasionally with adhering resin residue. Only 49 pieces in the sample of the Pavlov collection were identified as backed implements. These were predominantly blades (46 pieces), with only a few flakes (pieces). In addition, 393 pieces in the sample had the same retouch along one of the edges.

These were classified into such groups as blades (36), burins (4), endscrapers (2), points (12), debitage (9), scraper (1), and microliths (329). A subgroup was created within the microlithic category, identifying those pieces that had steep retouch along the edge as backed microblades, 329 out of 551 in the originally assigned group of microliths (59.71%). However, an additional 48 pieces meet the criteria of a microblade, and are backed; yet were not classified under either category. Among these are 29 backed blades, 6 blades, 6 points, and 5 debitage pieces, indicating the inconsistency in the application of the typological labels. In addition also the possibility that a single piece could potentially belong to several categories, and the final typological variability may, to a large extent, be as much a function of the classificatory system, as of the prehistoric behavior that left these artifacts behind.

Since the backed microliths were already discussed in the previous section — 231 backed microblades (70.21%) showed traces of use or hafting, I will focus here on the remaining 113 pieces that were modified with steep retouch along one edge. An overwhelming majority of these was made on blades (105), only a few flakes were represented (8). The size of the implements was from the range of the modified lithics — mean length 41.04 mm (∂ (st d) 10.32), and mean width 12.43 mm (∂ (st d) 7.2), which is slightly smaller than the points but larger than the endscrapers or burins.

The most common activity for which backed implements seem to have been used was cutting mainly plant material and fresh meat and/or hide. Typologically, there does not appear to be any difference, and all pieces, whether labeled as backed, blades, points, or debitage were potentially utilized. A small portion of this group was also possibly used for scraping the same kind of materials. (see table 5. 20 below)

Combinations

The last and the smallest group of the formal tools was labeled as combinations. These were all blades that had both ends, proximal and distal, modified into a different kind of a recognized tool type. There were 8 pieces included in this group, and all of them were a combination of a burin and an endscraper. I have already indicated the debatable nature of a burin as an intentional tool, and the pieces in the combination category further support this suggestion. Seven pieces had traces of use, and six of them had also discernible traces compatible with hafting. All six appeared on the burin end of the tool, suggesting that the sharp end, notwithstanding its origin as a manufacture byproduct or an intentional modification, was in this case used for wedging into a handle. The one piece that did not appear to have been hafted had traces of wear on both ends. Both could be interpreted as having been used for scraping the same type of material — fresh hide or meat. Due to the small size of the sample, however, statistical analysis does not offer any significant insight.

Debitage

The third largest category, after blades (39.7%) and burins (18.35%), was debitage, containing 660 pieces (16.17% of the sample). These were all the stone artifacts that did not fit in any obvious way into the other formal tool classes, and were presumed to have been byproducts of either manufacture or use. Some of them showed traces of retouch, others were edges of cores or reduced cores, and some flakes or blades had no obvious signs of modification. Although typologically they are problematic, the lithics in this group can add a great deal to our knowledge of the prehistoric activities that took place at the site, as well as the overall treatment of raw materials that were used to produce the tools. The group is also of particular significance due to its large size, thus potentially revealing more information than many of the formal tool categories that were numerically small, such as scrapers (1.49%), or backed implements (1.2%), that are usually discussed in detail.

Reduced cores, residue blades, and flakes are all included in the category. Both blades and flakes are about evenly represented, 270 and 276 pieces respectively, while 114 pieces are expended cores. A portion of each category showed traces of use, totaling 121 pieces, and additional 19 pieces were haft ends that broke either in use, or postdepositionally, and wear could not be determined.

Table 5.11. Pavlov: Debitage wear and haft by shape

shape	count	%	wear/haft	%
core	114	17.27	4	3.51
flake	276	41.82	43	15.58
blade	270	40.91	93	34.44
total	660	100.0	140	21.21

The debitage category offers the best example how classification highlights only certain aspects of the collection, focusing on whether a recovered artifact conforms to set typological standards. In the sample from the Pavlov collection, only 18 cores were recorded, and these had no traces of any use. On the other hand, the category debitage contained 114 expended cores, out of which 4 were subsequently used as wood scrapers. This detail is already absent in the usual presentation of typological variation in a collection, where only percentages of types are noted. Yet, if we were to consider the same sample from the perspective of manufacture stages, and possible utilization of the pieces, we are dealing with a much larger group of 243 pieces that started as cores, and were further modified by reduction, or transformed into other tools. Some were typologically assigned to the debitage category, some to burins, or scrapers.

Amidst the flakes belonging to the debitage category, only a small portion (15.58%) had traces of use. Yet even this small group provides additional behavioral information that is not immediately apparent from mere visual inspection of the artifacts, and complements information gained from microscopic analysis of other lithics in the whole sample. Only two of the debitage flakes were resharpening flakes, i.e. used as a part of a larger piece, and later removed either to sharpen the edge, or to create a new one. The rest of the flakes were used without having been shaped or modified in any extensive way. Blades appear to have been utilized more extensively (34.44%), but a large portion of them (47 out of 93) were in fact resharpening blades, indicating a greater intensity of use, or a change of form, while retaining the same tool.

The size of the debitage did not vary to any significant degree, with mean length for the flakes 38.29 mm (∂ (st d) 10.9), and 43.15 mm for the blades (∂ (st d) 9.3). This is consistent with the size of most of the lithics in the sample collection, further supporting the suggestion of a fairly standardized size of all stone pieces present at the site.

Cutting and scraping were again the activities for which these pieces were used most. However, considering that a large number of them were resharpening blades, it is somewhat surprising that the largest category of processed material was plant matter which leaves very light traces of wear, unless used extensively. Either these lithics were used over a protracted period of time, and eventually became worn out, or they were refashioned into another tool, once the task for which they were made was finished.

The second largest group was frozen or dry material, indicating that use of unmodified flakes and blades, and resharpening of tools were not restricted to one season only, but rather were common all year round. One of the implications of such an observation, I would suggest, is a stable supply of raw material that did not fluctuate seasonally. This could have been achieved in a number of ways, ranging from a stable access to raw material sources, to regular contacts with people from the area where the raw material originated, as the majority of minerals used at Pavlov were not local.

Pieces in the debitage category, mostly unmodified, were similar in their utility as unretouched blades, discussed earlier. Both types of lithics were useful for cutting and scraping of plant matter, most likely in the warmer months, and dry and frozen meat and/or hides during colder times. This similarity may have been the result of the non-specific nature of implements needed for general tasks. The different utility appears in processing wood, for which only very few unretouched blades were used, while a number of debitage pieces had wear traces consistent with wood polish, indicating not only wood shaving but also carving, cutting, and battering. Some of the debitage pieces with this type of wear were resharpening blades (6 out of 19), usually edges removed from larger pieces, or reduced cores (3 pieces), used secondarily. The majority were only partly modified, unilaterally, or at the proximal end, so as to allow for hafting. (see table 5. 21 below)

The width of the used debitage is from a similar range as the rest of the used pieces, particularly unretouched blades, regardless of the activity for which they were most likely used. Length however, tends to be smaller, hardly ever exceeding 60 mm. This may be the consequence of most of these pieces being retouch blades or flakes, removed from larger stones. Only those likely used in batter of harder materials were slightly larger, ranging from 39 to 61 mm. The results in this category point out the need to view lithic artifacts from more than one perspective, since typologically debitage is not an informative group. Yet, upon a closer examination valuable insight can be gained about the treatment of raw materials, tools, and the range of prehistoric activities.

Conclusions

The extended analysis presented here was conducted with several aims in mind. I will point out their inter-connectedness, and the ways that issues of function and style affect each stage of analysis. Since many tool types are defined functionally, I examined the possibility of studying prehistoric activities as related to individual tool categories.

The results of the microwear analysis of 4,081 stone pieces from the Pavlov site show that a correlation between form and function is far from straightforward, and that a tool type, such as a "scraper", could not be identified with only one particular activity. The majority of the typologically described pieces display evidence of having been used for a number of activities, and many of the types appear to have been interchangeable for several different, possibly repetitious, activities (such as cutting or scraping hide, meat or plant material). Burins, scrapers, blades, and microliths were all to various degrees used in scraping as well as cutting of processed materials. I conclude that organizing a collection into typologically described groups does not *in itself* greatly enhance our knowledge of activities or lifeways of prehistoric occupants of a site. Indeed, the evidence presented by use-wear study would caution against any quick assumption of function on the basis of observable form alone.

At the next level of analysis, we confront a basic premise of prehistoric archaeology: that a particular site represents an observable material manifestation of an identifiable cultural group. Like most Palaeolithic sites, Pavlov has primarily been described in terms of its lithic collection, that is to say its distinctive tool types, and their proportional representation. When regional comparisons are made, the variability of type percentages between different sites defines their degree of similarity or difference and, by extension, their inclusion or exclusion in a spatially and temporally defined cultural group.

The distinct character of the "Pavlovian" cultural group has been based on a high percentage of burins, relative to the other tool types in the representative collections. The foregoing study questions the coherence of the assembled stone pieces collected under the single name of "burin". It appears that a straight edge created by a single blow recognized as a characteristic feature of a "burin", could result from a number of different manufacture, use, and discard steps. A significant portion of burins in the sample examined were in fact reduced cores that had no traces of further use. In addition to being products of a range of techniques, the assembled "burins" also display evidence of having been used in a variety of ways, with different edges, and/or ends deployed. In addition, some appear to have been secondarily modified.

Thus I conclude that with regard to manufacture, potential function, and utility, a burin does not hold up well as a tool type, unless defined in the most general of terms. Rather than burdening a class of objects with a long list of necessary and sufficient qualifying conditions, I would suggest that a "burin" probably should not be considered an independent category. Rather more attention should be given to the use of different manufacture and modification techniques, including the burin blow technique. If this proposal is accepted then we must consequently question the basis upon which the Pavlovian has been described. If the very tool type that has stood as the primary feature of this prehistoric group appears in large part to represent debitage from manufacture or retouch, I would call for the need to redefine the basis on which this particular culture has been established. Furthermore we need to examine the tradition of cultural groups defined by stone tool types. Beyond burin's questioning, the study presented here shows that a large portion of each tool category only meets its classificatory criteria loosely, and many of the stone artifacts actually satisfy the criteria of several formal categories at once.

It is not my intention to suggest that Palaeolithic collections are simply not described rigorously enough, or that this particular site has been dealt with in a careless fashion. Quite to the contrary, I was able to carry out this analysis only due to the careful and conscientious manner in which the artifacts were recovered. Rather, this study demonstrates how few real groups of artifacts actually satisfy the abstract formal divisions invoked in their classification. Typologies by themselves are too often simply tautological exercises, assuming scientific guise through statistical analysis of relative distributions, yet not meeting basic criteria of classificatory systems in terms of exclusiveness and boundedness. In addition, by themselves, they provide only minimal insight into prehistoric behavior.

Tables:

Table 5.12. Pavlov: Traces of use-wear on retouched blades, suggested activities and worked materials (out of 552 pieces).

activity	soft	frozen	fresh/ tough	wood	bone/ antler	count	%
cutting	51	34	16	3	9	113	50.9
scraping	13	11	30	3	4	61	27.48
shaving	—	—	—	19	—	19	8.56
drill	7	2	1	1	—	11	4.96
chopping	—	2	—	4	4	10	4.51
mixed	8	—	—	—	—	8	3.6
carving	—	—	—	—	—	—	—
count	79	49	47	30	17	222	100
%	35.59	22.07	21.17	13.51	7.66	100.0	

Table 5.13: Pavlov: Traces of use-wear on unretouched blades: suggested activities and worked materials (out of 712 pieces)

activity	soft	frozen	fresh/tough	wood	bone/antler	count	%
cutting	20	26	9	—	2	57	54.29
scraping	11	—	11	—	7	29	27.62
shaving	—	—	—	9	—	9	8.57
drill	—	—	—	—	—	—	—
chopping	—	1	—	—	2	3	2.86
mixed	7	—	—	—	—	7	6.66
carving	—	—	—	—	—	—	—
count	38	27	20	9	11	105	100.0
%	36.19	25.71	19.05	8.57	10.48	100.0	

Table 5.14. Pavlov: Traces of use-wear on microburins suggested activities and worked materials (out of 312 pieces)

activity	soft	frozen	fresh/ tough	wood	bone/ antler	count	%
cutting	10	10	1	0	9	30	19.87
scraping	14	28	9	0	17	68	45.03
shaving	—	—	—	41	—	41	27.15
drill	—	1	—	1	1	3	1.99
chopping	—	—	1	—	—	1	0.66
mixed	—	—	—	—	—	—	—
carving	—	—	8	—	—	8	5.3
count	24	39	19	42	27	151	100
%	15.89	25.83	12.58	27.82	17.88	100	

Table 5.15. Pavlov: Traces of use-wear on burins, suggested activities, and worked materials (out of 383 pieces)

activity	soft	frozen	fresh/tough	wood	bone/antler	count	%
cutting	9	2	13	1	5	30	17.86
scraping	20	19	12	—	12	63	37.5
shaving	—	—	—	38	—	38	22.62
drill	—	—	—	—	—	—	—
chopping	—	2	1	2	3	8	4.76
mixed	—	—	—	—	—	—	—
carving	—	5	6	12	6	29	17.26
count	29	28	32	53	26	168	100.0
%	17.26	16.66	19.05	31.55	15.48	100.0	

Table 5.16. Pavlov: Traces of use-wear on microliths, suggested activities, and worked materials (out of 1192 pieces)

activity	soft	frozen	fresh/tough	wood	bone/antler	count	%
cutting	98	60	63	4	15	240	39.87
scraping	112	82	102	—	16	312	51.83
shaving	—	—	—	18	—	18	2.99
drill	11	1	12	—	2	26	4.32
chopping	—	—	—	1	2	3	0.5
mixed	—	—	—	—	—	—	—
carving	—	—	—	3	—	3	0.5
count	221	143	177	26	35	602	100.0
%	36.71	23.75	29.4	4.32	5.82	100.0	

Table 5.17: Pavlov: Traces of use-wear on points: suggested activities and worked materials (out of 367 pieces)

activity	soft	frozen	fresh/tough	wood	bone/antler	count	%
cutting	41	16	30	1	6	94	58.39
scraping	10	2	10	—	5	27	16.77
shaving	—	—	—	2	—	2	1.24
drill	12	4	5	—	3	24	14.91
chopping	—	—	—	—	—	—	—
mixed	—	—	—	—	—	—	—
carving	—	3	2	5	4	14	8.7
count	63	25	47	8	18	161	100.0
%	39.13	15.53	29.19	4.97	11.18	100.0	

Table 5.18. Pavlov: Traces of use-wear on endscrapers, suggested activities, and worked materials (out of 177 pieces)

activity	soft	frozen	fresh/ tough	wood	bone/ antler	count	%
cutting	2	—	—	—	—	2	1.33
scraping	28	38	38	—	21	125	82.78
shaving	—	—	—	21	—	21	13.91
drill	—	—	—	—	—	—	—
chopping	—	—	—	—	—	—	—
mixed	—	—	—	—	—	—	—
carving	—	—	(1)	2	1	3	1.99
count	30	38	38	23	22	151	100.0
%	19.87	25.17	25.17	15.23	14.57	100.0	

Table 5.19. Pavlov: Traces of use-wear on sidescrapers, suggested activities, and worked materials (out of 72 pieces)

activity	soft	frozen	fresh/tough	wood	bone/antler	count	%
cutting	2	1	—	—	—	3	9.38
scraping	8	4	7	2	6	27	84.38
shaving	—	—	—	—	—	—	—
drill	—	—	—	—	—	—	—
chopping	—	—	—	1	1	2	6.25
mixed	—	—	—	—	—	—	—
carving	—	—	—	—	—	—	—
count	10	5	7	3	7	32	100.0
%	31.25	15.63	21.88	9.38	21.88	100.0	

Table 5.20: Pavlov: Traces of use-wear on backed implements, suggested activities, and worked materials (out of 113 pieces)

activity	soft	frozen	fresh/tough	wood	bone/antler	count	%
cutting	21	14	16	1	7	59	75.64
scraping	7	1	4	—	—	12	15.39
shaving	—	—	—	2	—	2	2.56
drill	1	—	4	—	—	5	6.41
chopping	—	—	—	—	—	—	—
mixed	—	—	—	—	—	—	—
carving	—	—	—	—	—	—	—
count	29	15	24	3	7	78	100.0
%	37.18	19.23	30.77	3.85	8.97	100.0	

Table 5.21: Pavlov: Traces of use-wear on all debitage, suggested activities and worked materials (out of 660 pieces)

activity	soft	frozen	fresh/tough	wood	bone/antler	count	%
cutting	18	18	9	2	4	51	42.15
scraping	19	7	6	—	5	37	30.58
shaving	—	—	—	14	—	14	11.57
drill	—	1	1	—	1	3	2.48
chopping	—	3	1	2	8	14	11.57
mixed							
carving	—	—	—	1	1	2	1.65
count	37	29	17	19	19	121	100.0
%	30.58	23.97	14.05	15.7	15.7	100.0	

Chapter 6: Going to Austria

"I know Austria," my grandfather said, looking out the window at a small patch of grass, houses, and a street that he did not recognize.

"I am a man of two centuries, I was born in the last one, lived in this one, and hopefully won't see the next one." He was frail, and frequently forgot who we were, and where he was.

"I was in Austria during the war. It is a very poor country. I was in the Alps, and those people had barely any food, meat only on Sundays." He meant the First World War, although he was in the Second one too.

"And my mother sent me a letter: 'I had seven sons,' she wrote, 'and at one point I thought I had none...You have not written in three months.'" He paused, remembering.
I wondered how the mail got delivered during the war.

"The war was hell for mothers, they had no idea where their sons were, or if they were alive. You don't want to go to Austria; it's a very poor land"

My cousin Jozef taught tennis to any club member who showed up. He started at 6am, and lived in the club all week long, including the weekends. Viennese were quite generous to him, as my mother said "supported him", so that he would not have to work. Nobody in Eastern Europe thought of tennis as work, but Vienna was then not a part of our world. Austria was a place to which people from my homeland escaped during the Cold War, if they were not caught at the border.

(December 1993, Bratislava/Vienna, Slovakia/Austria).

Willendorf II: The Site and the Collection

Introduction

The archaeological location of Willendorf II, while perhaps even more renowned than Dolní Věstonice/Pavlov, has received limited attention in terms of full analysis and interpretation of the recovered materials. This is partly due to Austria's political and geographical position within the field of inquiry. Palaeolithic archaeology in Austria is placed within the realm of natural history and generally marginal. Consequently, beyond a larger-than-life reproduction of the famous "Venus of Willendorf" statuette [the original dated to 25 800 ± 800 B.P. (Otte 1990)] marking the location of the find, the site has received little recognition in Austria itself and has been the subject of only a limited number of studies since its discovery in 1908 (Broglio and Laplace 1966; Felgenhauer 1959; Haesaerts 1990, Haesaerts et. al. 1996, Kozlowski 1986; Montet-White 1988a, b; Otte 1990; Otte & Haesaerts 1989).

The Willendorf II collection has been covered and analyzed in detail in two major works (Felgenhauer 1959; Otte 1981), one lengthy typological study (Broglio and Laplace 1966), and a number of articles deal with Willendorf II marginally. All this work has been published in either German (Felgenhauer 1959; Hahn 1977; Bosinski 1989) or French (Broglio and Laplace 1966; Otte 1981), and no comprehensive treatment yet exists in English. The three major works (Felgenhauer 1959; Broglio and Laplace 1966; Otte 1981) are based on primary research, the remaining publications all rely on the published information contained in these studies. Although this system of secondary reference is an established — and necessary — practice of scientific research, it at the same time contributes to transformation of partial information into canonical truths, and their subsequent perpetuation as unquestioned facts. Thus while issues related to the typology of stone artifacts have been the center of significant discussion since the mid-1950's or 1960's, each new publication simply reproduces the received typological definitions of individual layers at Willendorf, if it includes Willendorf II at all. Considering the renown that the site has achieved (mainly due to the famous female statuette), it has received remarkably little direct attention in archaeological debates of the time period, or specific scrutiny.

The results I present here are based on primary research with the lithic collections stored at the Museum of Natural History in Vienna, Austria, chosen as a comparative sample against which to evaluate the Pavlov materials. For a period of six months I evaluated the accepted typological divisions at Willendorf II and conducted a use-wear study, to provide a alternative set of data derived from these stone tools. Only levels 5-9 were selected for this study, as these are the closest temporally to the Pavlov collection, with dates ranging from 30.000 B.P. in Layer 5, to 23.000 B.P. in Layer 9, and were characterized typologically in general terms as Aurignacian and Gravettian. I subsequently compared my results with both the Felgenhauer (1959) and Otte (1981) reports and have noted differences, which may in part be due to the differences in sample sizes studied. However, Felgenhauer's study is presented solely in percentages, making sample size estimates impossible. Otte's research involved a smaller overall sample than this work, yet some of the differences in our results cannot be solely attributed to the sample size, e.g. differences in raw material proportions in each layer. At all points my study followed the catalogue inventory at the Museum of Natural History in Vienna, Austria, and the formal classifications within it, set by Felgenhauer (1959).

Table 6.1: Willendorf II: Sample size (stone tools) in the studies discussed (in numbers).

W II layers	5	6	7	8	9	total
Otte 1981	382	220	166	374	995	2137
Tomášková 1995	600	199	283	518	1031	2631

Stratigraphy and dating

Willendorf, a loess site on the left bank of the Danube, is in reality a sequence of seven single sites (W I - W VII) nested next to each other in the south - north direction at the foot of the Nussberg mountain. A clear stratigraphic sequence was recorded only at Willendorf II where layers 1-9 were identified (Felgenhauer 1959: 91-4). Ten C-14 dates were originally obtained from the Aurignacian layers 1-4 and the Gravettian layers 5-9. The Aurignacian layers ranged from 41,500 to 30, 000 B.P. and, the Gravettian layer 8 had a date of 25,800 ± 800 B.P. (Otte 1990: 66; Otte & Haesaerts 1987). Layers 5-9 were identified as containing a "Gravettian culture of western as well as eastern types with specific Willendorf culture in Layer 9" (Felgenhauer 1959: 211; see also Broglio & Laplace 1966b; Kozlowski 1986; Montet-White 1988a).

In 1993 a new profile at Willendorf II was cleared for extensive samples aimed at palaeoenvironmental reconstruction, C-14 dating, TL stratigraphy, and TL dating (Haesaerts et. al. 1996). Cultural layers 6-9 were identified on the basis of pedological analysis with sediment unit B, and layer 5 with unit C. Charcoal identification showed a dominance of Picea from unit D up to unit B, and the authors of the study suggest that spruce was the essential tree cover, while pine trees probably existed in the region during the whole Interpleniglacial and the Upper Pleniglacial (Haesaerts et. al. 1996:33). The collected charcoal samples from layers 3-8 were dated with both traditional C-14 methods (GrN), and accelerator mass spectrometry (GrA). Layer 9 contained no charcoal suitable for dating, and mammoth bones and scapula, available at the Museum of Natural History in Vienna, were used for the purpose.

Table 6.2: Willendorf II: Radiocarbon dates

Sediment unit	Cultural layer	C-14 date	±	No. date	Material	Author
B1	9	23,860	270	GrN – 21898	bone	Damblon et. al
B1	9	24,370	290	GrN – 22208	bone	Damblon et. al
B1	9	24,910	150	GrA – 5006	bone	Damblon et. al
B1	9	23,180	120	GrA – 5005	scapula	Damblon et. al
B1	above 8	23,200	140	GrA – 893	charcoal	Damblon et. al
B1	above 8	23,400	190	GrA – 493	charcoal	Damblon et. al
B1	above 8	23,670	120	GrA – 494	charcoal	Damblon et. al
B2	8	24,710	180	GrA – 894	charcoal	Damblon et. al
B2	8	25,230	320	GrN – 17801	charcoal	Damblon et. al
B2	8	25,660	350	GrN – 17802	charcoal	Damblon et. al
B2	8	25,400	170	GrN – 21690	bone	Damblon et. al
B2	8	25,440	170	GrN – 20767	bone	Damblon et. al
B2	8	25,800	800	GrN – 11191	charcoal	Damblon et. al
B3	below 8	23,830	200	GrA – 491	charcoal	Damblon et. al
B3	below 8	23,990	130	GrA – 492	charcoal	Damblon et. al
B4	6	27,600	480	GrN – 17803	charcoal	Damblon et. al
B4	6	26,500	480	GrN – 20768	charcoal	Damblon et. al
B4	6	26,150	110	GrA – 1016	charcoal	Damblon et. al
B4	6	27,620	230	GrA – 895	charcoal	Damblon et. al
C1	below 6	28,560	520	GrN – 17804	charcoal	Damblon et. al
C2	5	30,500	900	GrN – 11193	charcoal	Haesaerts 1990
C2	5	32,000	3000	H – 246/231	charcoal	Felgenhauer 1959
C4	4	31,210	260	GrA – 501	charcoal	Damblon et. al
C8	3	37,930	750	GrA – 896	charcoal	Damblon et. al
C8	3	38,880	1530	GrN – 17805	charcoal	Damblon et. al

Authors: Damblon et. al. – c.f. Haesaerts 1990, 1996; Felgenhauer 1959 (after Haesaerts 1996:34, see detailed discussion of dating techniques and contamination issues)

What follows is a survey of each layer and a discussion of its associated materials, in ascending chronological order.

Layer 5

Charcoal samples collected from this layer in 1989 gave a radiocarbon date 30.500 ± 900 B.P. (GrN - 11193), and a humus sample was dated to 23.830 ± 190 B.P. (GrN - 11194), although the accuracy of the date is generally doubted (Haesaerts 1990: 211). Felegnhauer's 1959 charcoal sample was dated to 32,000 BP (H – 246/231) but with a very large deviation (3000).

The excavated area in this layer was fairly large (45 m by 5 m), and the cultural layer was 20 cm thick (Felgenhauer 1959: 93). Stone artifacts in this layer were mainly made out of gray chert (71.7%), and only marginally from other materials such as sandstone (11.5%), radiolarite (7.17%), jasper (4.5%), chalcedony (2%), and quartzite (0.67%). (see graph 6.1 below)

The majority of the artifacts from this layer were blades (442 pieces, 73.67%), only a small portion were made on flakes (144 pieces, 24%), or cores (14 pieces, 2.33%). The blade/flake ratio changed over time in the subsequent layers, and will be evaluated in the overall comparison.

The main focus of all currently available studies of the Willendorf material was based on typological analysis of the collection, using the following distribution:

Table 6.3. Willendorf II/5: typological distribution

typology	count	%
debitage	245	40.83
blade	201	33.5
microlith	64	10.67
burin	35	5.83
endscraper	35	5.83
scraper	10	1.67
point	7	1.17
combination	2	0.33
hammer	1	0.17
count	600	100.0

Debitage

The largest category in W II/layer 5 was debitage (245 pieces, 40.83%), a group of leftover artifacts deemed not to fit any specific typological category. In addition, upon examination I added 22 pieces to this category (10 blades, 5 scrapers, 4 burins, and 3 microliths), as they were unmodified, mostly a result of a reduction sequence. These pieces only stress the unclear nature of debitage, as no description gives clear guidelines as to the attributes sought in an individual stone piece in order to place it in, or exclude it from, this group. The number of broken blades or microliths that continue to be assigned to blade or microlith categories make it unclear whether modification might be the attribute that makes the difference where a lithic artifact is typologically grouped. A number of burins, particularly those described as nuclei form, are extended cores that do not show traces of use, could also be more appropriately categorized as debitage. The problem of debitage is not unique to Willendorf, but rather represents the unclear category of any typological system, including that applied to Pavlov discussed in the previous chapters.

The ratio of blades (66.53%), flakes (29.39%) and cores (4.08%) in the debitage group resembles the general distribution in the layer, and a larger percentage of blades show traces of wear or hafting. The size of the debitage was distinctly microlithic, with mean length 37.16 mm (∂ 18.66), and mean width 16.65 mm (∂ 9.96), while debitage with traces of possible use was even smaller: mean length 32.93 mm (∂ 17.3), and mean width 14.8 mm (∂ 8.86). This is compatible with the overall small size of the stone artifacts in the layer.

Table 6.4. Willendorf II/5: Debitage wear and haft by shape

shape	count	%	wear/haft	%
core	10	4.08	0	0
flake	72	29.39	26	31.32
blade	163	66.53	57	68.68
total	245	100.0	83	33.88

Only 33.88 % of the debitage had any discernible traces of wear or haft. Beyond the small portion of weathered and fire-cracked stones (11.7%), the majority of the group was byproducts of manufacture, rather than resharpening.

Table 6.5. Willendorf II/5, Debitage: traces of use-wear, suggested activities, and worked materials (numbers)

activity	fresh	soft	frozen	wood	total	%
cutting	21	4	6	—	31	47.69
scraping	19	4	1	—	24	36.92
shaving	—	—	—	5	5	7.69
carving	2	—	1	1	4	6.16
drilling	—	1	—	—	1	1.54
total	42	9	8	6	65	100.0
%	64.62	13.85	12.31	9.23	100.0	

Cutting and scraping were two activities most commonly attributed to the traces observed on the debitage. Fresh meat, hide, and/or bones were likely the dominant materials worked with these pieces, and size did not appear to have been of any major significance, as the length varied from very small to slightly above average (50 mm), still close to the microlithic definition. Only 15 of the pieces showed traces of heavy use, the rest (47 pieces.) had traces of moderate use. This finding does not support any simple hypothesis of "economic" behavior, with the expected stress on maximum use of each stone. A possible explanation could be a short term stay at the site, and/or secure access to raw materials, with a consequent lack of concern for the utilization of retouch blades or flakes.

A small number of debitage pieces showed traces of use compatible with that of working coarse, frozen materials. Such evidence would support a hypothesis of a winter occupation, or a short-term stay during a cold season.

Blades

The second largest category was composed of 201 blades. The numerous problems with this typological group discussed in the previous chapter on Pavlov, were only confirmed in the Willendorf II study. A small number of pieces (27 pieces, 13.43%) labeled as blades would fit more accurately the label "flake", since they did not satisfy even the basic criteria of length being twice the size the width. An additional 63 pieces (36.21% of the "true" blades) were the same size or smaller than the microliths. The question whether a piece classified as a blade should demonstrate retouch or not remained as unclear in the Willendorf II collection as it was in Pavlov. A number of blades were partly retouched, unilaterally or bilaterally, while some had backing along the entire edge, yet these were retained in the group of blades, while other pieces with similar retouch were assigned to other groups, such as points, backed knives, or microliths. Only 45 blank blades (22.39%) met all the criteria, and were of sufficient length that they could comfortably be labeled as blades.

All blades were included in the use-wear analysis, with the exception of those shorter than 40 mm, which were included in the microlithic category. Blades amounted to 138 pieces, out of which only 34 (24.64%) showed any traces that I would identify as having resulted from use, and one broken piece had traces consistent with hafting. The percentage of use (28.4%) is far below average, relative to the whole collection.

This could be a consequence of several possible factors related to the past behavior, although postdepositional changes should not be entirely excluded either. Some blades could have been used for only a short period of time. In that case the chances of use traces left on the edges are very small, particularly if the worked material was less resistant, particularly any plant material, or fresh fat or meat (bones or hide offer more resistance when either cut or scraped). That this could have been the case is further supported by the fact that blades with traces of use were dominated by fresh and plant material processing. Coarse materials (frozen bone, meat, and hide) that leave the most distinct and easily observable traces, did not seem to have been worked with blades at all.

Table 6.6. Willendorf II/5, Blades: traces of use-wear, suggested activities, and worked materials (numbers)

activity	fresh	soft	frozen	wood	count	%
cutting	16	3	—	—	19	55.88
scraping	8	—	—	—	8	23.53
drilling	4	—	—	—	4	11.77
shaving	—	—	—	2	2	5.88
carving	1	—	—	—	1	2.94
count	29	3	0	2	34	100.0
%	85.29	8.83	0	5.88	100.0	

Table 6.7. Willendorf II/5, Blades: Size of pieces used in different activities (mm)

activity	length	width
cutting	43-115	14-28
scraping	42-97	15-20
drilling	45-49	16-19
shaving	76-141	22-52

The choice of blades for different activities did not seem to vary significantly. Blades, most likely used for cutting, were not restricted to a specific size range, and were represented by all lengths present in the studied sample (mean length 66.68 mm, ∂ 21.99; width 19.15 mm, ∂ 8.08). Only pieces that were interpreted as having been used for drilling fresh hard material (most likely bone) were smaller, and those identified as possibly having been used for shaving wood were larger. Potentially either of these could have been a specialized activity, for which tools with specific features were made or selected. However, the small sample size should serve as a caution to over-interpreting the significance of the results.

Microliths

Only 64 pieces were identified as microliths, although a large number (166) of additional pieces met the length requirement of a microblade. A total of 230 lithics (38.33%) could be labeled as microliths (length < 40 mm). Since a number of the debitage blades (69 out of 245) could also be included in the microlith group, this would make the microliths the largest category.

Overall the pieces from W II/layer 5 had a very low average length 38.51 mm (∂ 19.56) and width 15.58 mm (∂ 9.25). The pieces that had discernible traces of use were even smaller (36.48 mm) than the overall mean length but the variation in size increased slightly (∂ 20.37), suggesting that very small and large pieces were most likely frequently used. Considering that a significant portion of each category could qualify as microliths, I analyzed only blades and microblades together, and all the other pieces were considered within their respective groups, so as to evaluate whether size was a significant factor in any particular activity.

As with the blades, cutting was the dominant activity for which the microblades were most likely used, followed by scraping. Traces of all other activities were represented only marginally, partly probably due to the overwhelming majority of fresh materials (meat, hide, bones), as the worked substance.

Table 6.8. Willendorf II/5, Microblades: Traces of use-wear, suggested activities, and worked materials

activity	fresh	soft	wood	frozen	count	%
cutting	41	2	—	—	43	63.24
scraping	15	3	—	2	20	29.41
shaving	—	—	3	—	3	4.41
drilling	2	—	—	—	2	2.94
count	58	5	3	2	68	100.0
%	85.29	7.36	4.41	2.94	100.0	

Size does not appear to have been a factor in the choice of individual microliths for any particular activity, since most of the used pieces ranged 10 to 30 mm in length, with slightly larger pieces having traces compatible with shaving wood. The two pieces that had traces at the tip of the blade, most likely resulting from rotation involved in drilling, were larger, but again the small sample size cautions against any further analysis.

Burins

Burins seemed to have formed only a small portion of the collection in W II/layer 5, and only 35 pieces were identified as such. On the other hand, 21 additional pieces were either burin spalls, or were retouched with the burin blow technique, yet typologically identified as blades, debitage, a point, or as combination tools. The burin edge was not utilized in any uniform fashion. Occasionally it appears to have facilitated hafting, at other times it seems to have been used for shaving wood or scraping bone, hide or some similar matter, yet in many cases it showed no traces of use at all. The multiple patterns of use, as well as the frequent absence of utilization of the edge created by the burin blow technique supports the suggestion made in the analysis of the Pavlov collection. Namely that the majority of lithics labeled as burins are likely manufacture byproducts that are occasionally put to secondary use.

A third of the burins (11 pieces.) were similar in size to the microliths, and only one had any additional retouch, indicating that these were all probably extended cores and/or flakes from microblade production.

Table 6.9. Willendorf II/5: Traces of use-wear on burins, suggested activities, and worked materials

activity	wood	fresh	count	%
shaving	10	—	10	45.46
scraping	—	9	9	40.91
batter	2	—	2	9.09
drilling	—	1	1	4.55
count	12	10	22	100.0
%	54.55	45.46	100.0	

The most common use of burins appears to have been woodworking, closely followed by scraping fresh material, probably bone and/or hides. I have not observed any traces of other activities or worked materials. Size did not seem to play a significant role in terms of actual use, as both microburins and regular burins showed traces of woodworking, as well as scarping fresh matter. However, within the groups themselves, the smaller tools were observed as having more often traces of scraping, while the larger pieces were more frequently associated with traces of wood-shaving.

This group of tools points out yet again the need to reconsider the assumption that all used tools were intentionally produced for a specific purpose. Many burins may have been an unintended consequence, or a byproduct resulting from one of the steps in a manufacturing process, a debitage that was eventually used. As noted elsewhere, the typological approach to lithic analysis does not address this question sufficiently, and commonly masks it by treating all tools as finished variations of an ideal type.

Endscrapers

W II/layer 5 had 35 endscrapers. Similarly to the Pavlov collection, these are one of the least problematic tool categories. Nevertheless some questions appear, particularly about the specifically functional definition of this tool type, when a wide range of other lithic pieces performed this function. When the endscraper is defined in terms of retouch of the distal or the proximal end, we are still left with a number of lithics (11) that were made on a blade or a flake, with a retouched end. Yet these were not included in the endscraper category. Only 11 of the endscrapers were made on blades, flakes were the more common shape (24 pieces). A half of the group was microlithic in size, complicating the distinction between microliths and the rest of the tools. However, the use-wear traces were quite uniform, and appear to have been the result of a fairly standard use in scraping fresh materials, and woodworking. Out of the 29 endscrapers that showed traces of use, 15 had also traces of hafting. Seven were broken and I could not determine whether they were hafted, and the remaining 7 did not appear to have had any traces resulting from some form of attachment to a handle.

Table 6.10. Willendorf II/5: Traces of use-wear on endscrapers, suggested activities, and worked materials

activity	wood	fresh	count	%
shaving	16	—	16	55.17
scraping	—	13	13	44.83
count	16	13	29	100.0
%	55.17	44.83	100.0	

Size did not appear to have been a significant factor, as endscrapers in both suggested activities were within very similar size ranges.

Scrapers

As noted earlier, scrapers were recently a focus of some interest in the analysis of stone implements from earlier time periods (Dibble 1984, 1987, 1988; Kuhn 1991, 1992), suggesting that at least some forms were more likely reduction stages than distinct types. The point may be well taken when dealing with the small group of scrapers in W II/layer 5. Only 10 pieces were identified as side-scrapers, and four of them were quite clearly extended cores that did not show any traces of wear. Two pieces had some traces that could be interpreted as a result of use, but one of them was an endscraper. On the other hand, a number of assorted pieces from various other typological categories had traces that were compatible with those observed in scraping with a side of a flake or a blade, either retouched or blank. This suggests that the typology grouping may be useful in the initial sorting of the material but additional information should be considered in the final analysis of a collection.

Points

Several pointed blades (7 pieces) were singled out into a separate category of points. Three of them were backed microliths (Gravette points), and one a resharpening blade taken off a micro-gravette. The rest were two unilaterally retouched blades, and a burin-blow retouched blade, forming a rather mixed group. Four of these blades had traces that could have been made by either cutting or scraping. On the other hand, 106 additional lithics were pointed blades, debitage, or microliths, indicating that the same feature was present in other pieces as well.

Summary: The lithic collection in W II/layer 5.

The preceding discussion focused on the problematic nature of individual typological categories, showing not only the inconsistency with which they are applied, but at the same time stressing how such inconsistency is inevitable once the typological approach is adopted. Many lithic categories are described in rather general terms (e.g. blades or points), while a large number of recovered lithics is not a finished ideal type that could easily be assigned to a particular typological category. We may gain a greater insight into the prehistoric lives of the makers of these stone implements if we analyze the collection in terms of use-wear, considering the tool type only as a secondary information, reversing the former approach, when tool typologies were the main concern driving the entire analysis of any stone tool collection. The analysis of the collection of stone tools from Layer 5 at Willendorf II would be as follows:

The assemblage that I examined contained 600 pieces, with 69 (11.5%) weathered stones where the traces of use could not be determined. Traces of hafting or use-wear were identified on 257 pieces (48.4%) but 29 of these were broken blades with only the hafting end preserved. thus allowing suggestions regarding use and possible worked materials only on 228 pieces. On the basis of the microscopic observations, the remaining 274 pieces (51.6%) could not be associated with any particular activity.

Table 6.11. Willendorf II/5: Traces of use-wear on all pieces, suggested activities, worked materials (multiple activities in brackets)

activity	fresh	wood	soft	frozen	count	%
cutting	80	—	9	6	95	40.95
scraping	67	—	8	3	78	33.62
shaving	—	38	—	—	38	16.38
drilling	8 (+4)	—	1	—	13	5.6
carving	3	2	—	1	6	2.59
batter	—	2	—	—	2	0.86
count	158	42	18	10	228	100.0
%	69.3	18.42	7.9	4.39	100.0	

The overwhelming majority of the stone pieces (69.3%) appear to have been used in processing fresh organic materials (e.g. bone, meat, hide), where cutting (40.95%) and scraping (33.62%) seem to have left the most observable traces. Four pieces may have been used for several activities, as they had both traces along the edge of what could be interpreted as cutting, and at the tip traces of possible drilling and/or piercing. Wood seems to have been the second most common worked material, though trailing far behind the first group, with only 18.42%. Shaving wood for either construction material, fuel, or some other purpose dominated. Judging from the use-wear traces, the wood was collected rather than cut, as no lithics appeared to have any traces of woodcutting. Only two pieces had traces that I would identify as results of carving, thus either this was a completely marginal activity, or it might have been a very short lasting activity that left few traces. Plant matter could have been worked to a limited degree (7.9%), and coarse and/or potentially frozen material was only marginally represented (4.39%) in the traces observed.

Table 6.12. Willendorf II/5: Size of all pieces used in different activities (mm)

activity	count	length	width
cutting	95	8-80	2-28
scraping	78	12-65	3-25
shaving	38	27-58	15-27
drilling	13	21-50	9-20
carving	6	25-65	17-37

Cutting, the most commonly observed activity, does not appear to have been restricted to a particular size, since associated lengths range from the minimal 8 mm to 80 mm. Wood working, on the other hand, appears to have been the one activity for which size requirements were quite specific, since shaving and carving appear to be the two activities for which larger pieces were used. This distinction could have been purely functional, and woodworking tools might have been specialized implements with specific requirements placed on them.

Layer 6

Four new C-14 dates for W II/6 were obtained from charcoal in 1993 (Haesaerts et. al. 1996), and all clustered around 27 – 26 K years BP (27,600 ± 460, 26,500 ±480, 26,150 ± 110, 27,620 ± 230). Three dates were taken from the same sample, suggesting a contamination due to solifluction (Haesaerts et. al. 1996: 36). The authors indicate that the date should be around 26,500 – 26,150 BP.

The original interpretation, based on sedimentology, suggested that climatic conditions were assumed to have been the same as they were during the occupation in W II/5: cold but relatively humid, tundra-like environment (Haesaerts 1990:209). The cultural horizon in this layer was 20 cm thick, and was found 70 cm above W II/5 (Felgenhauer 1959: 93). Although levels 5 and 6 were separated from each other by a distinct sterile loess layer, Otte (1990:220) considered them to be a part of the same "stage" 30.000-28.000 B.P., characterized by microliths, and represented not only by the French site La Gravette, but also Mauern, Geissenklösterle, and Dolní Věstonice. The new dates and a more detailed pedology associate the cultural layer with those superimposed above (layer 8), indicating a more recent date around 26,000 BP, and a cold and dry climate (Haesaerts et. al. 1996: 40-41)

Layer 6 contained the fewest artifacts of the layers under consideration, with a lithic assemblage of only 199 pieces. Raw material proportions differ slightly from W II/5, as flint decreased by 10% to 64.3%, while sandstone (up by 2% to 13.6%), quartzite (up by 10% to 10.6%), and radiolarite (up by 2% to 9.05%) all increased. However, flint remained the predominant raw material, both in terms of presence as well as use. (see graph 6.2 below)

Typologically W II/6 exhibited slight changes compared to W II/5. Blades and debitage were still the two dominant categories but blades now predominate (45.73% as opposed to 33.5%), and debitage ranks only second (41.71%). Microliths drop drastically from 10.67% in W II/5 to 1.01% in W II/6. The remaining typological categories remained relatively stable, with only slight changes. I will examine the assemblage in this layer using the same criteria as before, in order to evaluate whether the change is a result of a shifting patterns in the prehistoric record, or whether different classification of some stone pieces may have resulted in the distribution observed.

Table 6.13. Willendorf II/6: Typological distribution

typology	count	%
blade	91	45.73
debitage	83	41.71
endscraper	10	5.03
burin	8	4.02
point	3	1.51
microlith	2	1.01
scraper	1	0.5
hammer	1	0.5
count	199	100.0

The overall size of the assemblage in W II/5 was smaller (mean length 42.49 mm, ∂ 20.16, mean width 15.88 mm, ∂ 9.6) than the size in W II/6 (mean length 49.8 mm, ∂ 17.3, mean width 19.85 mm, ∂ 10.2). One has to remember that the difference of several mm is in most contexts rather minor.

Table 6.14. Willendorf II: Lithic size in layers 5, 6 (mm)

cult. layer	mean l.	∂	mean w	∂
Layer 5	42.49	20.16	15.88	9.6
Layer 6	49.8	17.3	19.85	10.2

Blades

Overall, the assemblage from W II/6 was dominated by blades, as 150 out of the 199 pieces were blades, while only 44 flakes were present, and cores were in the minority with 5 pieces.

The largest typological group in W II/6 consisted of 91 blades. Six of these did not meet the basic criteria (length < 2 width) and thus were actually flakes (11 broken blades that did not meet the criteria were not included in the count). An additional 24 pieces were the same size or smaller than microblades (length < 40), and out of the remaining 64 blades, 44 pieces had some form of retouch. Hence only 20 blank blades (21.98% of the total) satisfy all the criteria, and could not be included in any other category. This situation is similar to the breakdown in W II/5 where only 22.39% blades met all the requirements of the blank blade category.

Although the proportion of microliths appears to have dropped drastically in layer 6, only 2 pieces were identified as such, a number of blades (24 pieces) were smaller than 40 mm, suggesting that the reduction in the microlithic component may not have been as dramatic as the initial categorization suggested. Moreover, 68 pieces in the whole assemblage were below the 40 mm length, contributing to a revised figure of 34.17% for the microlithic component of the collection. This is lower than the microlithic component in W II/5 by 4% which suggests only a slight, insignificant reduction.

Only about a quarter of the analyzed blades showed any traces of hafting or use (23 pieces, 25.28%). This is a very similar low percentage of use-wear traces on blades as in the previous W II/5 (24.64%), but W II/6 had an overall low rate of use marks. A majority of the blades (68%) showed traces that could be interpreted as having resulted from cutting fresh and frozen organic materials (such as meat, bone, hides). A few other pieces had any traces of wear, and a number of lithics might have been used for several activities, showing multiple traces on one piece. Particularly three blades that had parallel cutting marks, appeared to have had also rounded tips that could have resulted from drilling, rotating action. One blade appeared to have been used for cutting as well as for bone carving. Only two blades were observed to have traces of woodworking, and plant processing traces seem to have been entirely absent.

Table 6.15. Willendorf II/6: Traces of use-wear on blades, suggested activities, and worked materials (multiple activities in brackets)

activity	fresh	frozen	wood	count	%
cutting	8	9	—	17	68.0
drilling	(1)	(2)	—	3	12.0
scraping	2	—	—	2	8.0
shaving	—	—	2	2	8.0
carving	(1)	—	—	1	4.0
total	10	9	2	21	100.0
%	47.62	42.86	9.52	100.0	

Debitage

Eighty-three pieces from W II/6 were identified as debitage. The majority was blades (60.24%), while flakes comprised 34.94% of the group, and four extended cores made up the rest. Some of the debitage had no retouch (26 pieces, 31.33%), most of these (22 pieces) were blank blades and could have been classified together with the blades in the previous category. The remaining pieces that had some form of modification were mainly retouch blades or flakes, removed as a part of the manufacture process, as none of them were resharpening blades.

The cores in this group did not show any traces of use beyond the initial blade and/or flake detachment. The remaining lithics had a few traces of use but maximum utilization of material does not appear to have been a major concern, as the majority of the pieces had no observable traces of wear. However, a short-term use that left no visible traces should not be entirely excluded. Size of the debitage in this layer was larger than in W II/5, with mean length 42.25 mm (∂ 13.2), and mean width 23 mm (∂ 11.8).

Table 6.16. Willendorf II/6: Debitage wear and haft by shape

shape	count	%	wear/haft	%
core	4	4.82	0	0
flake	29	34.94	5	17.24
blade	50	60.24	5	10.0
total	83	100.0	10	12.05

The group of lithics that had any traces of use was very small, making any statistical comparison relatively meaningless. The one noteworthy observation gained in the analysis of this group, just as in the former blade group, is that no traces consistent with the processing of plant material were present in this layer.

Table 6.17. Willendorf II/6: Traces of use-wear on debitage: suggested activities and worked materials (numbers)

activity	fresh	wood	frozen	total	%
shaving	—	3		3	33.33
scraping	1	—	1	2	22.22
cutting	2	—	—	2	22.22
drilling	1	—	—	1	11.11
carving	—	(1)	—	1	11.11
total	4	3	1	8	99.99
%	50.0	37.5	12.5	100.0	

Endscrapers

A total of 10 endscrapers were found in W II/6, seven made on flakes, and three on blades, one of them a bipolar scraper. Three of these were the same size as the microliths (length < 40 mm). On the other hand, the scraping of various materials and shaving wood left discernible traces on a total of 19 lithic pieces in the whole assemblage from this layer, almost twice the number of those typologically associated with this function. Within the endscraper category, six showed some marks of wear, the majority most likely left by wood-working, and one endscraper had traces that were compatible with scraping fresh bone or similar hard material.

Burins

A total of eight thicker blades was modified by the burin blow, and were thus assigned into the burin category. None of these had any further modification traces, and could have been debitage from blade production, used later for other purposes. Only two of the burins could be labeled as microburins. This small number further supports the suggestion that microburins were extended microblade cores, as W II/6 had a very low count of microblades; therefore we would not expect a larger number of microburins than present.

Five of the burins had traces that could have resulted from use, three of them possibly from scraping fresh harder material such as bones, and two appeared to have been used in wood-working. The location of use marks along the edges varied with each piece, suggesting that the resulting burin shape was not necessarily a premeditated product. The size of the pieces varied slightly. The regular burins all ranged in length from 60 to 80 mm and 20 to 30 mm in width. This is the upper end of the common size of burins in most of the layers at Willendorf II (mean burin length for all layers is 51.67 mm, ∂ 17.67, mean width 20.74 mm, ∂ 5.85). Therefore I would conclude that criteria of size do not appear to reveal any specific information that would provide any further insight into the prehistoric activities at Willendorf.

Points

A small number of blades were labeled as points (3 pieces). This category remains unclear in many Palaeolithic contexts, since typologically characteristic points have certain distinctive features that are not always present in all blades labeled as points. Gravettian points have a typical backing along one edge and a modified ventral face of the base of the point. Font-Robert points have a distinct bilaterally retouched stem. Solutrean points have a flat retouch on both sides of the blade, Azillian points have retouch along the convex edge of the blade. Shouldered points have a concave retouch at one edge of the base. All of these subcategories are described in detail and are easily recognizable in most collections.

However, a number of pointed blades are frequently labeled as "points", even if they lack any additional retouch that would justify the creation of a separate group. The layer 6 at Willendorf II is a just such a case, since the three points do not feature any additional attribute that would distinguish them from all the other pointed blades in the assemblage. There were five additional pointed blades in the blade group, and one of them was a backed point (Gravette type), yet it was classified as a blade. Further five debitage pieces were pointed blades, and the two identified microliths were pointed as well. Consequently, I would suggest that points should either be included in the general blade group, if we were to use such a system, or only those with any additional completed retouch should be grouped together.

Two out of the three identified points had traces that could be interpreted as results of use. Both parallel striae along edges suggest cutting, in one case of a fresh organic material, in the other frozen hard substance. These proposed activities further question the explanatory value of a separate category for these particular pointed blades, as they appear to have been used for cutting like many other blades, rather than piercing as is usually assumed with points.

Microliths

Only two pieces were identified as microliths, yet as mentioned earlier the dramatic decrease in the microlithic component of W II/6 compared to Layer 5 is more a result of the classification than a function of prehistoric behavior. Applying the length as a major criteria of a microlith, I have concluded that 68 microliths were recovered from this layer, contributing 34.17% of the total assemblage, a proportion only slightly lower than the 38.33% in Layer 5.

Table 6.18. Willendorf II/6: Traces of use-wear on microblades suggested activities and worked materials (out of total 68)

activity	fresh	wood	frozen	total	%
cutting	4	—	1	5	71.43
shaving	—	2	—	2	28.57
total	4	2	1	7	100.0
%	57.14	28.57	14.29	100.0	

Only 7 of the microblades had any traces that could be interpreted as having been produced by cutting or scraping materials. Four of them seem to have been a result of cutting fresh softer material (e.g. hides or meat), while two had traces compatible with wood shaving, and one microblade appeared to have been used for cutting frozen hard bone or any similar organic material.

Summary: The lithic collection in W II/6.

The assemblage from layer 6 at Willendorf II contained 199 pieces. It was the smallest of the five examined Gravettian layers (W II / 5-9). Although I would suggest that the typological change attributed to the sequence of levels 5 and 6 was not as distinct as some have proposed (contra Otte 1981), it appears that some shifts did occur between the two layers, particularly when judged in the light of the results of the use-wear analysis.

Out of the 199 stone artifacts recovered from this level, 30 pieces (15.08%) could not be evaluated for traces of use-wear as they were weathered (23 pieces) or fire-cracked (7 pieces), processes that usually obliterate any visible marks. Only 48 pieces (28.4%) had any discernible traces of hafting or use, which is a far lower percentage of use than that in W II/5 (48.4%). Four blades were broken, with only the hafting end preserved, thus not allowing any proposition of possible activities or worked materials. Additional differences were observed in terms of the type of activities and worked materials that I proposed for each layer on the basis of the use-wear results. These will be described in the following chapter. In terms of the possible activities that could have taken place at the site during the occupation of W II/6, the observed traces of use could be interpreted as follows:

Table 6.19. Willendorf II/6: Traces of use-wear on all pieces suggested activities and worked materials (multiple use in brackets)

activity	fresh	wood	frozen	total	%
cutting	13	—	10	23	46.94
shaving	—	11	—	11	22.45
scraping	7	—	1	8	16.33
drilling	1+(1)	—	(2)	4	8.16
carving	(2)	1	—	3	6.12
total	21	12	11	44	100.0
%	47.73	27.27	25.0	100.0	

On the basis of use-wear, I would suggest that most common apparent activity in this layer was processing fresh organic materials (such as meat, bone, hides). Woodworking and cutting frozen hard materials seems also to have been practiced, but to a lesser degree. Judging from the use-wear analysis, plant material may not have been processed at the time at all, or for such a short period of time, and to such a limited degree, that this activity may have left no traces. Hence, I would propose that this assemblage might have been produced over a cold time period when little fresh plant material was available, but frozen hard matter was present. Judging from the small number of artifacts, and the low percentage of used artifacts, I would suggest a short stay oriented towards a particular hunting activity. While a comprehensive overview of faunal evidence extends beyond this study, I note that the faunal remains from this layer include bones of reindeer, ibex, cave lion, bear, and wolf. Which of these were hunted or scavenged cannot be determined on the basis of the data available at the moment. The mammoth bones that were abundant in Layer 5 were absent in Layer 6.

The overall mean size of lithics that had observable traces of use was 51.29 mm in length (∂ 15.7), and width 19.97 mm (∂ 9). Pieces that were interpreted as having been used for cutting ranged from the smallest ones to those quite large in size, suggesting that specific requirements were not necessarily in place, and any cutting blade was suitable. The largest pieces appeared to have been used for scraping, rather than wood-shaving, as was the case in W II/5. The few pieces that had any marks left possibly by carving or drilling tended to be from the middle range rather than small.

Layer 7

This layer does not have a C-14 date, and has been evaluated only typologically (Felgenhauer 1959, Otte 1981, 1990). Furthermore, the existence of a separate cultural horizon in W II/7 has been questioned by researchers who attempted reconstructions of the sequence (Felgenhauer 1959:93, Haesaerts 1990:207). They noted that W II/7 was very poorly delineated and absent in the photographic documentation, and that the new profile created in 1981 also failed to provide a clear definition of it.

The time frame proposed for levels 7 and 8 was between 28.000 to 26.000 BP (Otte 1990: 220), but the new dates obtained in 1993 shifted towards 25-26,000 BP (Haesaerts et. al. 1996). New evaluation of the pedogenesis allowed a more detailed knowledge of the palaeoclimate, and the authors suggest that starting with W II/6, the climate became drier, colder, and more tundra like (Haesaerts et. al. 1996:40-41).

Raw material distribution changed to a high degree in W II/7. Flint contributed only 44.5%, while radiolarite increased markedly to 38.16% (from 9.05% in W II/6 and 7.17% in W II/5), and sandstone remained at about the same level at 13.07%. Quartzite, on the other hand, dropped from 10.55% in W II/6, to 2.47% in W II/7. The remaining chalcedony and jasper stayed at about the same low percentage as before. (see graph 6.3 below)

The assemblage recovered from W II/7 had the highest proportion of blades (82.33%) among the layers analyzed, and the lowest share of cores with only 2 pieces (0.71%). The typological distribution was also noteworthy in the predominance of blades and lower amount of debitage than in any of the other layers.

Table 6.20. Willendorf II/7: Typological distribution

typology	count	%
blade	211	74.56
debitage	51	18.02
endscraper	6	2.12
scraper	5	1.77
burin	4	1.41
microlith	4	1.41
point	2	0.71
total	283	100.0

Blades

The largest group was composed of 211 blades. Nineteen of these did not meet the basic criteria of a blade (length ≥ 2 width). An additional 40 blades were the same length as microblades present in the assemblage (broken blades were not judged by the same measure), and were therefore analyzed together with that group, leaving only 152 blades in the original typological group. Only 17 of these were blank blades, and the remaining 135 (88.82%) had some form of modification that would enable their inclusion in some other typological category. Thus only 8.06% of the blades in the largest category were actually blank blades that met all the criteria for their separation into an exclusive group.

Table 6.21. Willendorf II/7:
Traces of use-wear on blades suggested activities and worked materials (out of 163) (multiple activities in brackets)

activity	fresh	frozen	wood	soft	total	%
cutting	45	3	1	1	50	71.43
drilling	3 (7)	—	—	—	10	14.29
scraping	8	1	—	—	9	12.86
shaving	—	—	1	—	1	1.43
total	56	4	2	1	63	100.0
%	88.89	6.35	3.18	1.59	100.0	

Out of the 163 blades considered, 69 pieces (42.33%) had marks of either hafting or use. Only 63 blades could be evaluated as to the possible material that might have been processed with the stone tools, as the additional 6 pieces were broken and only the hafting ends of these blades were recovered. The use-wear traces observed suggest that an overwhelming majority (71.43%) of blades in W II/7 were used for cutting fresh organic material (e.g. meat or hide). Only a few blades had traces compatible with scraping, possibly indicating that processing of the materials took place elsewhere. Seven blades with cutting marks, had also traces that would suggest rotating action, such as drilling. Only a few other traces were determined as having been the result of cutting coarse, frozen materials and wood. One blade had a polished round edge, compatible with traces left by cutting plant material over an extended period of time.

The mean length of all blades in this layer was 54.91 mm (∂ 17.2), and the mean width was 19.56 mm (∂ 7.2). The blades with traces of use were only a slightly longer (57.3mm), and the majority of those with traces that could be interpreted as having resulted from cutting clustered around the median size. The few blades with traces of scraping tended to be at the lower end of the range, and most of them appeared to have been hafted (6 out of 9), possibly in a composite tool.

Debitage

The second largest group consisted of 51 pieces (18.02%) labeled as debitage. The position of broken pieces is unclear, as it is difficult to assess whether these broke before, in the process, or after use, or postdepositionally. As noted before, debitage is generally ignored in typological studies, partly due to the lack of any definition of attributes that would have to be present for any one piece to be classified as debitage, the broken blades being one such a case in point. Furthermore, 14 burin spalls were classified as blades although they were most likely manufacture and/or resharpening byproducts.

A large number of the debitage pieces were the same size as the microliths in the layer, reminding us of the porous boundaries that exist between all the typological categories (21 pieces excluding broken blades). The mean size of the pieces with traces of use was 31.8 mm in length (∂ = 13.4), and 17 mm in width (∂ = 10.14). Both blades and flakes were equally represented in the group, as well as two extended cores. Considering that the overwhelming character of the assemblage in this layer was that of blades (82.33%), the flakes seem disproportionately high in the debitage group. On the other hand, adding up all the burin spalls and broken blades increases the whole group to 105 pieces, and the blade component of it to 82 pieces (78.1%). This proportion is then much more in line with the other layers, suggesting a less dramatic shift than the typological division proposed.

Table 6.22. Willendorf II/7: Debitage wear and haft by shape

shape	count	%	wear/haft	%
core	2	3.92	0	0
flake	21	41.18	4	19.05
blade	28	54.9	8	28.57
total	51	100.0	12	23.53

Table 6.23. Willendorf II/7: Traces of use-wear on debitage: suggested activities and worked materials

activity	fresh	wood	total	%
scraping	3	—	3	42.86
carving	2	—	2	28.57
shaving	—	1	1	14.29
cutting	1	—	1	14.29
total	6	1	7	100.0
%	85.71	14.29	100.0	

Only seven pieces out of the 51 showed traces of use. An additional six appear to have been hafted, yet the missing broken segment does not allow for interpretation of use. Considering the low instance of use or hafting marks, it is possible to suggest that the majority of debitage was most likely manufacture byproducts, and that resharpening dulled, or used tools was not a major concern. Only some traces of fresh material working could be suggested; other traces were not detected in this particular group.

Endscrapers

Six pieces were identified as endscrapers, although additional six blades showed traces of scraping on either end. Moreover a number of blades (10 pieces), and debitage (4 pieces) were used for scraping as well. Out of the six typologically described endscrapers, five had discernible traces of use. Fresh meat or bone (3 pieces), and wood (2 pieces) were the two kinds of materials that could be inferred as possibly having been processed. All the endscrapers were made on radiolarite or flint blades.

Scrapers

Five pieces were identified as side-scrapers. All of them were made on a flake but only one piece had any retouch along lateral edges, the remaining were either endscrapers on a flake (2 pieces), or reduced cores (2 pieces) that should have been assigned to the debitage group. In terms of size, the stones ranged in length from 20 mm to 43 mm, with only one piece exceeding 40 mm, casting all the other in the microlith category. Three of the scrapers had traces of use, two possibly having been used on fresh material, and one on wood.

Burins

Four pieces were designated as burins. Two of them were reduced cores that had traces compatible with wood-working, shaving and carving. One additional piece had traces that could be interpreted as resulting from scraping fresh softer material (e.g. hides) along the edge adjacent to the burin end. This piece was actually a microburin (length 32 mm, width 18 mm), most likely a core reduced in the production of microblades. The rest ranged in length from 50 mm to 80 mm, a size that seems to have been fairly common in implements that might have been used in wood processing.

Microliths

W II/7 appears to have had an equally low number of microliths as W II/6, since only four pieces were identified as such. However, as in the previous level, a large number of additional pieces met the length requirement to qualify for the status of a microlith, additional 84 pieces were smaller than 40 mm in length (the 29 broken pieces were not included in the category). Even with the increased number of microliths, this layer appears to have had a lower microlithic component than the previous two levels, with only 29.68%.

Table 6.24. Willendorf II/7: Traces of use-wear on microblades suggested activities and worked materials (out of 47)

activity	fresh	soft	total	%
cutting	9	2	11	44.0
scraping	7	—	7	28.0
drilling	2 (+5)	—	7	28.0
total	18	2	20	100.0
%	90.0	10.0	100.0	

Among the microblades 20 pieces out of 47 (42.55%) had traces that could have resulted from use. A great portion of them (16 pieces) appears to have been used for processing fresh softer materials (e.g. hide, or meat), as both cutting (9 pieces) and scraping (7 pieces) seem to have left discernible marks. In addition five of these pieces may have been used for drilling as well, as they had traces not only along the edges but also at the tip. Two more pieces had traces only at the tip, that may suggest rotating action such as drilling. The remaining two pieces were resharpening blades, that appear to have been removed from a stone that was likely used for cutting plant matter, and became dull from extensive use. Both edges had parallel striae and scratches, and were rounded with distinct glossy polish, all patterns compatible with cutting grass or similar vegetal substance.

Summary: The lithic collection in W II/7

The lithic assemblage from W II/7 consisted of 283 pieces, which makes it the second smallest of the Gravettian layers, after W II/6. This layer has traditionally been evaluated together with W II/8, as there is no C-14 date, and it is not entirely clear from the stratigraphic sequence what the exact location of W II/7 was. However, I will first describe the collection from this layer on its own, in terms of use-wear, and later will consider if the results obtained support any suggestion of links with the subsequent layers in the sequence

Table 6.25. Willendorf II/7: Traces of use-wear on all pieces, suggested activities and worked materials

activity	fresh	wood	frozen	soft	count	%
cutting	55	1	3	3	62	53.45
scraping	24	—	1	—	25	21.55
drilling	5(+12)	—	—	—	17	14.66
shaving	—	6	—	—	6	5.17
carving	4 (+1)	1	—	—	6	5.17
count	88	8	4	3	103	100.0
%	85.44	7.77	3.88	2.91	100.0	

Out of the total of 283 stone pieces, 25 pieces were weathered and one blade was burnt to such a degree that use-wear was not discernible. In the remaining sample, 116 stones (44.96%) had traces that could be interpreted as results of hafting or use. Only 103 of these were sufficiently complete to allow suggestions of materials that might have been worked with the stone tools, the 13 additional pieces were broken with only the hafting end preserved. Cutting appears to have predominated as an activity (53.45%), followed by a much lower degree of scraping (21.55%), and drilling (14.66%). Other activities (wood working and carving) seem to have been quite peripheral. Fresh organic material appears to have been the dominant worked matter (85.44%), as all the other materials were represented in the analysis only marginally.

Size did not seem to have played a major role, although a sense of "the right size" seems to have existed, as all the stone pieces that had traces of wear appear to have been from a very similar size range (20 to 80 mm in length). The only difference could be suggested for the cutting implements that included narrower pieces, as well as a standard upper limit of 30 mm. The mean length of all the lithics from this layer was 46.95 mm (∂ = 17.88), those that had possible traces of use were longer: mean length 48.82 mm, ∂ = 21.45, but included also smaller pieces (judging by the greater ∂). Further evaluation will follow in the comparative analysis of the whole sequence.

Layer 8

Layer 8, excavated in 1908, 1909, and 1927, was located only 50 cm above W II/6 (Felgenhauer 1959: 93). Clear definition of the cultural horizon was originally visible only in the middle section of the sequence, achieving 50 cm in thickness, disappearing in all other directions (Felgenhauer 1959: 93). A new profile cut in 1981 allowed a much better distinction of the layer, as well as charcoal samples for C-14 dating: GrN-11191: 25.800 ± 800 BP. (Haesaerts 1990: 211). Haesaerts suggested the onset of cold, dry climate, relative to the warmer humid environmental setting of layers 5 and 6 (Haesaerts 1990: 209). This interpretation was improved by the pedology and malacology results from 1993, indicating a generally cold and dry climate developing into the Upper Pleniglacial in the previous layers as well, with W II/8 represented a brief period of stabilization under less cold conditions (Haesaerts et. al. 1996: 40-41).

The research in 1993 unveiled a new hearth-like structure with surrounding stones, and an abundance of charcoal for dating. The samples taken allowed for conventional dates and cross dating, confirming the previous C-14 date (GrN-17801: 25,230 ± 320 BP; GrN-17802: 25,660 ± 350 BP). Furthermore, dates on bones from the middle horizon B2 fit in the time range (GrN-21690: 25,400 ± 170 BP; GrN-20767: 25,440 ± 170 BP) (Haesaerts et. al. 1996: 36-37).

As in W II/7, flint and radiolarite continue to be the dominant raw materials, but the proportion of flint increases by 15% (to 59.85%), while radiolarite drops by 11% (to 27.03%). Sandstone decreases by 8% (to 5.02%), and the proportions of the remaining raw materials change only slightly. (see graph 6.4 below) However, as I will mention later, these percentages should be considered within the wider context of the Willendorf II sequence, since there is some discussion (Haesaerts 1990: 209) whether W II/7 represents a separate cultural layer, or whether it would be better treated as a part of the W II/8.

Typologically, the lithic assemblage from W II/8 is distinct from the preceding layers in the significantly higher representation of endscrapers (8.49%), and an increased proportion of microliths (7.14%). On the other hand, the proportion of blades and debitage is more in line with W II/6, than with W II/7. A more detailed examination of the individual typological categories is as follows:

Table 6.26. Willendorf II/8: Typological distribution

typology	count	%
blade	222	42.86
debitage	200	38.61
endscraper	44	8.49
microlith	37	7.14
burin	8	1.55
scraper	5	0.97
point	1	0.19
hammer	1	0.19
total	518	100.0

Blades

The largest group in the assemblage consisted of 222 blades. Eleven of these did not meet the basic blade criteria (length ≥ 2 width); 63 blades were the same size or smaller as the microblades (length < 40 mm), and only 20 pieces had no traces of modification, while all the other blades (128 pieces) had various forms of retouch. Consequently, only 9% of the blades were blanks that could not be assigned to any other typological group.

The proportion of microliths seems to have increased in this layer compared to the previous one, yet a number of blades (63 pieces) were again not classified as such but instead retained in the blade group. Furthermore, 100 additional pieces were of the same size or smaller as the microliths, increasing the group to 198 pieces (38.22% of the whole assemblage).

I have included all the blades in the analysis, except for the microblades (i.e. blades shorter than 40 mm). Out of the 154 blades considered, 67 (43.5%) had traces that could be interpreted as having resulted from either hafting or use. The cutting and scraping of fresh material appears to have the primary activity for which the stone tools were used, indicating processing of the materials, besides just the primary cutting. A number of tools (7 pieces) had traces on disparate edges that indicated different activities in which the same stone piece could have been used, particularly cutting and scraping with same piece, and scraping and drilling. Traces related to the working of soft plant matter comprises the second largest suggested group, represented in a fairly substantial proportion considering the low visibility of plant processing activities in the archaeological record. Woodworking was also marginally represented in the observed traces, but the use of frozen coarse materials seems to have been entirely absent.

Table 6.27. Willendorf II/8: Traces of use-wear on blades suggested activities and worked materials (multiple use in brackets)

activity	fresh	soft	wood	count	%
cutting	22 (+7)	8	2	39	51.32
scraping	22	7	—	29	38.16
drilling	1 (+4)	—	—	5	6.58
shaving	—	—	2	2	2.63
carving	1	—	—	1	1.32
count	46	15	4	65	100.0
%	70.77	23.08	6.15	100.0	

Size did not appear to have played a significant role in the choice of blades for any one particular activity, as all the ranges of blades associated with cutting, scraping, or drilling fell within similar limits. The mean length of the used blades was 51.39 mm (∂ = 13.0), and the mean width 17.76 mm (∂ = 5.0).

Debitage

The second group of lithics in W II/8 consisted of 200 pieces of debitage. The debitage in this layer differed significantly from all the previous levels in the high proportion of cores present (38%), and a relatively low proportion of blades (40%), which ranged between 50-70% in the previous levels. In addition 26 blades, burins, and scrapers were also either resharpening pieces, reduced cores, or burin spalls that could be added to this group. The cores, despite their high presence, did not appear to have any marks of further utilization, suggesting an overall abundance of raw material, and/or little sense of a need to economize and make use of any available cobble.

Table 6.28. Willendorf II/8: Debitage wear and haft by shape

shape	count	%	wear/haft	%
flake	44	22.0	10	22.73
core	76	38.0	0	0
blade	80	40.0	32	40.0
total	200	100.0	42	21.0

Thirty-one pieces displayed some traces of use, and additional 11 pieces were broken hafting ends where use could not be estimated. A significant number of these were resharpening blades (13 pieces), the edges of dulled cutting tools. Fresh material, such as meat, hides, or bones, seems to have been processed most frequently, although plants and wood were also cut or scraped to a limited degree.

Table 6.29. Willendorf II/8: Traces of use-wear on debitage, suggested activities, and worked materials

activity	fresh	soft	wood	count	%
cutting	16	2	—	18	58.07
scraping	9	1	—	10	32.26
shaving	—	—	2	2	6.45
drilling	1	—	—	1	3.23
count	26	3	2	31	100.0
%	83.87	9.68	6.45	100.0	

The size of the debitage pieces that had traces of use-wear did not vary widely in length, regardless of suggested activity, although wider blades were more often linked with interpretations of scraping, and narrow blades were associated more frequently with cutting.

Endscrapers

The 44 endscrapers (8.49% of the collection) represent a significant proportional increase of this type in W II/8, relative to W II/7 (2.12%), and some increase compared to W II/6 (5.03%). The majority of them were made on blades (32 pieces), but a few flakes appeared as well (12 pieces). Although some smaller endscrapers were present (18 pieces were the size of microliths), the majority were only slightly smaller than the blades in the assemblage. The mean size of the scrapers with traces interpreted as results of use was 42.23 mm in length (∂ = 17.74), and 20.14 mm in width (∂ = 5.7).

Table 6.30. Willendorf II/8: Traces of use-wear on endscrapers, suggested activities, and worked materials

activity	fresh	soft	wood	total	%
scraping	24	9	—	33	97.06
shaving	—	—	1	1	2.94
total	24	9	1	34	100.0
%	70.59	26.47	2.94	100.0	

A majority of the scrapers (70.59%) had traces that could be interpreted as resulting from hide scraping, while plant working accounted for another quarter (26.47%). Only one scraper had marks that I would attribute to wood shaving. Frozen coarse material marks were not detected in this group of tools at all. Since all the activity that left traces on the tool edges appeared to have been of the same kind, it is interesting to notice some difference in size depending on the suggested material that might have been processed with the tools. It appears that the larger scrapers (25 - 82 mm long) were more frequently associated with hide scraping, while marks associated with plant materials were more often found on the edges of the smaller tools (18 - 46 mm long).

Microliths

Only 37 blades were identified as microliths, but additional 80 pieces of debitage (63 blades, 18 endscrapers, one point, a burin, and a scraper) were from the same small size range (below 40 mm). The total microlithic component of the assemblage from this layer then amounted to 201 pieces (38.8%), a higher proportion than in the previous layer 7 (29.68%).

I considered 100 pieces as a part of the use-wear study, 37 microliths and 63 blades, since the rest were analyzed and discussed within the respective categories where the size was discussed (see debitage and endscrapers).

Table 6.31. Willendorf II/8: Traces of use-wear on microliths, suggested activities, and worked materials

activity	fresh	soft	wood	count	%
cutting	16 (+4)	1	2	23	71.875
scraping	5	3	—	8	25.0
shaving	—	—	1	1	3.125
count	21	4	3	28	100.0
%	75.0	14.29	10.71	100.0	

Cutting fresh softer materials was the most common activity interpreted from the observed traces. Several microblades (4 pieces) had traces that were compatible with both cutting and scraping. The remaining microblades had marks suggesting plant and woodworking. Only nine of the blades with traces appeared to have been used heavily; the great majority of the traces seem to have been a result of moderate activity. Size did not suggest any variation in terms of the type of activity, but some differences were noted as to the materials processed. The traces associated with cutting or scraping fresh hides were observed on blades ranging from the smallest (9 mm) to the largest (39 mm long). Those associated with plant working were from the middle range (17 to 31 mm), and wood processing was linked to microblades in the 35 to 37 mm range.

Burins

Only eight pieces in W II/8 were identified as burins, and one of these was a microburin (length < 40 mm). On the other hand, eight additional pieces were burin spalls, one blade and seven debitage pieces. This amount is consistent with the low number of lithics with burin blow retouch in levels 7 and 6, but differs significantly from levels 5 and 9 where burins are represented in much higher numbers.

Three of the burins had traces that could be interpreted as a result of use or hafting, all of them appear to have been hafted, and possibly used in woodworking. The microburin had traces compatible with hide scraping. Only one of the burin spalls had any traces of use, being a resharpening blade removed from a dulled tool, apparently used in cutting bone or a similar hard fresh organic material. The remaining pieces were all reduced cores, flakes, or retouch blades with no obvious marks that would indicate utilization. The small sample size renders any statistical analysis meaningless.

Table 6.32. Willendorf II/8: Traces of use-wear on burins, suggested activities, and worked materials

activity	wood	fresh	count
shaving	2	—	2
scraping	—	1	1
carving	—	(1)	1
count	2	1	3 (4)

Scrapers

Five pieces were identified as side scrapers, yet three of them were actually reduced cores with no traces of utilization, and another thick flake with unilateral retouch was heavily weathered and its use-wear could not be determined. Only one flake had any traces that could have been interpreted as results of scraping fresh hide at a very high intensity and/or length, as the marks were remarkably pronounced, and the entire edge was rather rounded from use.

Points

Only one point in this layer was separated into a category of its own, but an additional 109 blades, 32 pieces of debitage, and 12 microliths were shaped into a point and could have been grouped together, unless some other criteria was imposed to separate them. However, no such explicit statement of typological logic was made in any of the studies that involved typological analysis of the Willendorf II collections.

Considering its size (length 34 mm width 7 mm), the one identified point should belong to the microlithic category. On the basis of retouch this object was identified as a Gravette point, yet another nine out of the 12 pointed microliths had identical retouch and were not grouped with it. Thus I would conclude that the category point is not only poorly defined but also redundant in the assemblage of this layer, as it overlaps with lithic pieces in a number of other categories in several respects. This particular point was backed and had a very round, polished opposite cutting edge that I would interpret as a result of scraping soft, possibly vegetal matter.

Hammers

One item was classified as a hammer due to its large size (length 210 mm, width 87 mm), and polished, slightly battered appearance. The traces seem to have been more likely caused by polishing, smoothing of leather, hide or some similar softer material. The sides of the piece had traces of bashing, so we may suggest that the stone might have been used in several steps of leather processing.

Summary: The lithic assemblage in W II/8.

The collection from W II/8 discussed here contained 518 stone pieces. A small number of them (59 pieces) could not be evaluated for use, as they were either heavily weathered (47 pieces) or fire-cracked (12 pieces). In the remaining group, 181 lithics (39.43%) had traces that I could interpret as having resulted from either hafting or use. A number of these (15 pieces) were broken blades, where only the segment that had marks compatible with hafting was preserved. One blade and an endscraper had potential hafting traces but no sign of use could be observed. These two pieces could have either never been used (although made and hafted), or they could have been used over a very short time period, possibly on softer plant matter that leaves traces of wear only after an extended use. Consequently, a total of 164 stone pieces could be discussed in terms of possible activities of use, and materials worked.

The lithics from W II/8 that I observed as having use-wear marks, seem to have been used equally for cutting (45.95%) and scraping (43.78%), mainly of fresh organic materials, such as hide, bones, or meat. A number of pieces had traces along several edges that appeared to have been caused by diverse activities, suggesting the possibility that the same tool might have been used for multiple purposes, or was reused later after the completion of the first activity. Whether this involved the same person using a multipurpose implement, or multiple users employing the tool during the same occupation, or its reuse during subsequent occupations cannot be determined from the data at hand. However, this phenomenon of using stone implements for multiple tasks was rare in W II/layers 5 and 6, and appeared only to a limited degree in W II/7. The multiple activities suggested in these layers included scraping, and either drilling or carving. The combined tasks in W II/8 were cutting and scraping. This may indicate some shift in practices, if we were to assume that cutting may have represented a primary activity, and scraping, drilling, or carving secondary, processing activities.

Other activities appeared to a lesser degree: use of plant materials (19.51%), and a minor amount of wood working (7.32%). Coarse frozen material seems to have been completely absent from the materials processed in W II/8, suggesting that the occupation may have taken place over a warmer time period (or periods), without extending into colder months when at least some frozen materials would presumably have been used.

Table 6.33. Willendorf II/8: Traces of use-wear on all pieces, suggested activities, and worked materials (out of 459)

activity	fresh	soft	wood	total	%
scraping	51 (+12)	21	—	85	45.95
cutting	66	11	4	81	43.78
shaving	—	—	8	8	4.32
drilling	2 (+6)	(1)	—	9	4.87
carving	1 (+1)	—	—	2	1.08
total	120	32	12	164	100.0
%	73.17	19.51	7.32	100.0	

Table 6.34. Willendorf II/8: Size of all pieces used in different activities (mm)

activity	count	length	width
scraping	85	20-80	9-30
cutting	81	17-60	3-29
drilling	9	16-58	12-30
shaving	8	20-70	7-23
carving	2	35-38	16-26

The mean length of all the unbroken stone pieces in W II/8 was 43.95 mm (∂ = 17), and the mean width was 19.06 mm (∂ = 9.9). The pieces that had identifiable marks of use were on average slightly longer and narrower, with a mean length of 45.58 mm (∂ = 21.5), and a mean width of 16.46 mm (∂ = 9.3). The standard deviation of length indicates that the range was quite broad, with small as well as very long pieces used. Pieces interpreted as having been used for either scraping fresh materials or woodworking tended to be slightly longer, and carving implements appear to have been smaller.

Layer 9

W II/9 was excavated during the 1908, 1909, and 1927 seasons, and new profiles were created in 1955, 1981, and again in 1993 (Felgenhauer 1959, Haesaerts 1990, Haesaerts et. al. 1996). Some discrepancy exists in the literature as to the exact position of this layer in relation to the preceding W II/8. Felgenhauer states, on the basis of Obermaier's and Bauer's plans, that it was located 1.40 m above W II/8 (Felgenhauer 1959: 94). Yet illustrations of profiles that Haesaerts presented in 1990, seem to suggest a much smaller distance between the two layers, 0.3-0.5 m (Haesaerts 1990: 210), indicating some recent postdepositional shift. This was further supported by the recent pottery mixed in the top of the layer when excavated in 1993, and the lack of success in attempt to recover some charcoal from the layer further into the wall of the profile (Haesaerts et. al. 1996). Consequently, the final samples that were dated were taken from a bone located in the Museum of Natural History in Vienna (GrN-21898: 23,860± 270 BP; GrN-22208: 24,370 ± 290 BP; GrA-5006: 24,910 ± 150 BP; GrA-5005: 23,180 ± 120 BP). The authors consider the younger dates to be a result of contamination by conservation of the bone in the museum, and view the dates in the 24,000 BP range to be valid (Haesaerts et. al. 1996: 38).

As in W II/5, flint is again the dominant raw material (74.3%), while the presence of radiolarite drops to 14.26%. Chalcedony, sandstone, and quartzite together contribute only 10%, other minerals are present only as few individual pieces. (see graph 6.5 below)

Typologically the assemblage was assigned to the central European variant of the Gravettian. Otte recently re-evaluated the Willendorf II sequence, and considered W II/9 as a proof of a complete break between western and eastern European regions based on typological analysis (Otte 1990: 221). The majority of the assemblage was composed of blades (69.16%),a proportion similar to that of the preceding layer. Cores, on the other hand, were not very abundant (3.1%). Typologically, shifts were noted mainly in terms of higher presence of debitage, and a dramatic increase in the number of burins (Otte 1981: 289).

Table 6.35. Willendorf II/9: Typological distribution

typology	count	%
debitage	440	42.68
blade	213	20.66
burin	116	11.25
endscraper	116	11.25
microlith	68	6.6
point	45	4.37
scraper	16	1.55
combination	8	0.78
hammer	7	0.68
borer	2	0.19
total	1031	100.0

Debitage

The largest group (440 pieces) was debitage, containing a variety of blades, flakes, and a few cores. The proportions of raw materials used in the debitage group match those for the whole assemblage, with flint dominating. Both the debitage blades and flakes appear to have had only a small amount of use-wear traces (13.15% and 16.36% respectively), and the cores had no discernible traces of utilization at all. The whole debitage group was microlithic in size, with mean length 36.74 mm for blades (∂ = 14.44), 27.97 mm for flakes (∂ = 15.72), and 43.42 mm for cores. An additional 87 pieces grouped in different categories could also be identified as debitage. These included blades, burins, endscrapers, and scrapers.

Table 6.36. Willendorf II/9: Debitage wear/haft by shape

shape	count	%	wear/haft	%
core	24	5.46	0	0
flake	165	37.5	27	16.36
blade	251	57.05	33	13.15
total	440	100.0	60	13.64

Out of the group of 440 pieces, 91 could not be evaluated for traces of use, as they were either weathered (14 pieces) or fire-cracked (77 pieces). The remaining group (349 pieces) contained mainly retouch blades and flakes, as only 60 pieces (13.64%) showed any marks of hafting or use. A number of broken blades were preserved that had only traces compatible with hafting (22 pieces), but their use could not be determined. Those pieces that were observed as having traces of use, seem to have been mainly employed in cutting and scraping fresh materials such as meat, hides, bones. Some of these (15 pieces out of 30) were resharpening blades/flakes, as the traces appeared very pronounced, accompanying rounded edges, polished dulled surface. Only a few stone pieces could be associated with processing other materials — plants (4 pieces), wood (3 pieces), or frozen materials (1 piece). Seven lithics had traces of possible multiple activities, all of them involved scraping, and either carving or drilling, most likely of bone.

Table 6.37. Willendorf II/9: Traces of use-wear on debitage, suggested activities, and worked materials (multiple activities in brackets)

activity	fresh	soft	wood	frozen	count	%
cutting	21	1	1	1	24	53.33
scraping	8	3	—	—	11	24.44
drilling	(5)	—	—	—	5	11.11
carving	1+(2)	—	—	—	3	6.67
shaving	—	—	2	—	2	4.44
count	30	4	3	1	38	99.99
%	78.95	10.53	7.9	2.63	100.0	

Blades

The second largest group consisted of 213 blades. A portion of them (19 pieces, 8.92%) did not meet the blade definition (length < 2 width). Another 58 blades were in fact microblades, as they were the same length, or smaller than the typologically defined microblades in the assemblage from W II/9. Only 15 remaining blades were blank with no modification, while 121 blades were either unilaterally or bilaterally retouched, a number were partly or completely backed. Thus only 15 blank blades (7.04%) could not be included in any other typological group.

Although the mean length of the blades was 53.63 mm, and of those with traces of use slightly higher (56.45 mm), the variation among them was extremely great (∂ = 23.99 and 28.36 respectively), and a significant number of them (58 pieces, 27.23%) were microblades.

Only blades longer than 40 mm will be considered here, as those falling within the range of microblades were included in that typological group, reducing the original 213 pieces to 167. Within this group 53 blades (31.74%) were weathered beyond recognition, and 63 blades (37.73%) showed discernible traces of either hafting or use. The remaining pieces did not have any marks that could be viewed at the magnification used for the analysis.

Judging from the observed use-wear traces, processing fresh materials was the most common activity for which blades were used. Scraping seems to have predominated, even over cutting (41.18% and 32.35% respectively). A number of blades (13 pieces) had traces at the tip that would be compatible with a rotating action, such as drilling. The majority of these latter objects (10 pieces) had also marks of scraping, indicating that these may have been used for multiple activities, or reused sequentially. Only 5 pieces showed traces that could be interpreted as having resulted from contact with wood, and frozen and plant materials appear to have been touched by blades only minimally, as only 3 blades showed any traces of working of either of these materials. Out of the 57 blades that were determined as having some traces of use-wear, 33 pieces (57.9%) also showed signs of having been possibly hafted. Seventeen blades were broken, and the possibility of hafting could not be established, and 7 blades did not appear to have been attached to any handle.

Table 6.38. Willendorf II/9: Traces of use-wear on blades suggested activities and worked materials (out of 167) (multiple activities in brackets)

activity	fresh	wood	frozen	soft	count	%
scraping	27	—	1	—	28	41.18
cutting	20	1	1	—	22	32.35
drilling	2(10)	—	—	1	13	19.12
shaving	—	4	—	—	4	5.88
carving	—	—	(1)	—	1	1.47
count	49	5	2	1	57	100.0
%	85.97	8.77	3.51	1.75	100.0	

Size did not seem to have played a major role in selecting tools for particular activities, since blades with traces linked to both scraping and cutting, were from the entire length range (50 mm to the long blades over 100 mm). Implements that could have been used for woodworking fall mainly in the mid-range rather than the higher end of the scale, unlike the pattern in the previous layers. However, given the small number involved, this result should not be over-interpreted.

Burins

The number of burins increased dramatically in W II/9, reaching 116 pieces as compared to the 8 pieces in the previous layer. The majority of the group was made on blades (92 pieces, 79.31%), and only a small part were flakes (24 pieces, 20.69%). Size did not range widely, with mean length of 52.09 mm (∂ = 18.5), and width of 21.03 mm (∂ = 4.8). A number of these (43 pieces) were the same size as the microliths, being possibly extended microblade cores. Burins that were larger than the microliths were of a fairly standardized size, measuring 61.7 mm in length (∂ = 16.8), and 21.74 mm in width (∂ = 4.9).

Within the whole group of 116 burins, 49 (42.24%) had traces that could be interpreted as having been caused by either hafting or use, while 11 pieces were weathered and of indeterminate use. Variation in portions of used edges and hafting seems to be fairly substantial, supporting the argument that most burins were likely reduced from larger pieces that were at times used for a number of purposes. Among the 43 burins that had discernible traces of use, 23 appear to have been hafted. A half of the hafting (12 pieces) seems to have been at the end that had the burin retouch, while the other half had traces of use along the burin edge.

Table 6.39. Willendorf II/9: Traces of use-wear on burins, suggested activities and worked materials (multiple activities in brackets)

activity	fresh	wood	frozen	soft	count	%
scraping	27	—	2	1	30	60.0
shaving	—	9	—	—	9	18.0
carving	1+(5)	(2)	—	—	8	16.0
cutting	3	—	—	—	3	6.0
count	31	9	2	1	43	100.0
%	72.09	20.93	4.65	2.33	100.0	

The majority of traces observed on the burins in this layer corresponds with the scraping either fresh materials, such as bone or hides, or, to a lesser degree, wood. Only three pieces had traces that could have been caused by contact with plant or frozen material. This pattern of activity appears fairly consistently throughout the layers where burins were present. Seven pieces had traces of multiple activities, mainly scraping and carving, where the burin shaped tip may have been used.

Burins that had traces which could be interpreted as having been caused by scraping were longer and wider, and cutting traces were more frequently associated with shorter and narrower lithics; other activities did not seem to have been tied to any particular implement size.

Endscrapers

The number of endscrapers increased slightly in W II/9, contributing 11.95% to the total assemblage. Most of them were made on blades (65 pieces, 56.03%) but flakes were common to a higher degree than in the previous levels (51 pieces, 43.97%). A number of the endscrapers (58 pieces, 50%) were microliths in size (below 40 mm) with mean length 41.5 mm (∂ = 18.06), and mean width 21.27 mm (∂ = 5.65). Those that had traces of use were slightly larger than endscrapers in the previous W II/8 (mean length 46.06 mm, mean width 22.32 mm) but hardly reaching the length of some of the longer blades present in the assemblage.

Table 6.40. Willendorf II/9: Traces of use-wear on endscrapers, suggested activities, and worked materials

activity	fresh	frozen	wood	soft	total	%
scraping	28	15	—	4	47	78.34
shaving	—	—	11	—	11	18.33
cutting	2	—	—	—	2	3.33
total	30	15	11	4	60	100.0
%	50.0	25.0	18.33	6.67	100.0	

A high proportion of the group (61 pieces, 52.59%) showed traces of either hafting, use, or occasionally of both (30 pieces). The traces observed on most of the endscrapers corresponded with the activity that is commonly associated with these tools — scraping. Fresh materials (mostly hides and bones) could be suggested as the most frequently worked, followed by scraping frozen, hard materials, and wood working. Plant materials did not seem to leave many traces on these tools. Hafting traces did not appear to have been associated with any particular worked material, and endscrapers that had hafting marks seem to have been used for the same types of materials as those that did not have marks interpreted as resulting from hafting. Size did not seem to have been a concern either when different materials were processed, as traces related to any plausible worked material were not associated with any particular size range of stone tools.

As discussed previously, although endscrapers are generally considered as the most indisputable, straight forward stone tools that most often have use-wear marks, it needs to be remembered that a large number of other blades and flakes may have been used for exactly the same kind of activity. This leaves us with a question why was a specific "endscraper" implement needed when other lithics served the same general purpose.

Microliths

The group of microliths in this layer consisted of 68 identified pieces (6.6%), about the same proportion as in the previous W II/8 (7.14%). An additional 299 pieces were from the same size range (less than 40 mm). These additional pieces were from such typologically diverse groups as 166 pieces of debitage, 57 endscrapers, 46 blades, 10 burins and points each, nine scrapers and a borer. Taken as a whole the increased group represented 35.6% of lithic materials from the layer, a slightly lower share than in the previous W II/8, where microliths contributed 38.8%.

I have analyzed all typological groups by themselves, without separating the microliths. The only exception were the blades where I only considered those over 40 mm in length and broken pieces, while microblades were added to the following analysis. Thereby the presented results included the study of 68 microliths and 46 microblades. Out of these 114 lithic pieces, 54 had traces that could be interpreted as having been caused by either hafting or use. Three of these were haft ends of blades with the rest broken, not allowing any suggestion of possible activities for which these could have been used. The remaining 51 microblades were interpreted in the following fashion:

Table 6.41. Willendorf II/9: Traces of use-wear on microliths, suggested activities, and worked materials (multiple activities in brackets)

activity	fresh	soft	count	%
cutting	28	2	30	57.69
scraping	15	5	20	38.46
drilling	(1)	1	2	3.85
count	43	8	51	100.0
%	84.3	15.7	100.0	

In keeping with the pattern observed in the previous lithic groups of this layer, the microliths seem to have been mainly used for processing of fresh materials (84.3%). Only a small portion of the microblades (15.7%) had traces that could have been caused by contact with plant materials. No other traces were observed under the magnification used, suggesting that neither wood nor frozen hard materials were worked with microliths. Size did not appear to be a factor in either different activities, nor in worked materials, as microblades with plausible traces of use-wear were from the same length range (mean length 28.47 mm, $\partial = 8.45$, mean width 8.74 mm, $\partial = 4.8$).

Points

A small number of blades were classified as points (45 pieces, 4.37%), a group that appears in a higher degree than in W II/8 (0.19%). However, just as in the previous layer, a number of additional pointed blades might be added to the group if the basis of the definition were clearly stated and strictly applied. Thus if we were to account for all modified pointed blades, the group would consist of 192 stone pieces (18.62% of the whole assemblage), a proportion lower than that in W II/8 (29.73%). Caution should be therefore exercised when evaluating any shifts and changes based solely on typological categories. The analysis presented here included the original 45 typologically defined points, as all the other blades were analyzed within their respective categories.

Table 6.42. Willendorf II/9: Traces of use-wear on points, suggested activities, and worked materials

activity	fresh	soft	count	%
cutting	19	—	19	86.36
scraping	1	1	2	9.09
drilling	1	—	1	4.55
count	21	1	22	100.0
%	95.45	4.55	100.0	

Out of the 45 points, 29 (64.44%) had traces that could have been caused either by hafting or by use. Seven of these were in fact broken, and only the hafting end was preserved, precluding any interpretation of the activity for which they may have been used. Twenty-two points had rather distinct marks parallel with their edges, most likely caused by cutting fresh material, such as meat, hide, or bones. Only one piece had traces that could have resulted from processing plant materials.

The size of the points should be noted mainly because it was even smaller than the size of the previous group of microliths. The mean length was 27.78 mm (∂ = 8.4), and the mean width 12.27 mm (∂ = 4.98). The fact that the points were shorter than the microliths (with the same ∂) makes the creation of this category the more puzzling, particularly since the other pointed blades that were larger were not included in it. If a consistent typological analysis were the aim of the study, then either this group should be merged with the microliths, or other points should have been included.

Scrapers

Sixteen flakes were identified as side-scrapers but eight of them were actually debitage pieces, and four of these were extended cores that had no additional traces of modification. Two of the scrapers had traces of use, both of them appeared to have been used for bone scraping. Only one of them was bilaterally modified at the proximal end, while the other had no apparent retouch and appeared to have been a residue flake with cortex still present on part of the dorsal side of the piece.

Combinations

W II/layers 5 and 9 had a few pieces that were identified as combination tools. All eight pieces in this layer combined endscrapers with a burin retouch at the opposite end. In five cases the burin ends seem to have been used for hafting. A total of seven pieces had traces of wear, and the one remaining stone was too weathered to allow any observation. Traces of scraping fresh material could be suggested in four cases, and traces of other materials (wood, plants, frozen hard substance) appeared in one case each.

Hammers

Seven larger pebbles were identified as hammers. All of them were extremely smooth, polished stones with some scratches, pitted surfaces, and rounded edges. Five of them had identifiable marks, while the two additional cobbles were too weathered to enable any interpretation of use. In all cases the smooth surface and gloss suggested polishing, smoothing type of action rather than a heavy impact usually induced by hammering. On the other hand, the smooth surface might have resulted from carrying the implement in a leather pouch, where constant contact with that material over an extended period of time could have contributed to a fairly uniform gloss similar to that observed.

Borers

The smallest typological group contained two borers, both bilaterally retouched pointed implements that could easily have fitted the description of the points in the collection. One of them was a longer blade (98 mm) with a retouched distant end, while the other was a short flake (38 mm) with a sharp point at proximal end. The two pieces had traces that could be interpreted as the result of drilling fresh harder material, such as bone.

Summary: The lithic collection in W II/9

The lithic assemblage from this layer consisted of 1031 pieces. Among these 205 pieces (19.88%) could not be evaluated in terms of possible use, as they were either heavily weathered (114 pieces, 11.06%), or burnt in fire (91 pieces, 8.82%), whereby any use-wear traces were obliterated. In the remaining group (826 pcs), 332 pieces (40.2%) showed traces of either use, hafting, or both. Four pieces appeared to have been hafted but did not seem to have any traces of use. These could either have been imperfect results, reserve tools made for future use and then abandoned, or tools actively employed for such a short time period that no use-wear marks were produced (particularly if soft plant matter was the material processed).

Table 6.43. Willendorf II/9: Traces of use-wear on all pieces, suggested activities, and worked materials (multiple use in brackets)

activity	fresh	wood	frozen	soft	**count**	%
scraping	110	—	19	15	144	45.0
cutting	95	2	2	3	102	31.88
shaving	—	26	—	—	26	8.13
drilling	5(16)	—	—	2	23	7.19
carving	2(14)	1+(2)	(1)	—	20	6.25
polishing	3	2	—	—	5	1.56
count	215	31	21	20	287	100.0
%	74.91	10.8	7.32	6.97	100.0	

Judging from the results of the use-wear analysis of the lithic collection in W II/9, processing fresh materials (meat, hide, bone) appears to have been the primary activity that took place at the location (215 pieces, 74.91%). A number of implements (23 pieces) that had traces interpreted as the result of scraping, had also additional processing marks resulting from drilling or carving. Not one stone piece had multiple traces that could be interpreted as having involved cutting and scraping, a pattern observed in the previous layer. Therefore, one of the possible interpretations that could be proposed would involve the separation of primary activities (cutting up an animal or a piece of meat), and secondary activities that involved processing of the acquired material (scraping, carving, drilling).

Marks consistent with shaving and working wood make up 10.8% of the use-wear observed. Although far smaller than the first group in terms of numbers (31 pieces), this was the category where the traces were most pronounced (58.07% heavy marks), suggesting heavy use of the same stone tools over a longer period of time. Overall moderate use of lithic implements seems to have predominated in this layer, since 57.14% traces suggested moderate length and/or intensity, and 39.72% indicated heavy, intensive use.

A small number of stone tools appear to have been involved in processing either frozen (7.32%) or plant materials (6.97%), suggesting that presence at the site occurred also during colder times when frozen meat, bones or hide were relied upon. However, from the current data it is not possible to infer whether this was a continuous occupation, or a return to the same site during a cold time season. Plant resources appear to have played a small role, although we always have to keep in mind the feeble traces that plant working leaves on stone tools (particularly if the additional factor of handling the artifacts for almost a century in the museum is taken into account).

Table 6.44. Willendorf II/9: Size of all pieces used in different activities (mm)

activity	count	length	width
scraping	144	20-120	5-35
cutting	102	14-65	3-30
shaving	26	35-85	16-30
drilling	23	24-66	12-35
carving	20	24-85	16-33

The overall size of the stones in this layer tended to be smaller than in other layers, with a mean length of 43.34 mm (∂ = 22.15) and a mean width of 17.04 mm (∂ = 9.4). However, those pieces observed to have some traces of use-wear appeared slightly larger than this average, and exhibited a greater degree of variation, with a mean length of 48.13 mm (∂ = 25.5), and a mean width of 18.61 mm (∂ = 9.78). In this instance size does seem to have been relevant in choice of tools for particular activities. Shaving and scraping traces were most frequently observed on longer pieces, and wood working marks on wider stones, frequently in the company of hafting traces (42.3% of wood working implements had possible hafting traces). Cutting marks appeared most often on smaller and narrower lithics, and were also frequently associated with hafting marks (64.71% of cutting implements appeared to have had also hafting marks).

Conclusion

Since the mid-1980's Austrian Palaeolithic research has emerged as a vibrant research field, and hopefully new studies will provide materials amenable to questions that go beyond the typological attribution of sites. Czech sites close to the Austrian locations are plentiful, but excavated and studied with stone tool typology as the sole focus, which makes any inferences as to the prehistoric human behavior very difficult. Furthermore, analysis of recovered materials in most cases seriously lags behind excavation.

An example at hand is Milovice, a southern Moravian site that could provide a potentially interesting comparison for the artifacts from layers 5 and 6 at Willendorf II, since it may be contemporaneous (Milovice: GrN 14826: 29,200 ± 950 B.P. No C-14 date exists for Willendorf II/Layer 6, but Layer 5 below has a C-14 date GrN 11193: 30,500 ± 900 B.P.). Geographically close, it has been excavated since 1988. However very few details have been released, beyond the statement that it is an Aurignacian site with an evidence of a large mammoth bone dwelling (Haesaerts 1990: 205; Oliva 1989, Svoboda 1994: 124). General statements abound about "Aurignacian" cultures developing, changing, and locally differentiating, but few suggest what changes in lithic morphology might mean in terms of the experience of prehistoric populations.

Graph 6.1. Willendorf II, Layer 5: Raw material distribution (%)

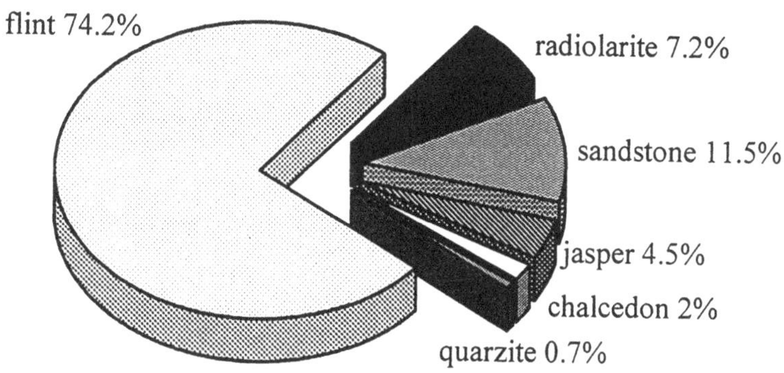

Graph 6.2. Willendorf II, Layer 6: Raw material distribution (%)

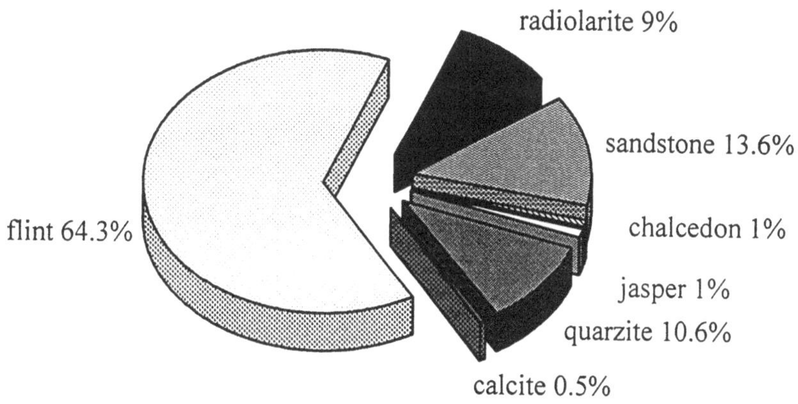

Graph 6.3. Willendorf II, Layer 7: Raw material distribution (%)

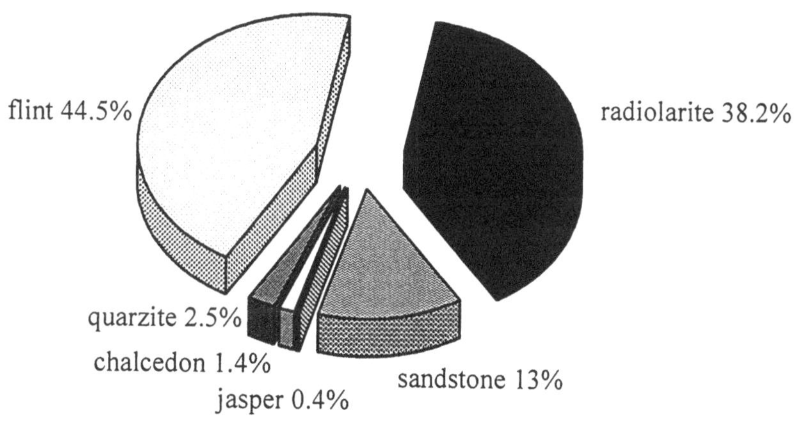

Graph 6.4. Willendorf II, Layer 8: Raw material distribution (%)

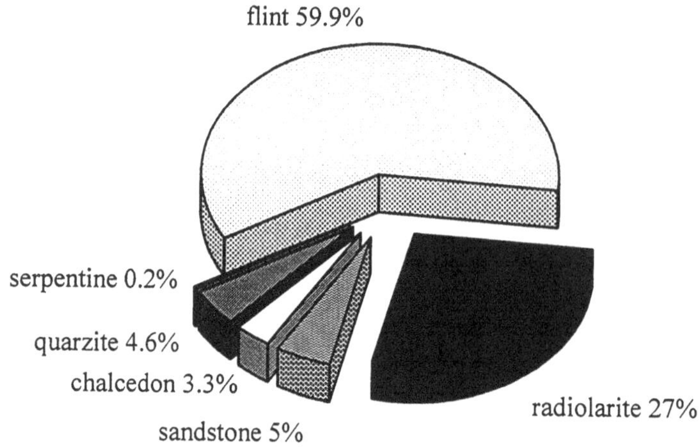

Graph 6.5. Willendorf II, layers 7+8: Raw material distribution (%)

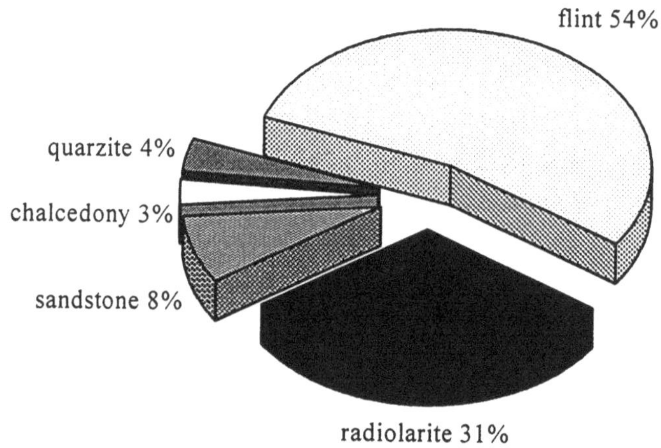

Graph 6.6. Willendorf II, Layer 9: Raw material distribution (%)

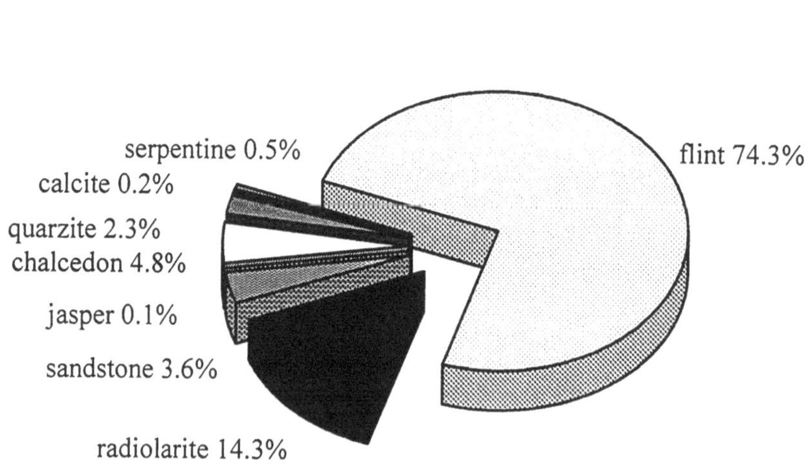

Graph 6.7. Willendorf II, Layer 5: Breakdown of suggested processed materials (%)

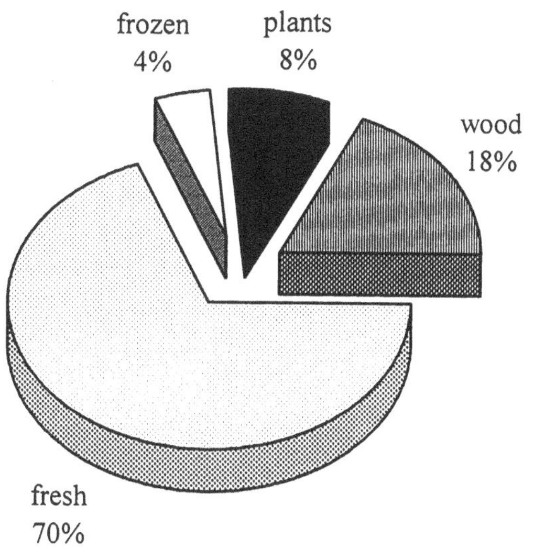

Graph 6.8. Willendorf II, Layer 6: Breakdown of suggested processed materials (%)

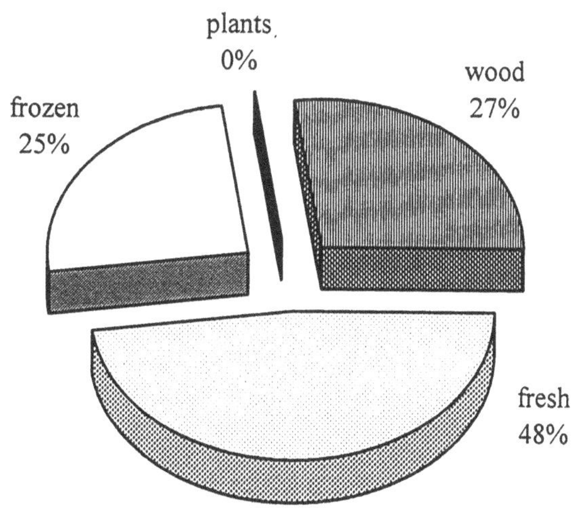

Graph 6.9. Willendorf II, Layer 7: Breakdown of suggested processed materials (%)

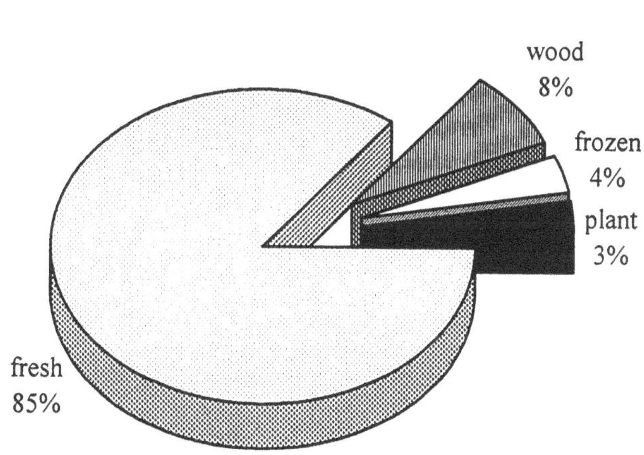

Graph 6.10. Willendorf II, Layer 8: Breakdown of suggested processed materials (%)

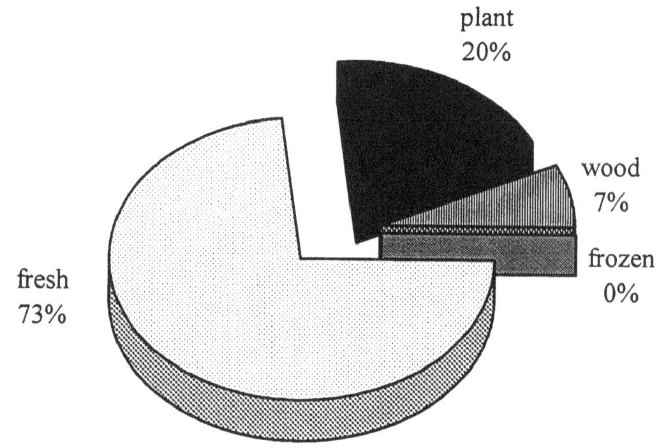

Graph 6.11. Willendorf II, Layer 9: Breakdown of suggested processed materials (%)

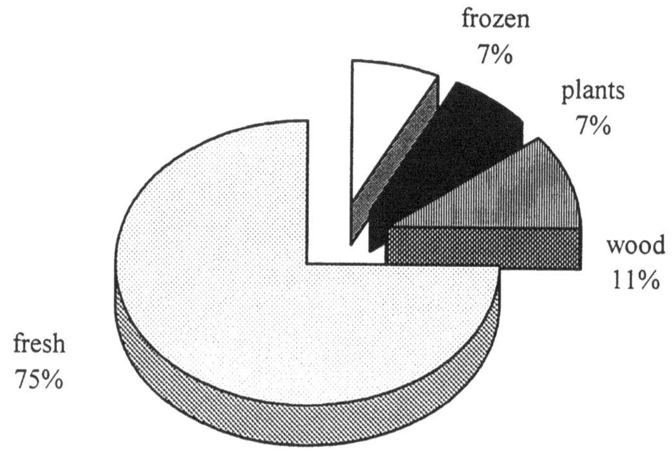

Graph 6.12. Willendorf II, Layer 5: Breakdown of suggested activities (%)

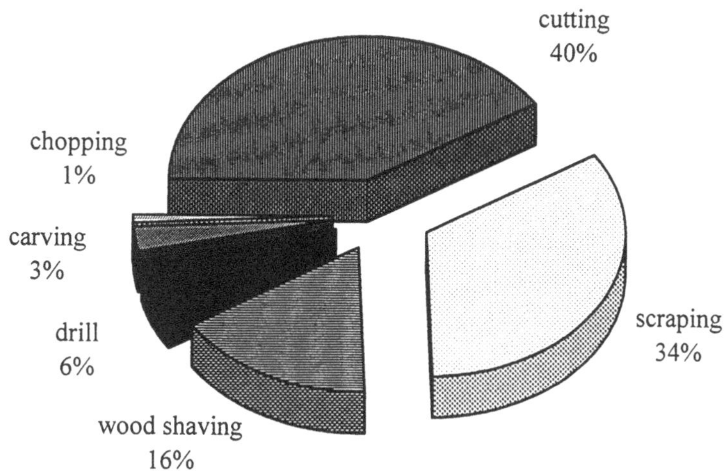

Graph 6.13. Willendorf II, Layer 6: Breakdown of suggested activities (%)

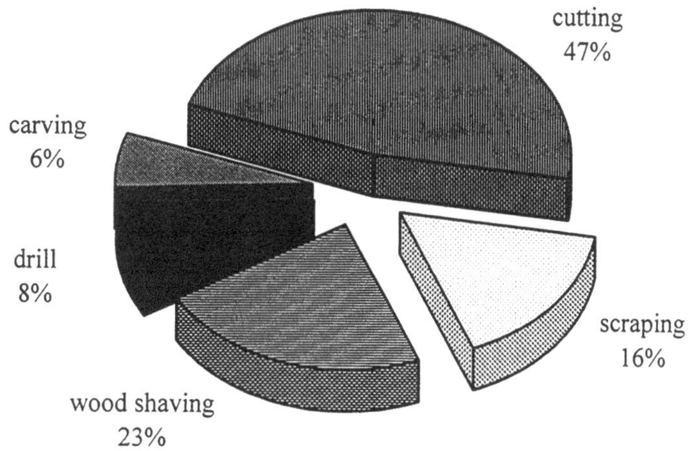

Graph 6.14. Willendorf, Layer 7: Breakdown of suggested activities (%)

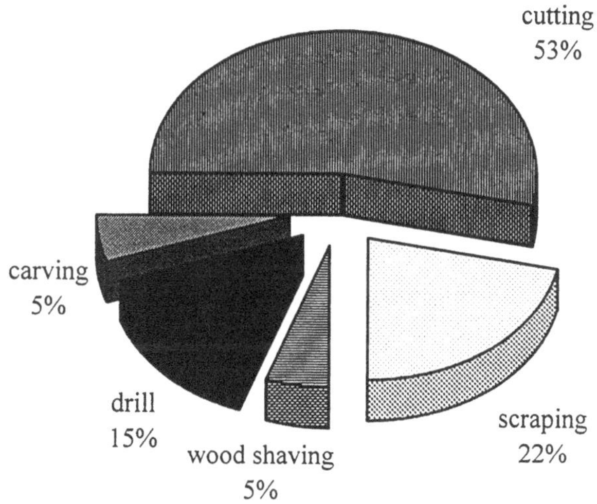

Graph 6.15. Willendorf II, Layer 8: Breakdown of suggested activities (%)

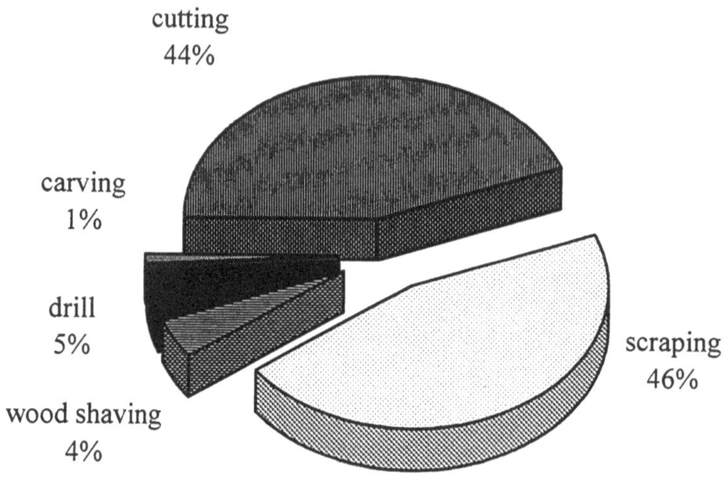

Graph 6.16. Willendorf II: Layer 9: Breakdown of suggested activities (%)

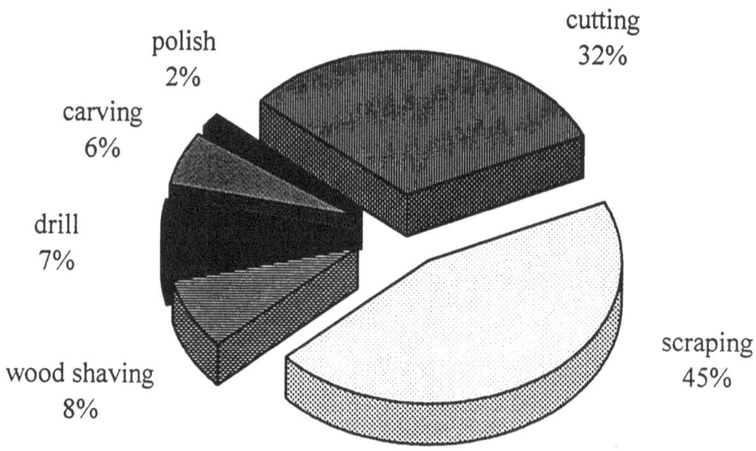

Chapter 7: Seasons Of Difference: Interpretations Of Pavlov And Willendorf II

Twelve months: A fairy tale

The stepmother ordered Maruska to go out and get strawberries, as she and her daughter Holena desired them.

"But it is winter, everything is covered in deep snow; the forest is fast asleep, and there are no strawberries at this time of the year," poor Maruska replied, to no avail. She wrapped herself in a woolen scarf, and off she went, looking for strawberries. The wind was howling the snow was up to her waist, and she went further and further into the forest. Suddenly she spotted a light in the distance, and as she came close, she saw a large bonfire with twelve men sitting around it.

"May God send good fortune upon you," she greeted them politely. "May I warm my frozen feet at your fire just for a few moments?"

"Why are you in the forest in the dead of the winter?" asked the oldest man who was tending the fire.

"My stepmother and stepsister wanted strawberries and they sent me to fetch some, although there is snow everywhere," Maruska replied.

The oldest man looked around and called out: "Brother August, come and tend the fire for a while."

A younger man stood up stirred the fire, and suddenly everything around changed. Trees were green, flowers were blooming, and wild strawberries appeared in large patches all around. Maruska thanked them from the bottom of her heart, and picked as fast as she could. A short while later, when her basket was full, the old man stood up again, and took over tending the fire. The forest turned white and quiet again, just as it should be in the dead of the winter

(After "Twelve Moons", a Czech/Slovak folktale).

Interpretations of Pavlov and Willendorf II: Seasons of Difference.

Raw materials: uses and meanings

The sample of the Pavlov collection discussed here contained 4081 stone pieces. It was divided into three groups in terms of raw material used: flint, radiolarite, and calcite. An overwhelming majority was flint (4005 or 98.14%), with only a minor representation of radiolarite (50 or 1.23%) and calcite (26 or 0.64%). Even if the sources are not known, it appears clear that there was a focus on one type of material: flint. This fact could be interpreted in an ecological, an evolutionary, or a social framework, the most commonly evoked explanatory devices in prehistoric studies, or, most productively, a combination of insights that these approaches offer.

I. Environmental constraints, such as availability of a particular raw material, and a shortage of other materials, combined with concern for energy expenditure, were in most general terms the primary motives for any behavior accounted for in ecological approaches of the 1970's and 1980's (e.g. Cronk 1991; Kirch 1980; Mithen 1989; Speth and Spielmann 1983). Despite the need to take environmental conditions into account, however, most historically recorded relationships between human behavior and natural settings are not as straightforward that one could be directly deduced from the other.[1] Ecological reasoning assumes rationality to be a universal and timeless characteristic, through which people are always able to evaluate their condition and make choices in terms of energy. Furthermore, ecological approaches tend to deal with archaeological sites and their environmental conditions without necessarily comparing them to contemporaneous or preceding occupations in the area. On the other hand, when dealing with the distant past, such as the Palaeolithic, environmental information is often one of the few reliable data sources, and every archaeologist is only wise to make the most use of out them. The number of innovative studies that have recently focused on prehistoric landscapes have given ecological approaches a new incentive, motivation, and energy to carry it much further than "mini-max" models could have ever gone (e.g. Jochim 1998; Bender 1997).

II. Evolutionary models, on the other hand, rely on comparison of archaeological materials within a geographical area, with time as the crucial variable through which any record is interpreted. Preceding and subsequent time periods are viewed as different due to an evolutionary force that moves the adaptation process. This explanatory model is generally favored in east European archaeology, where differences in the use of raw materials over time are considered to have been a part of a general unidirectional change (e.g. Adamenko and Gladilin 1989; Kozlowski 1986; Svoboda 1991).

Comparison with other sites shows that flint is the dominant raw material in most Aurignacian and early Gravettian sites in Moravia, but radiolarite sharply increases with time (Svoboda 1994). Supporting evidence is drawn from a similar trend in the sequence of Willendorf II layers. However, it needs to be pointed out that radiolarite appears already in Middle Palaeolithic sites (Kulna, Předmostí), and an Aurignacian site Tvarozna in the same southern region is almost entirely composed of radiolarite (Prichystal 1994:46). Within the framework of the evolutionary theory as applied in archaeology, change in the use of raw materials tends to be interpreted as an indication of greater economy, and development of increasingly more rational behavior, as evidenced by "risk avoidance" (e.g. Bamforth and Bleed 1997; Tainter and Tainter eds. 1996.). Similarly to traditional ecological approaches of the 1980's, evolution of rational behavior seems to be an assumed universal trend. While this may be the case in some contexts, it has to be remembered that this still remains only an assumption that needs to be shown more convincingly that any studies within the evolutionary approach have managed to do so far.

III. Special cultural significance attached to different sources, materials, or their use, lack of access to a source, either due to a limited mobility range, or exchange contacts with other groups, or a reliable long-term source, are among the most common cited reasons for preference for a particular material from a cultural/social perspective. In addition, one could suggest an association of a certain raw material with a segment of the inhabitants of the settlement, who were obtaining the material, or who were involved in the daily routine of using the stone tools, whether this division was along the lines of age, gender, or some other attribute unknown to us.

However, a cautionary note should be voiced, to ground the interpretive flight in observed facts. It is plausible that the vast accumulation of materials at Pavlov is a result of multiple occupations, a palimpsest, rather than a single event. The time difference between individual occupations may be in the order of one year, or they may be 300 years. Collapsing a potentially long time period into a one-event picture of the past only provides a narrative that projects the present onto the past background without much evidential support.

On the other hand, accepting the long time span, and interpreting the archaeological materials with one frame of explanation inevitably leads to the conclusion of either static, unchanging cultural values, or universal human attitudes. The fact that the lithic collection at Pavlov was predominantly composed of one raw material, flint, is by itself an insufficient base for interpretation. Whether the reasons for selecting mainly flint were environmental, social, a part of a more general evolutionary trend, or a combination of these, it cannot be evaluated on the basis of the number alone, and independent criteria would have to be used to strengthen any of the interpretive models. As I will discuss later, any credible story of the past ultimately has to draw on numerous lines of evidence that complement and support each other, even if that means being able to state only a modest claim.

Differences in raw materials at Pavlov and in the individual Willendorf II layers are striking at first sight, but much less so upon closer examination. The Pavlov assemblage is heavily dominated by flint (98.14%) with only a minor portion of radiolarite (1.23%). As I discussed in the Pavlov chapter, a part of the problem is terminological. The flint in the assemblage was treated on the most general level and no differentiation was made between varieties or colors of flint.

The Willendorf II collection, since it was a part of a natural museum research, was analyzed with a much higher degree of attention to the varieties of flint, and these were identified and separated from the general category flint. Consequently jasper, commonly identified as an iron tinted variety of flint, was treated separately at Willendorf II but not at Pavlov. Nevertheless, statistically these differences are minor and do not alter significantly the overall picture of raw material distribution in the collections.

The variation of raw materials at Willendorf II is indisputably higher than at Pavlov. Although general claims about the local origin of all these materials have been made (Felgenhauer 1959; Otte 1981), on examination these remain hypotheses, leaving this aspect of the study open for future research. I present here only a sketch comparison of the raw materials, and their use, suggesting that the different sources of the materials should be investigated in the future, as it would add an important dimension to the study of the prehistoric peoples that inhabited the area. However, judging solely from the results of the use-wear analysis, the actual treatment of the raw materials in the collections in terms of intensity and economy of use, does not appear to have been significantly different.

The lithic collections at Willendorf II consisted mainly of flint, i.e. more than a half of an assemblage in each layer was made out of this material. The only exception was W II/layer 7 where flint contributed only 44.5%. The ranges of the other layers varied between 60-75%. No clear pattern can be observed over time, as the oldest, W II/layer 5, had as much flint (74%) as the most recent, W II/layer 9 (also 74%). The use of the material seems also consistent over time (40-50%) except for W II/layer 6, where only 27% of lithics showed traces of wear. A possible link of this pattern with the interpreted activities will be discussed later. (see graph 7.1 below)

If Pavlov is assigned chronologically to the same time span as cultural layer 8 (25,800 - 25,200 BP), differences in raw material variability in these assemblages (98% flint at Pavlov, 60% in W II/layer 8) could be attributed to possible variations in length of stay at the location (i.e. mobility), permanence of a settlement, and/or reliance on regular sources of raw materials that could have resulted from specific social contexts — contacts, interaction with inhabitants in other areas, as opposed to opportunistic use of available materials, that could possibly result in a mix of local minerals. This appears to be the case in the Willendorf II layers, where a greater variety of local materials may correspond with shorter, special purpose visits to the location. Nevertheless, one has to be cautious in offering a single explanation for variability in several occupations that occurred over a 5000 year time period.

The overall pattern, based on the presence and use of the major raw materials, shows W II/layer 6 to have been anomalous, since flint predominated, but was used to a lower degree than in all the other Willendorf II and Pavlov layers, mere 27%. The other raw materials in this layer (radiolarite, sandstone) were used only marginally, 11% and 7.4% respectively. Quartzite was represented minimally in all layers (2-4%), but in W II/layer 6 it was the third most common mineral, 10.55% (after flint and sandstone) used to an even higher degree than flint — 28.6%. However, we should keep in mind the small size of the assemblage from W II/layer 6 (199 lithics, of which only 48 showed traces of use). Considering that no information is available on the sources of the raw materials, beyond the general suggestion of local origin, I can evaluate the differences only in the context of the overall interpretation presented here.

The earliest layers, 5 and 6, differ from the general raw material pattern, since radiolarite played only a minor role, and sandstone and quartzite were instead proportionately higher. During the occupation of these two layers, radiolarite was used, minor as it was, to very different degrees: significantly in W II/layer 5 (46.5%), and only marginally in W II/layer 6 (11%). Some correlation between the raw materials and the processed materials appears in both W II/layers 5 and 6. (see graph 7.2 below)

Frozen materials do not seem to have been worked with radiolarite implements during the occupation of either layer, only flint tools showed traces that I would interpret as having resulted from cutting and/or scraping frozen meat, bones, or hide. Since the pattern of radiolarite use appeared in two separate occupation layers, most likely separate causes produced the same end result, particularly considering the different pattern of radiolarite use in the two layers in every other respect.

During the occupation of W II/layer 5, radiolarite tools appear to have been used for a wide range of activities — shaving wood, cutting and scraping hides and meat, and cutting plants. On the other hand, during the occupation of W II/layer 6, the few radiolarite tools that showed any traces of wear, seem to have been used in woodworking only. All other activities were more often carried out with either flint or quartz tools.

Radiolarite appears in W II/layers 7, 8, and 9 in declining proportions. The correlation observed in the previous layers – no working of frozen material traces on radiolarite tools - did not seem to occur in W II/layer 7, where radiolarite tools show traces of use on frozen materials but the same traces were not observed on flint tools. W II/layer 8 does not have any lithics with traces of processing frozen materials, and all other materials seem to have been worked with both flint and radiolarite tools. The assemblage from W II/layer 9 resembles the earlier layers again, since radiolarite tools, unlike flint tools, do not show any traces of having processed frozen materials.

Hence I conclude that, except for W II/layer 8, there appears to have been correlation between tools, the raw materials out of which they were made, and the materials for which they were used. This pattern seems specific to each layer, suggesting that it was not directly related to any particular physical characteristic of the stone, or inherent to the raw material itself.

Seasons of being

At the beginning of this section I wish to stress a cautionary note that the interpretations proposed are strictly tentative and subject to refutation, since they are all currently based on only one line of evidence — the use-wear analysis. Thus they should be understood not as final claims, but rather as propositions to evaluate against further research.

The interpretive focus of my research was to evaluate potential differences in seasonal activities that could have taken place during the occupation of the locations. On the basis of the experimental work, I felt confident that I could distinguish traces produced on the stone tools by contact with frozen meat, bones, or hide from those produced by fresh resistant materials, plant materials, or wood. Considering that it is possible to find some, limited as it may be, vegetal matter in the cold winter season, but that it is impossible to come across frozen meat in warm summer months, I view the presence of traces left by the frozen materials as a suggestive clue about seasonal occupations of the sites. Since I am aware of the cultural construction of seasons and divisions of time (Fabian 1983; Ingold 1988), I am not automatically assuming the perception of four seasons as we know them; rather I only propose activities related to warmer or colder times, indicating variations in occupations.

Willendorf II, layer 5: Leaves of late fall.

The assemblage from the oldest Gravettian cultural layer at Willendorf II, layer 5, contains stone tools with traces produced by processing a mix of materials, predominantly fresh meat, hides and bones (69%). This was followed by wood (18%), plant materials (7.9%), and a small portion of frozen materials (4.4%). (see graph 7.3 below) The most common activities were cutting (41%) and scraping (33.6%), while woodworking played a smaller role (16%), and carving and drilling were only marginal (2.6% and 5.6% respectively). (see graph 7.10 below)

In interpreting the patterns, I took a step beyond the immediate evidence. I divided the activities into a hypothetical sequence composed of the preliminary activity of cutting (assuming that it was done as a first step in any hunting event, or acquiring any materials), and subsequent processing activities such as scraping, woodworking, carving, drilling, that could have been postponed or done at different times. This is not meant as a hierarchical sequence but rather as a temporal one.

Thus the scenario during the occupation of W II/layer 5 that I propose would have concentrated on hunting, and the initial division of the acquired meat, but also included processing of the materials, such as scraping hides/bones. This would either require a group of people working together, a longer stay, or both. The occupation possibly occurred during a colder time period, suggesting the transition between late fall into winter since the processing of plant materials, rare as they might have been, was still more common (7.9%) than the cutting or scraping of frozen materials (4.4%). It would appear that the location had been seldom visited in summer for a longer stay, as plant remains were not processed to any major degree, and only rarely visited in cold winter months, given the lack of evidence of significant processing of frozen materials.

Woodworking, whether for shelter construction, or heating seems to have featured prominently (18.4%), strengthening the suggestion of colder time of occupation, when such provisions would be needed. Shaving wood for construction material, fuel, or some other purpose predominated, and judging from the use-wear traces, the wood was most likely collected rather than cut, as no lithics appear to have any traces of woodcutting. Only two pieces had traces that I would identify as results of carving; thus either this was a completely marginal activity, or it might have been a quick and brief one, leaving a few traces.

Woodworking appears to have been the one activity for which size requirements of the lithic implements were quite specific, since shaving and carving seem to be the two activities for which larger pieces were used. Tools for shaving wood might have been specialized implements with specific requirements placed on them. Cutting, whether meat, hide, bones, or plants, was the most commonly observed activity, and does not appear to have been restricted to a particular lithic size, since length of the lithics associated with such traces ranges from the minimal 8 mm to 80 mm.

A fairly high portion of the lithics (28%) showed traces of hafting. These were not restricted to any particular activity, or processing of any specific material, as all were evenly represented: cutting as well as scraping, plants as well as hides, or meat. This pattern changed in the subsequent layer where only 17% of the artifacts had traces left by hafting. The investment of time and effort into hafting, as well as formation of specific tools rather than use of generic blades, may indicate either a longer occupation, or a repeated visit with tools brought for planned activities. Considering the low presence of cores (2.33%) and burins (5.4%) in this cultural layer, it appears more likely that tool production took place elsewhere.

Only four lithic pieces in this cultural layer had traces of multiple activities, all of them involving a combination of cutting and either drilling or carving with the tip of the same tool. This could be interpreted as a possible sign of a division of tasks and activities that were not carried out with the same tools, and, with the four exceptions, the implements would not have been used for multiple purposes. This division did not appear to be typologically defined, as several tool types, blades, scrapers, endscrapers, burins, were used for the same kind of task, e.g. scraping or cutting. I would be less inclined to interpret this pattern as a sign of raw material abundance, since this layer also has the highest proportion of used pieces among all the Willendorf II and Pavlov assemblages as well (48.4%), indicating a fairly intensive use of the lithic materials at the site.

Thus in conclusion, I would propose that the assemblage from cultural layer 5 at Willendorf II could be interpreted as a fall/early winter hunting camp, where people came for a specific purpose, brought their tool kits with them, participated in divided/differentiated tasks, and most likely moved on shortly.

Willendorf II, layer 6: In the dead of the winter

Cultural layer 6 appears to have been distinct in several respects. Typologically it was primarily composed of blades and débitage, which together accounted for 174 lithic pieces out of the total 199 (87.4%). Raw material distribution and use, as described earlier, indicated that the occupation during the deposition of W II/layer 6 differed from the previous as well as the following layers in a number of ways. Most striking were the high proportion of sandstone (13%) and quartz (11%) among the raw materials, and a significantly higher use of quartz than in any other layer (29%). Judging only from the currently available information, it appears that W II/layer 6 was occupied, or visited, mainly in a cold season, and that the inhabitants relied on a somewhat specific choice of raw materials (although flint still predominated with 65%), performing different activities than those inferred from the artifacts recovered from the other cultural layers at Willendorf II.

The determination of whether we are dealing with a distinct "cultural" group is a much more complex issue. Considering the time differences between the individual layers at Willendorf II, it seems almost inevitable that each layer was occupied by a different "cultural" group, if we understand by that particular cultural practices, their meanings, and social relations among people limited to the historical memory of several generations. The meaning of the space, the material culture, and the natural context probably differed among the successive prehistoric inhabitants of Willendorf II.

Judging from the use-wear traces, it appears that fresh meat, hide, and bones were again the dominant processed materials, but in a smaller proportion than during the occupation of the previous W II/layer 5 (only 48% compared to 69%). Woodworking increased quite significantly to 27%, but the most dramatic increase occurred in traces left by processing frozen materials (to 25%), with a complete absence of plant processing traces. (see graph 7.4 and 7.16 below)

Cutting materials seem to have been the main activity (47%), followed by woodworking (23%). Scraping, or processing hides, appeared only as distant third (16%) occupation. (see graph 7.11 below) This pattern could be interpreted as related to winter seasonal activities when some hunting occurred but a significant portion of food came from frozen meat. The high proportion of woodworking traces (27%) suggests that heating and wooden shelters were probably needed more than during other time. (see graph 7.17 below)

This pattern of traces could have been the result of several types of activities. On one hand, this occupation might have been brief, as only a part of the acquired meat and hides seem to have been processed at the site, judging from the small share of scraping in the traces (16% as opposed to 34% in the previous layer). In addition, W II/layer 6 had the lowest portion of hafted tools (17%), suggesting that little additional effort and time was put into producing stone tool implements on the location. Those tools that displayed traces of having been hafted, were not related to any specific traces of use in terms of performed activities, cutting and scraping traces were present, as were all processed material.

The other possibility that may have led to the same pattern could be a longer winter occupation with only some hunting, or less than successful hunting, and reliance on previously frozen, already processed meat. This could have been either meat brought along, or meat previously stored in anticipation of repeated return. This scenario could be supported by the evidence of wood and bone-working, as carving and drilling wood, as well as bone, appears to have been practiced to a higher degree than in the previous layer (14% as opposed to 8%). Considering the small size of the assemblage, it may be suggested that carving and engraving objects could have been a part of a seasonal activity. Only lithics recovered from the subsequent W II/layer 7 showed a higher portion of traces consistent with carving and drilling. Since no carved wooden objects were preserved, and only one drilled tooth, and no carved bone were recorded in W II/layer 6, this suggestion remains speculative.

An interpretation of W II/layer 6 as a short term occupation oriented towards animal resources (the most significant share in the processed materials), could be further supported by the fairly standardized artifact size. The lithics in this collection were larger than most in the collections from the other cultural layers at Willendorf II, but the standard deviation was significantly smaller (∂ = 15), indicating a standardized size, a possible evidence of activities that were planned and known ahead of time. I suggest that this could have been a site to which a group(s) returned, storing frozen supplies, and bringing only a limited collection of finished tools. This difference, however, was not immediately obvious from the typological variability alone.

Willendorf II, layer 7: Early spring.

The study of the lithic assemblage from W II/layer 7 provided a number of new insights into the possible interpretation of the prehistoric occupation of the location. The assemblage differed yet again from those of the previous and subsequent layers. Typologically, this was a layer with the highest share of blades (75%), and the lowest share of débitage (only 18%, compared to 41% in the two previous layers, and a similar percentage in the following sequences). On the other hand, the typological categories, blades, débitage, or microliths, were not strictly applied and only 8% of all blades were blanks. The previous discussion of the lithic assemblage from W II/layer 7 identified problems with such an approach, and in the following passages I will focus on other aspects of the collection, asking the reader to keep in mind the possibility that layer 7 might not represent a separate layer.

The first visible variation was in the choice of the raw materials. The assemblage from this layer had the lowest portion of flint among all the studied collections (44.5%), and the highest share of radiolarite (38%). (see graphs 7.1 and 7.2 below) Some degree of correlation between the choice of the raw material and the processed materials appeared in the use-wear analysis. While in the previous two layers only flint tools seemed to have traces of working frozen materials, in W II/layer 7 it appeared to have been the exact opposite, only radiolarite implements showed traces that would be consistent with cutting and scraping frozen meat, bones, or hide.

Additional information would be necessary to support a detailed interpretation of the preference for a particular raw material in tool making for a specific activity. However, it seems that the reversal of the material of choice over time supports the suggestion that it was a decision mediated by availability or cultural choice, and not one tied to an inherently superior physical quality of either flint or radiolarite (e.g. cutting ability, or hardness). Had practical, material related choices were the sole reason for shift in the raw material, we would most likely witness some changes in the proportions of lithic sources but hardly a complete reversal, as is the case in the cultural layer 7 at Willendorf II.

The lithics in W II/layer 7 displayed a high rate of use-wear traces (45%) and the highest rate (30%) of traces suggesting hafting among all the layers at Willendorf II. Although I offered an interpretation of the low degree of hafting traces in the previous W II/layer 6 as possible evidence of a short term occupation when no additional time or effort was invested, I would not come to the immediate opposite conclusion in W II/layer 7 where the high rate of hafting appeared. We have to remember that W II/layer 7 seems to have been the most ephemeral cultural layer in the Willendorf II sequence (Haesaerts 1990: 207, Haesaerts et. al. 1996), not strengthening the suggestion of a longer occupational event. Moreover, the use-wear traces observed on the stone tools from this layer appear to have been rather "mono-thematic" - specifically fresh meat, bone, and hides, which does not lend particular credence to the notion of a longer stay at the location. The stone implements seemed to have been used primarily for processing meat products (85%), while all other materials played a minor role. Wood appeared as the second category (8%), and frozen materials and plants seemed quite marginal, represented only by 4% and 3% respectively. (see graph 7.5 below) Judging from these results, I would suggest a late winter, early spring hunting occupation, when some frost was still present.

A plausible interpretation would be that of a specialized hunting camp, or a repeated occupation for that purpose, where other activities either did not take place, or were carried out only marginally. In this scenario the camp could have been occupied almost during any season, although winter season alone would be unlikely, compared with the evidence from the previous layer which I interpreted as a possible winter occupation, and given the low degree of any wood working in this assemblage. As discussed earlier, the seasonality of the hunting would have to be ascertained with the help of faunal remains that at Willendorf II are unfortunately not available for analysis. The suggestion of the hunting camp could be supported by the high proportion of use traces on the stone tools that appear to have resulted from cutting (54%). (see graph 7.12 below) This proportion is higher than in any other layer under study, although overall it appeared to have been the most common activity in all cultural layers that left traces on the stone tools.

The presence of a very high portion of blades (82%), compared to flakes (17%) or cores (7%), could also be interpreted as an additional evidence of a specialized activity for which tools of a particular shape were produced. Size seems to have been a concern only in terms of the general tool size, as most of the lithics were from a size range of 30-60 mm in length, and 10-30 mm in width. No correlation was noted between any particular activity, or processed material and the implement size.

We may also consider the fairly standardized tool kit, composed for the most part of blades, many of them hafted, chosen from a narrow size range, and combined with a specific activity (as interpreted from the use-wear traces). This could be interpreted as an evidence of a higher social conformity on part of the particular group that inhabited the location. Whether this would have been voluntary or enforced cannot be suggested from the evidence available.

One of the interpretations of an increase in hafting (in this particular case in Lower Illinois Valley) was proposed by Odell (1994b). He suggested that hafting and mobility are negatively correlated, arguing that as a consequence of sedentism, lithic resources become depleted, risks in obtaining food increase, leading to improvements of stone tools in the form of various curation techniques, and securing the tool longevity by such means as hafting. While this is an interesting proposition that may be used to explain a change in lithic curation on a macroscale over a long time period, it appears to me that it is not a particularly appealing model on a microscale. The model is based on an assumed individualism where a person (or a group) can get out of a messy situation only by inventing something to secure self-reliance. This ignores the possibility of solving the problem by social means through relations with other people in the area, e.g. through altered patterns of trade, exchange, or raiding.

Besides the dominant activities, cutting and scraping, the lithics from W II/layer 7 differed from all the other assemblages in the high proportion of stone tools that showed traces of drilling (15%) and carving (5%), that even exceeded woodworking. However, only a few drilled teeth and one stone plaque with a hole bored in the middle of it were recorded (Felgenhauer 1959). Otte recognizes only a low number of engraved and drilled bone items in the assemblage from this layer (Otte 1981: 293).

One of the explanation for such an occurrence could be that most of the carved objects were not kept at the location but were carried away, exchanged, or disposed of even nearby (considering that the spatial limits of the excavation may not necessarily have been the limits of the occupation itself). Any proposed interpretation could only be on the level of a hypothesis that cannot be strengthened or refuted with the currently available data, and could range from a production of hunting or fishing implements, to a manufacture of perishable bone, wooden, or hide items.

Yet it is important to note that a change in the pattern that appeared over time involved the carved or drilled medium. While in the earlier W II/layers 5 and 6 bone and wood, as well as frozen bone, seem to have been used, in the later W II/layers 7 and 8 the lithics that showed traces of carving or drilling appeared on bone only but not on wood. This pattern could indicate a change in activities and practices, and possibly a result of different cultural norms.

Willendorf II, layer 8[2]: Sharing summer work

Layer 8 is an occupation level at Willendorf II for which the largest number of secure C-14 dates (11) has been obtained (Haesaerts 1990: 211, Haesaerts et. al. 1996: 34 - see previous chapter). Although this gives the cultural layer a fairly precise location in time (ranging from 25,800 BP to 23,200 BP above the layer), and provides a better ground for comparison with other sites in the area, it does not directly help in interpretations of the artifacts themselves. Our understanding of this layer is not inherently better than that of the Willendorf II layers for which there are no C-14 dates, since this analysis does not rely as much on the macroscale of the region as the microscale of the site. Thus the following interpretation of the use-wear analysis relies on the same evidence as those offered in the previous discussion.

Flint was the dominant raw material (60%) but radiolarite was also represented to a significant degree (27%). No particular preference appeared in the treatment of either raw material, and traces of all activities, as well as worked materials, were observed on flint and radiolarite tools. The only exception were frozen materials that did not seem to leave any traces on any tools in this layer, suggesting that the occupation (or repeated occupations) occurred at times other than winter. This suggestion is further supported by the high percentage of tools with traces of plant processing (20%), the highest percentage among all the Willendorf II layers. Nevertheless, meat, bone and hide working were still the dominant activity (73%). (see graph 7.6 below)

Judging from this evidence, I would propose that the location was occupied during the late summer, early fall. At this time both animal and plant resources would have been plentiful, accepting the possibility that the activities did not necessarily had to have taken place simultaneously. Woodworking appears to have been only marginal (7%), the smallest proportion in the Willendorf II sequence. If wood were connected to shelter construction and heating, then the occupation would most likely have been short term, perhaps in a warmer time period, when substantial constructions or heating would not have been needed. (see graphs 7.15, 7.16 below)

Activities inferred from the use-wear traces seemed fairly balanced between cutting (44%) and scraping (46%). This was the only layer where more tools were observed with traces of scraping than cutting. (see graph 7. 13 below) Throughout the discussion of the Willendorf II sequence, I interpreted the evidence of scraping, carving, and drilling as a suggestion of processing activities at the location, rather than the mere acquisition of food inferred from cutting only. Thus, the evidence of a fairly high degree of scraping hides, bones, but also plant materials could be taken to illustrate either a larger group of people who participated in different activities, a longer stay at the location, a repeated visit for distinct purposes, or all of these together.

This suggestion is further strengthened by the fact that a number of lithics had multiple traces of use, having been employed for more than one task. This phenomenon gradually increased through the Willendorf II sequence, from only a few pieces in W II/layer 5 to 35 pieces in W II/layer 9. Not only the number of lithics with multiple traces of use increased over time but also the type of activities that left the traces shifted over time. While in the earlier W II/layers 5, 6, and 7 a few of the blades that had traces of cutting, appear to have been used for drilling or carving, in W II/layer 8 the multiple activities involved cutting and scraping. This could mean a shift in the division of tasks, or a more individual processing of the acquired meat and hides. I am less inclined to propose a shortage of raw materials that would require maximizing utility of any usable blade, since only 40% of the lithic collection had traces of use. Furthermore, this assemblage contained the highest proportion of cores (15%) in the Willendorf II sequence that was even higher than the share of flakes (14%). Raw material was therefore available even in unused core forms, refuting the notion of a general need to utilize blades to their utmost potential. The high percentage of cores could also strengthen the suggestion of a longer stay with multiple activities that involved planning ahead, or having the raw materials available for longer than an immediate need, or even for a future reuse, if repeated visits were the case.

Another possibility that could be suggested for the observed pattern would be a minimal sharing of the raw materials, mainly the remaining cores, within the group. In such a case some members of the group would not have enough, and would have had to use their tools for cutting as well as scraping. This scenario however is merely hypothetical, as I have no other evidence to support or refute it.

One activity that seems quite marginal, particularly if compared with the previous layer, is carving. Only 1% of the lithics has traces that I would interpret as having resulted from carving bone. This was the lowest percentage in the Willendorf II sequence, and my initial reaction was that it must mean that layers 8 and 7 (where carving was significantly higher) were probably one cultural layer, and I was only looking at the skewed result of a divided collection. However, although this remains a very real possibility (which will be examined below), it could also be a phenomenon that had its roots in the prehistoric behavior of the people at the location.

Was carving a specifically winter-related activity? Or was it tied to a specific place, therefore done only to a limited degree during this occupation, and any engraved objects would have been brought from other places. Obviously any of these propositions could be valid, but none can be strengthened with any of the currently available data or analyses. A number of engraved bone artifacts were recorded, as well as three drilled Tertiary shark teeth (Felgenhauer 1959: 68). Otte reports more engraved bone artifacts from Layer 8 than from the preceding layers (Otte 1981: 293). This fact is not necessarily incompatible with my observations, as the engraved items could have been brought in from elsewhere, and not necessarily produced on location.

In conclusion I suggest that on the basis of the use-wear analysis the patterns observed in cultural layer 8 could be interpreted as a summer/early fall occupation of a possibly longer duration where hunting, as well as food processing occurred. The people who occupied the site did not seem to construct any substantial wooden structures, judging from the low representation of wood shaving, but made some items carving and drilling bone.

Willendorf II, layers 7 and 8: Early spring through late fall

The stratigraphic problems mentioned by Felgenhauer (Felgenhauer 1959: 93) and discussed recently in more detail by Haesaerts (Haesaerts 1990:207) open the possibility that W II/layer 7 could have been a part of the occupation that continued in layer 8. The lack of clear boundary between them has been discussed in terms of possible depositional problems, suggesting that it might have been a case of repeated occupation over a longer period of time. Here I will consider the two layers together, so as to examine whether the interpretations offered earlier would change to any significant degree.

Raw material representation in the combined assemblage would shift only slightly without any immediate change in the previous suggestions: flint would still be the dominant material (54%), followed by radiolarite (31%), and all the other minerals — sandstone (8%), quartzite (4%), chalcedony (3%), would contribute only marginally to the total. Since in both Layers 7 and 8 radiolarite appears to have had more traces of use-wear than flint, the effect of combining the results would be minimal, indicating a slightly higher use of radiolarite than flint, without any correlation with a specific activity.

The small number of lithics with traces produced by cutting frozen meat and bones (4 pieces) that were recovered from W II/layer 7 were all made from radiolarite. However, in the combined assemblage of 801 lithics, the number is so small that it would not present a very convincing basis on which to argue for an exclusive use of radiolarite for frozen materials.
The tool kit from the two combined layers (801 pieces) consisted mainly of blades (596 pieces, 74.4%); flakes were present to a small degree (121 pieces, 15.12%), and cores contributed 84 pieces, 10.5% - the majority of those (80 pieces) from cultural layer 8. Although the share of cores decreased compared to the assemblage from W II/layer 8 alone, it still remained the highest in the Willendorf II sequence, and was higher than the core component in the Pavlov collection. I interpret this as possible evidence of planning, and preparation for further stone tool manufacture. The additional suggestion could be anticipation of unusual hunting success or failure, if we accept traces of meat processing as an indication of such a focus at the site, leading to an uncertainty of the length of stay in one place. Expectations of success would motivate an accumulation of cores for processing tools, while failure might lead to a potentially longer stay. Yet both would archaeologically leave the same traces. Faunal remains would be needed to improve this hypothesis.

Since in both layers meat, bone, and hide processing appear to have been the dominant activity (85% in W II/layer 7, and 73% in W II/layer 8), it is not surprising that the combined effect is very similar (78%). Plant materials seem to still hold the second place (13%) but not to as high a degree as in W II/layer 8 (20%). Wood working traces were observed on 7.5% lithics, and traces of cutting frozen materials, entirely on implements from layer 7, in 1.5% lithics. (see graph 7.7 below) This proportion is still fairly small, and does not change dramatically the interpretation of the seasonal occupation that would have been probably from warmer times, late spring, through early fall. I have relied on the available information, from which I infer that no mammoth, or any other large bones were found that could be interpreted as an alternative source of fuel. Therefore I consider the wood working traces as a possible sign of both shelter construction and accumulation of heating material. No traces of carving could be linked to wooden materials; all seemed to have been produced from bone working.

The balance of inferred activities changes slightly. In the combined assemblage lithics with traces of cutting appeared more frequently (47.7%) than lithics that I interpreted as having traces of scraping (36%). If we were to assume contemporaneity for the activities involved, a plausible interpretation would be an occupation with activities focused on acquisition and processing of meat, hides and bones, supplemented with some plant gathering. In this presentation not only is time congruence assumed but also an equal representation of meat working and plant processing in the use-wear traces. However, as I have mentioned in the discussion of methods, plant materials have to be worked for at least twice as long to leave any traces. It is plausible that the practice of plant working was higher, particularly if it already was observed on 13% of lithics.

Carving and drilling, prominent activities according to the traces observed in the assemblage from W II/layer 7, decrease slightly to 3% and 9% of a collection that contained 801 pieces. On the other hand, the representation of these activities in the small assemblage from W II/layer 6 remains the most intensive. I can only guess whether it was related to hunting, seasonal activity, meeting grounds, or entirely common part of everyday practice for the group that inhabited the location at the time. It would be necessary to compare the assemblages with other contemporaneous collections in the area, so as to assess whether this was an exception or a norm.

Willendorf II, layer 9: Getting ready for winter

The largest assemblage at the top of the Willendorf II sequence, cultural layer 9, contained artifacts primarily made out of flint (74%, the same portion as the collection in layer 5). Radiolarite was represented by 14%, and all other minerals were marginal. Both major raw materials were used to a lesser degree than was the case in the studied collections from the previous layers, except for layer 6 (28% overall use-wear rate). Radiolarite in the collection from W II/layer 9 appeared to have had a somewhat lower rate of use (29%) than the flint lithics of the same collection (33%). Overall it appears that access to raw material was regular and secure, while relying on specific sources. Judging from the remains the occupants chose mainly flint and radiolarite but, considering the low use-wear rate, not using it exhaustively or with a sense of economy.

The use did not seem to have been linked to any specific activity but was correlated with processed materials. Twenty flint tools had traces left by working frozen materials. While flint tools had traces from cutting and scraping frozen bones, hide, and meat, such traces were not observed on the radiolarite implements. This correlation between flint and the specific processed materials was observed also in layers 5 and 6, but in reverse, i.e. radiolarite in these layers was used for cutting frozen materials. It is also plausible that the winter seasonal tasks were not carried out by the same group of people as were the activities related to the warmer times of occupation, leading to several occupational events that may appear in the same archaeological level.

Blades contributed only 69% of the assemblage, the lowest share in the examined Willendorf II sequence. A similarly low percentage appeared already in the preceding layer 8 (70%). While in the previous layer the number of cores was significantly high (15%), in the assemblage from layer 9 flakes were represented more than in any other of the studied collections (28%)[3]. Both blades and flakes had traces of use-wear, although the rate of use in blades was slightly higher (35% as opposed to 27%), while only five cores appeared to have been used. These were all large pebbles that resembled grinding stones, with very smooth polished surfaces, and had traces that could have been created by polishing or softening hides. Two additional cores appeared to have been used for a similar task but involving wood.

The focus of the occupation appears to have been the acquisition of animal resources, and subsequent processing, as suggested by the high percentage of use-wear traces formed by working fresh materials. Meat, hides, and bone were the most commonly treated materials (75%), followed by wood working (11%), cutting and scraping of frozen materials (7.3%) and plants (7%). Examining the activities in detail reveals that cutting may have been slightly less frequent (32%) than scraping (45%). This pattern may be suggestive about the division and sequence of performed tasks. The initial cutting of the meat could have been carried out away from the site where only the subsequent processing would have taken place. The lower portion of blades in the collection may substantiate the claim, as these may have been used away from the site. This scenario could be further supported by an interpretation of a number of lithics that had traces of multiple activities. (see graphs 7.8, 7.14 below)

While the increase in the use of the same implements for several tasks was noted in earlier Willendorf II layers, the combination of the activities involved varied in the assemblages from the different occupations. I have noted cutting and scraping traces on single blades in W II/layer 8, and cutting associated with carving or drilling in the earlier W II/layers 5 and 6. Lithics from W II/layer 9 showed a combination of scraping and drilling or carving. The pattern was fairly common: 33 pieces (11.5%) showed traces of such multiple activities that mainly involved bone as the medium.

If we accept the assumption that scraping, carving, and drilling could be considered processing activities that would most likely be carried out at a camp, while cutting up the kill could be done on the spot, then the observed pattern of use-wear traces could be interpreted as evidence of such a division of tasks. The earlier cases, in W II/layers 5 and 6, of cutting marks associated with carving and drilling traces could be interpreted as a use of the same tool for a number of activities. These may not have necessarily taken place at the same location but involved the use of the same implement for the sequence of primary and secondary tasks, presumably carried out by the same person. Implements from W II/layer 9, on the other hand, seem to indicate the use of the same tool for secondary tasks only - scraping, drilling and carving, possibly suggesting a different division of work. The person that carried out the primary cutting of the meat did not use the same tools for the secondary processing, suggesting specialization either in terms of different person (or a group of people) doing separate tasks, or specialized tools used for different tasks.

Specialization of any activity may lead to a smaller tool kit used by a group of people who posses a particular skill. In this case it would have been hunting, while the rest of the group would be involved in the more general processing stage, which would result in a greater number of tools in the archaeological record. On the other hand, a more involved, complex processing, consisting of a number of stages, tasks, and procedures could call for a more expansive tool kit, leaving a larger number of tools with such traces. The end result could be achieved either by a small group of people using a larger tool kit, or a bigger group participating in the activities.

The intensity of use did not seem to have played a significant role, as most of the lithics (57%) showed only moderate degree of use. Woodworking implements seemed to have been most frequently associated with heavy use. Moreover the assemblage from W II/layer 9 contained a significant portion of unused lithics, only 40% had observable traces of use, indicating that a shortage of raw material, or concern with economy could not have been very acute.

Judging from the presence of traces left by working frozen materials (7.3%), at least some of the activities occurred in wintertime. The evidence of wood working (11%) could be used to support such a claim, especially considering the traces of heavy use of tools possibly in shaving wood. Unfortunately, the stone tools from Willendorf II have not only been recovered almost a century ago but they have been carefully washed removing any possible residue that could provide any information on plant resources. Lithics may not be the best indicators of plant use in general, and judging from the use-wear traces alone, plant resources did not seem to have played a major role (7%) during the occupation of layer 9 at Willendorf II. However, the traditional problem with plant cutting traces on stone tools, as well as the possibility of relying on plants that can be picked or collected should be remembered. Furthermore, as Mason, Hather and Hillman (1994) have shown in their study of hearth remains from Dolní Věstonice, plant resources were a part of the diet, although it may not have been in a form palatable to our imagination.

The above discussed evidence suggests that the cultural layer 9 was occupied for a longer time, or by a greater number of people, during the late summer and fall season, processing materials at the location, possibly preparing for a winter season. Some division of labor existed, and the possibility of specialized tasks can be seriously considered.

I conclude the analysis of the Willendorf II sequence by suggesting that a number of changes seemed to have occurred over time. I would hesitate to interpret them as unidirectional development towards complexity. Most of these changes visible here involved possible shifts in seasonal activities, divisions of tasks, and the organization of practices related to subsistence activities.

Pavlov: the possibility of a long-term settlement

The extensive collection from the Pavlov site presents a challenge in a number of ways. First of all, although only recently analyzed in any detail (Svoboda 1994), it has been the eponymous site for the entire Pavlovian culture. The typological grounds upon which this claim has stood appear less than solid that they did at the start of this project. I feel that detailed analysis, typological as well as use-wear, has shown that although a unique "Pavlovian" culture may have existed, it should not be based on the typological variation of the collection.

Judging from the analysis of the lithic collection, and relying on the studies of the associated materials from Pavlov, I would suggest that this was a more permanent settlement than any of the occupations in the Willendorf II sequence, possibly a year round settlement. The starting point of this claim is the sheer number of materials recovered from the site (over 39, 000 lithic artifacts just in one season, summer 1953, and close to 1 million artifacts in 21 years of excavation). It is likely, however, that this accumulation represents a series of occupations, and I would not want to treat this vast amount of material as one uniform settlement. In dealing with the assemblage from one excavation season, I tried to evaluate the possibility of year round occupation, either continuous or temporary. The following interpretation will view the question from the angle of the seasonal nature of activities, their potential sequence, and task division.

The lithic implements at Pavlov were made almost exclusively on flint (98%), and the current state of sourcing in the area suggests that all of the material was imported (Prichystal in Svoboda 1994). The suggested sources lie in the north, in the area of Krakow, Poland, but neither Austrian sources to the south, nor German sources to the west have been fully considered. It appears that wherever origin of the raw material was, it was a secure source as the abundant material was not treated with any particular sense of economy; 44% of the lithics had traces of use and the majority indicated moderate use. A number of possibilities could be proposed as ways of securing a regular access to raw materials.

The inhabitants might have made regular journeys through the area, obtaining the materials first-hand. Since studies of sources and mining in Central Europe are fairly recent, no results are currently available to suggest any evidence of southern Moravian materials in other areas so as to indicate direct contact, or trade. The one possible lead may be the presence of fired clay pieces, identical to the numerous clay objects from both Pavlov and Dolní Věstonice, recovered at northern Moravian sites that had no evidence of kilns where such clay would be fired (Soffer and Vandiver in Svoboda ed. 1994). Tenuous as this evidence may be, it does suggest the possibility of contact, whether direct or trade, between these two areas, and potentially further to the north. The hope is that similar attention will be given in the future to recovered materials as possible links between sites in wider Central Europe. A discussion of direct access to the raw material sources should also involve a consideration of consequences of such model in social terms, and how we would recognize different scenarios in the archaeological record. Whether only some members or the whole group travel to obtain raw materials would not only require different explanatory models, but also would have different implications for social organization.

The Pavlov assemblage that I studied consisted mainly of blades (77%), and only marginally of cores (4%). The proportions of the sample resembled the composition of

Willendorf II layers 5 and 6, while being much more recent, indicating that different reasons for the similar effect should be evoked. Blades were identified slightly more often with traces of use-wear (42%) than flakes (32%) but the difference does not appear significant, and resembles the general pattern in the collections from the Willendorf II sequence.
Due to the fact that the Pavlov artifacts were recovered much more recently than those from Willendorf II, and that they were not carefully cleaned or washed, they were amenable to a more thorough study in terms of the worked materials. The traces left by prehistoric activities were more clear and distinct than those of the Willendorf II collections. As a consequence I was able to distinguish between bone and antler as a fresh hard material, and meat and hides as fresh, resistant but softer material. The more detailed distinction between the processed materials gives a more fine tuned picture of the possible activities at the site.

Activities inferred from the use-wear traces on the Pavlov lithics present a fairly balanced picture between scraping (42%) and cutting (38%). All other tasks seem to have been marginal. The third most common activity was wood working (10%), carving and drilling together left traces on less than 8% lithics. A number of stone tools had traces of multiple activities. The most frequent combination was cutting and scraping traces on the same tools; a few blades had traces of cutting with the tip possibly used for drilling or carving. Thus the overall impression is of mixed activities, with very few specific tools designed for one task only. Considering that both cutting and scraping were equally represented in the use-wear traces, and often present on the same tool, together with the interpretation of the larger portion of the traces as having been left by working plant materials, it may be suggested that these tasks may have been carried out by the same person(s). A similar interpretation may be offered with regard to the large number of faunal remains from smaller or medium size animals — rabbits, foxes and wolves — recovered from the vicinity of the hearths.

Judging from the use-wear traces, it may be suggested that the division and processing of the meat might have taken place at the site, rather than at the location of the kill. Musil (in Svoboda 1994) discusses spatial variation in species distribution throughout the site, occasionally even at the same hearth, indicating the possibility of some division of the prey, which then was processed by members of that particular group individually, as opposed to a sequence of collective hunting, cutting up the meat, and subsequent processing. This is a different picture than the one discussed earlier in cultural layer 9 at Willendorf II, where cutting and scraping were activities that were not carried out by the same tool.

The overall picture that appears from the study is a fairly balanced processing of different materials. Plants appeared to have been worked quite frequently (30%) - W II/layer 8 that had the highest share of plant material processing among the Willendorf II analyzed cultural layers, and was interpreted as a summer camp, reached only 19.5%. Considering the slow rate of attrition that vegetal matter has on stone tools, it would not be unrealistic to estimate an even higher role of plant processing. The usual assumption is that the processing of plants must have been for consumption in the form of food. However, as evidence from Pavlov has shown, some of the utilitarian clay may have been used in the form of baskets made out of twigs or reeds covered with clay, and either fired or used in an unbaked state (Soffer and Vandiver in Svoboda ed. 1994). Thus the observed high degree of plant processing traces on the lithics is not surprising, and might have been a consequence of the combined requirements of food consumption, as well as utilitarian uses, such as basketry, or even dwelling construction.

The second largest category is composed of frozen materials (23.5%). The size of this category is comparable to the collection from Willendorf II, layer 6 (25%) that I interpreted as a winter occupation. However, that assemblage did not contain any stone implements with traces of plant working. The presence, and a high percentage of both worked materials — plant and frozen — suggest occupation during times when both materials would have been in abundance, indicating winter as well as summer and fall occupations. Fresh meat and hides were represented in the use-wear traces to the same degree as the frozen materials (23%), indicating that hunting and processing of the acquired resources were taking place, possibly quite regularly, and that some of the obtained food may have been kept for the winter. The results of the faunal analysis show that the ages of the hunted animals varied from the very young to old individuals, with rabbits, foxes, wolves and reindeer having been the most common; mammoth being represented in the faunal remains to a lesser degree than at Dolní Věstonice (Musil 1959, in Svoboda 1994).

Surviving winter may not have been such an impossible task as is sometimes portrayed, particularly if sufficient resources were prepared ahead of time, and if, as it appears, substantial dwellings were built not only out of wood but also mammoth bones, plastered with clay (Soffer and Vandiver in Svoboda 1994). Furthermore, recent palaeoenvironmental reconstruction questions the common image of a universally harsh, inhospitable environment, suggesting that localized warmer zones may have existed, consisting of mixed forests, plant resources, and ecotypes that do not coexist at present (Opravil in Svoboda ed. 1994:165). Consequently, it seems quite plausible that the occupants of Pavlov could have stayed at the location over a period of time that extended over warm and cold seasons.

Traces of woodworking appeared on only 10% of the lithics, a similar percentage as observed at Willendorf II in layer 9. A number of wooden implements were recovered from Pavlov, including implements that resembled very closely stone tools, and that could have been used in a similar fashion (Klíma pers. comm.). None of these could be examined in terms of use-wear, as they were covered with a thick layer of varnish applied immediately upon their recovery to assure preservation. Thus it is possible that a wide range of wooden tools existed and the few traces of woodworking were created in the process of making them. If the dwellings were made out of other materials then wood, the low percentage of wood working traces could be explained as a consequence of a shift in construction materials from wood, such as may have been extant at Willendorf II in layers 5 and 6, to bone, hides, clay and mixed materials.

Notes:

[1]For example, even hunger - the most obvious example of environmental stress - is not simply universal experience (e.g. Arnold 1988; Garnsey 1988; Sen 1981)

[2] Following the interpretation of the results of the analysis of the assemblage from cultural layer 8, I will consider later in the chapter the possibility of W II/layer 7 having been a part of the same occupation as the subsequent W II/layer 8.

[3] I examined only a total of 32 cores, a small number considering the size of the collection (1031 pieces), but the 63 cores reported by Felgenhauer (1959: 71) would still not represent that significantly higher a proportion of cores in the whole assemblage.

Graph 7.1. Willendorf II and Pavlov: Flint presence and use in individual assemblages (% presence = light pattern, % use = dark pattern).

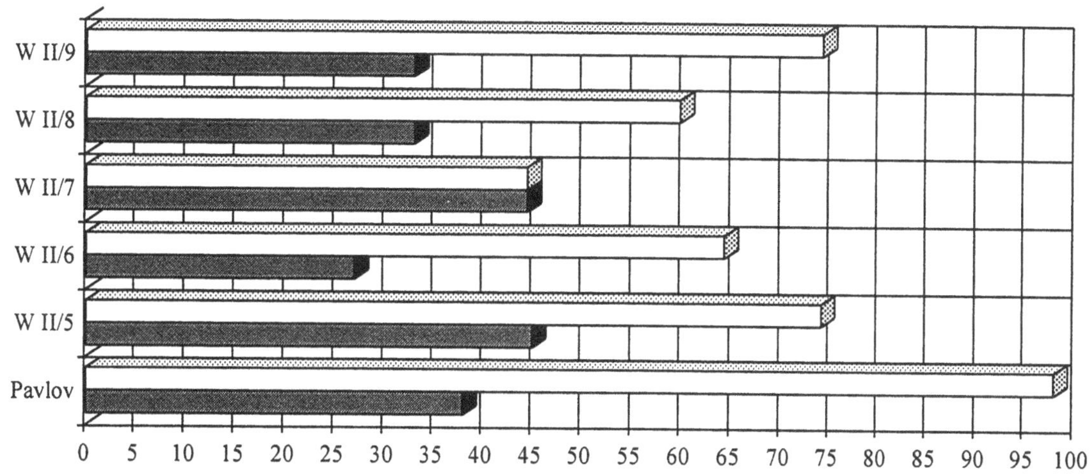

Graph 7.2. Willendorf II and Pavlov: Radiolarite presence and use in individual assemblages (% presence = light pattern, % use = dark pattern).

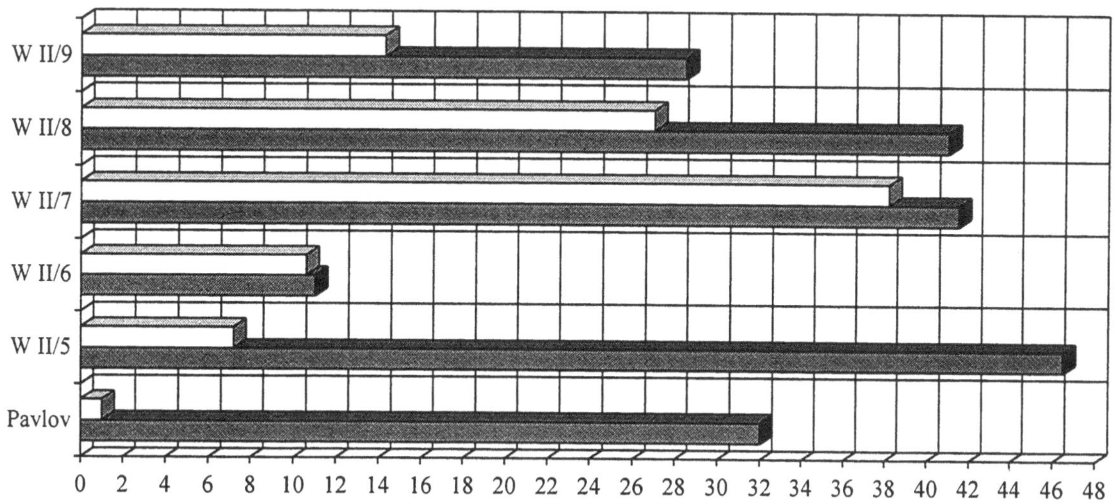

Graph 7.3. Willendorf II, layer 5: Seasonal division of processed materials (%).

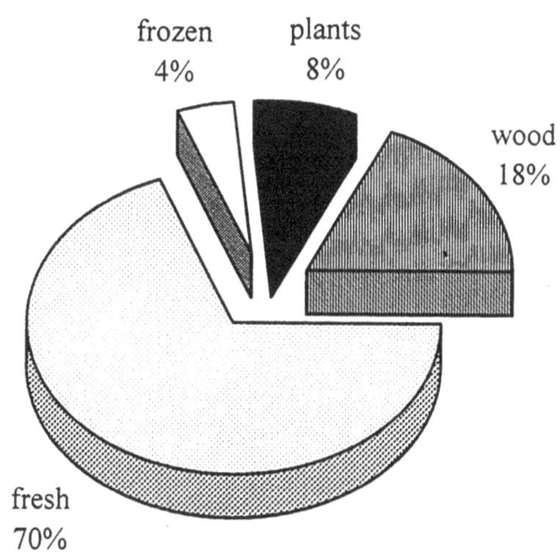

Graph 7.4. Willendorf II, layer 6: Seasonal division of processed materials (%).

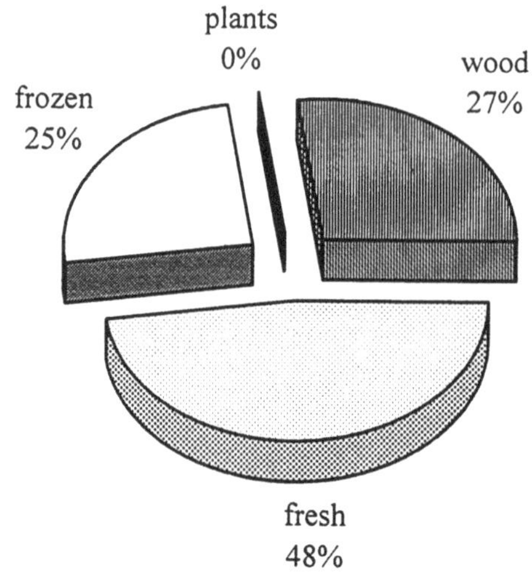

Graph 7.5. Willendorf II, layer 7: Seasonal division of processed materials (%)

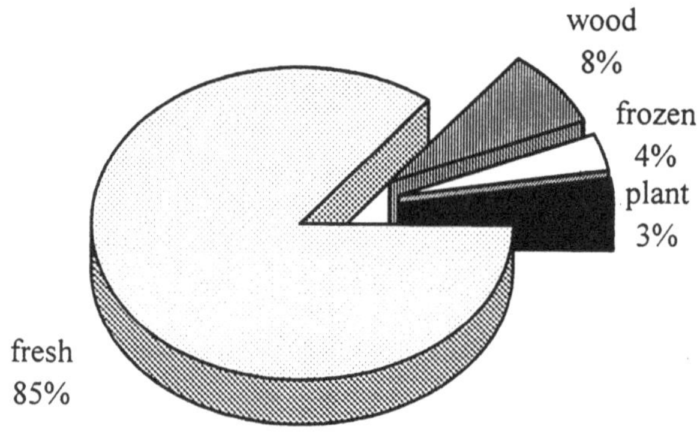

Graph 7.6. Willendorf II, layer 8: Seasonal division of processed materials (%)

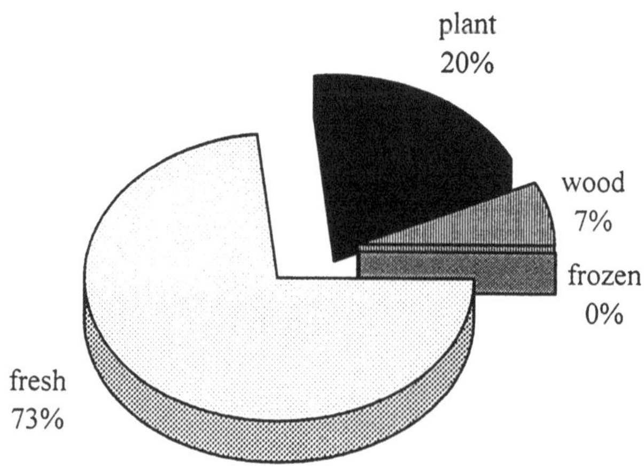

Graph 7.7. Willendorf II, layers 7+8: Seasonal division of processed materials (%)

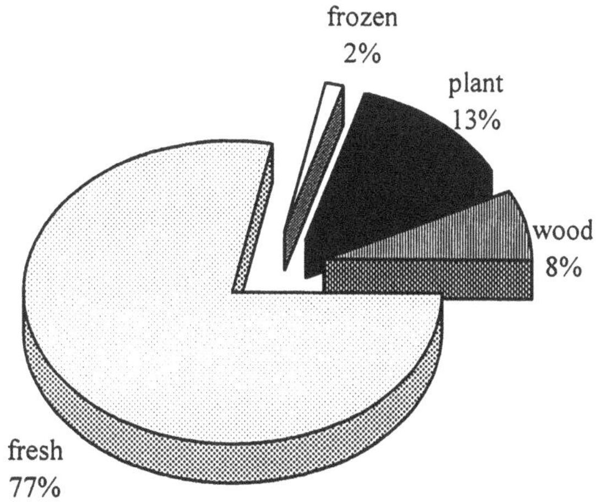

Graph 7.8. Willendorf II, layer 9: Seasonal division of processed materials (%)

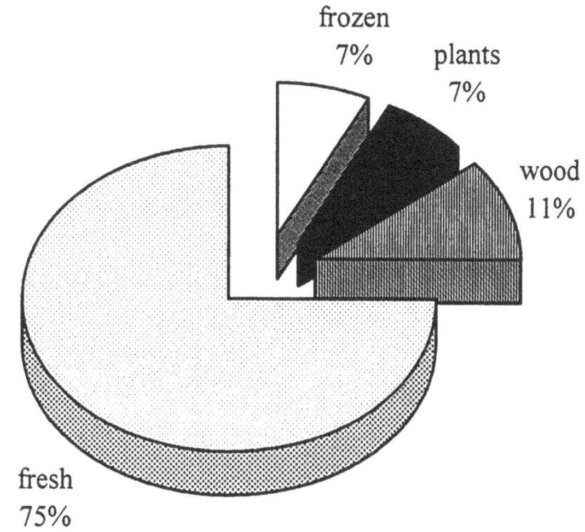

Graph 7.9. Pavlov: Seasonal division of processed materials (%)

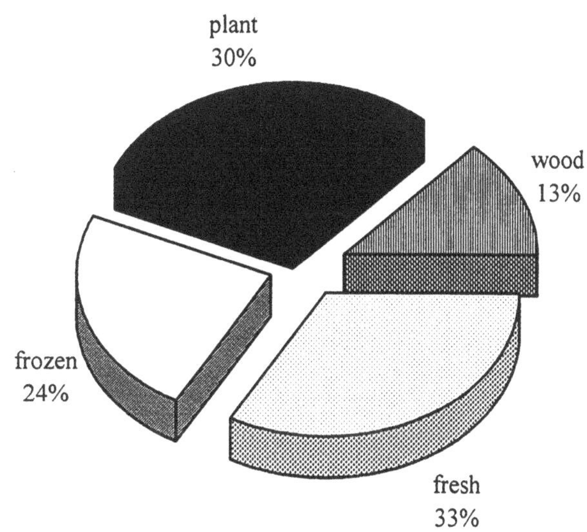

Graph 7.10. Willendorf II, layer 5: Breakdown of suggested activities (%)

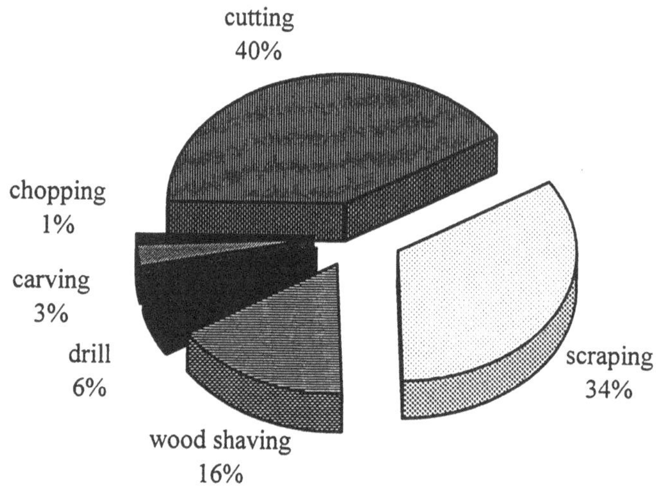

Graph 7.11. Willendorf II, layer 6: Breakdown of suggested activities (%)

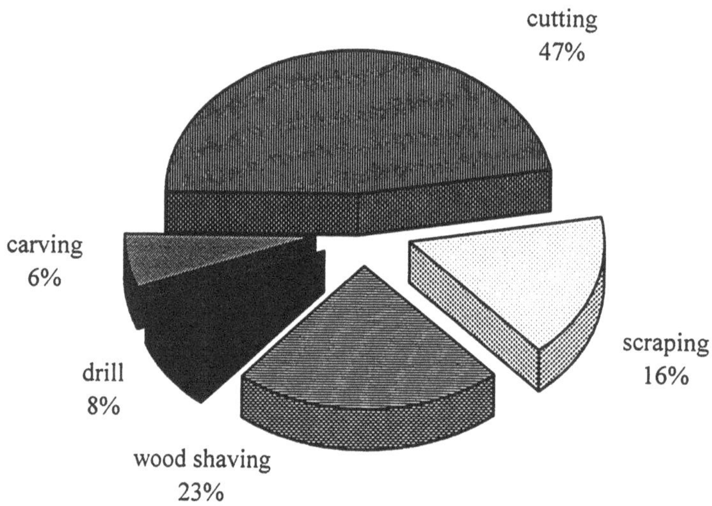

Graph 7.12. Willendorf, layer 7: Breakdown of suggested activities (%)

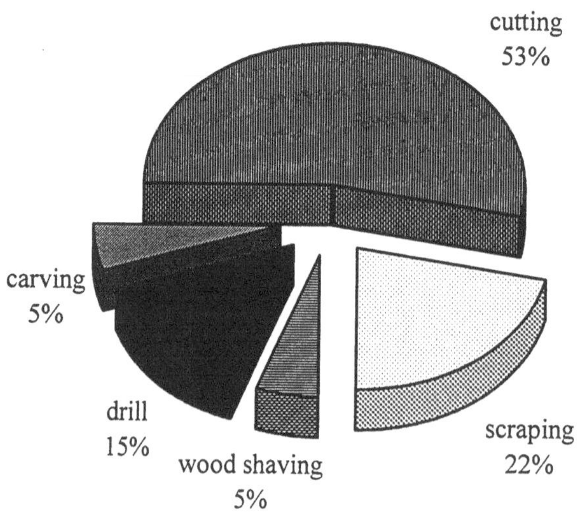

Graph 7.13. Willendorf II, layer 8: Breakdown of suggested activities (%)

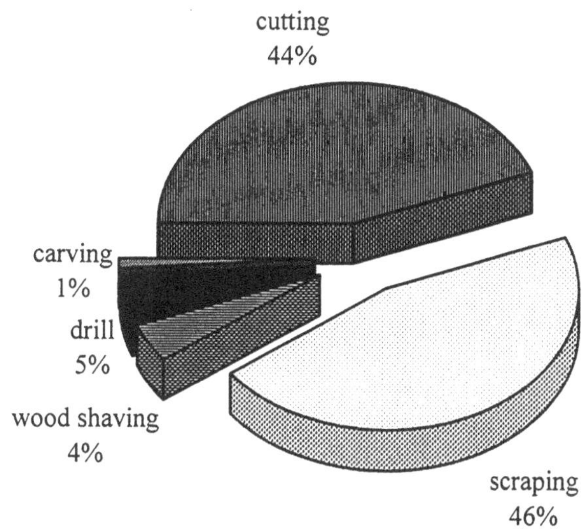

Graph 7.14. Willendorf II, layer 9: Breakdown of suggested activities (%)

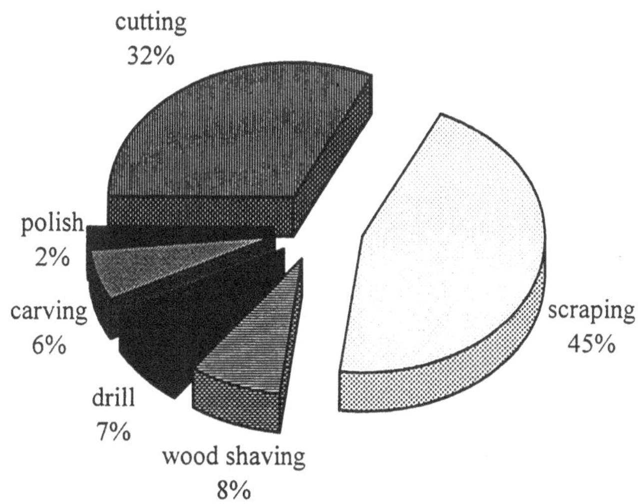

Graph 7.15. Willendorf II and Pavlov: Plant-processing traces (%)

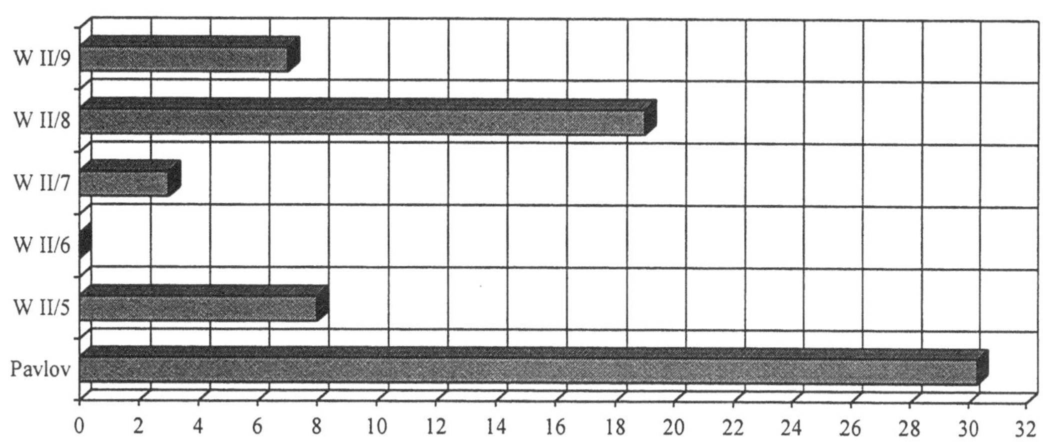

Graph 7.16. Willendorf II and Pavlov: Traces of processing frozen material (%).
(W II/8 – the lowest, interpreted as summer/fall occupation, W II/r 6 - the highest, interpreted as winter occupation).

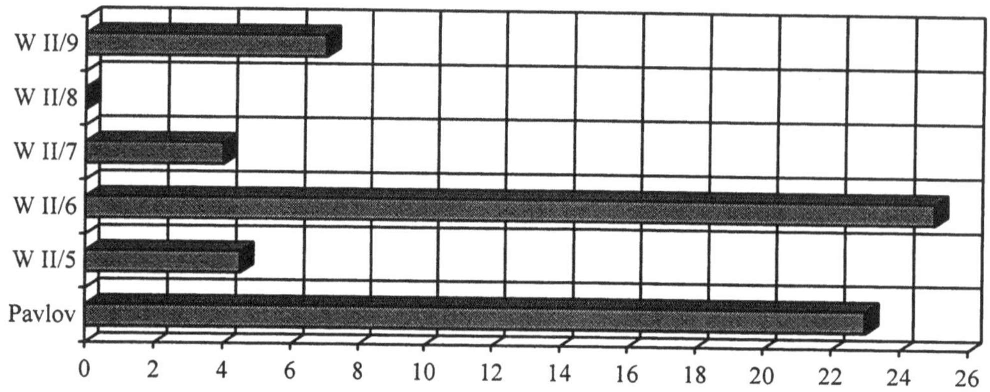

Graph 7.17. Willendorf II and Pavlov: Woodworking traces (%).
(W II/6 – the highest %, interpreted as winter occupation).

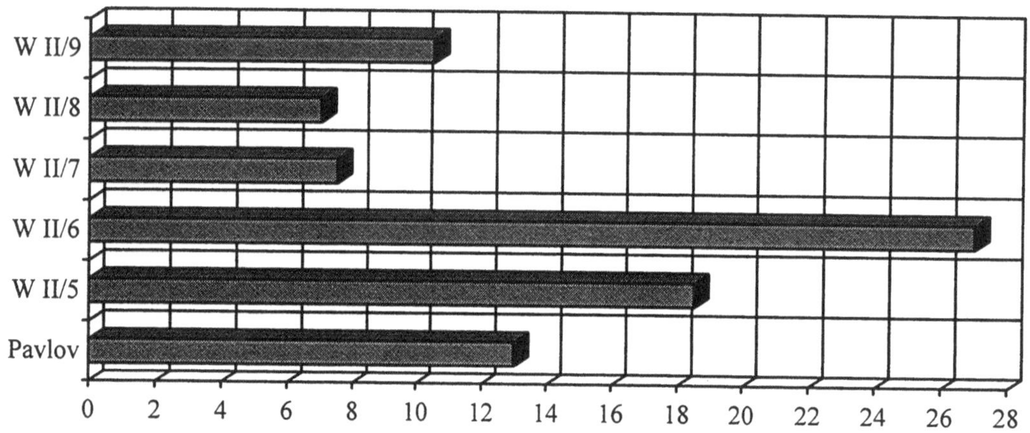

Table 7.1. Pavlov: Detailed breakdown of suggested processed materials
(multiple activities in brackets).

material/activity	plants	frozen	fresh meat/hide	wood	bone/antler	**total**	%
scraping	198 (+28)	162 (+5)	176 (+8)	—	58 (+3)	638	41.78
cutting	227	159	135	12	50	583	38.18
shaving	—	—	—	149	—	149	9.76
carving	—	13	9	22 (+1)	16 (+4)	65	4.26
drilling	20	6 (+2)	20	2	4	54	3.54
chopping	—	6	2	10	18 (+2)	38	2.49
total	445	346	342	195	146	1474	100.0
%	30.19	23.47	23.2	13.23	9.91	100.0	

Chapter 8: Conclusion: Meanings and Differences

The following chapter summarizes and compares the results of the preceding analysis of materials from each site, in order to evaluate the similarities and differences in the archaeological record. A detailed comparison serves as the basis for statements about the potentially different nature of the Pavlov complex and the Willendorf II sequence. Although the starting point of my study was to question the claims of difference between the sites, I have not made the immediate reactive leap of claiming their similarity or, even more, their common identity. Rather I tried to look for similarities and differences between the collections from the two locations, as well as within them, and the differences and similarities in approaches that the archaeologists working with material (including myself) have taken.

One of the issues that I was hoping to raise with the examination of the artifacts and the histories of the two sites, is the problem of comparison as a case of polarization, the notion that phenomena, in this case archaeological sites and by extension cultures, can only be either of the same kind, or different. Once the question is posed in this fashion — are they the same or are they different — we are forced to eliminate all uncertainty, all ambiguous cases, all gray areas in between, and are left facing only options of striking similarity or radical difference. Therefore, I would like to stress that my intention is to question the grounds upon which distinctions are made, and to shed more light on areas between the two extremes — same and different — rather than to simply suggest that if the claim of utter difference is false, then its opposite — pure identity — should be affirmed. In this chapter I try to point out overlapping areas of similarity in the assemblages and associated materials as well as their differences, and to identify these similarities and differences in terms of degree rather than absolute contrast. I have used this approach in both the formal analysis of the typological categories at both Pavlov and Willendorf II, and in the use-wear analysis of the stone tool assemblages from these sites; here I will expand it onto the level of site interpretation.

Classification and consideration of problems

Judging from the study of the two assemblages, I perceive two major problems with the typological approach, one methodological and one interpretive. First, from a methodological standpoint, the artifacts are dealt with as if they were idealized, finished end products, in some way (this step usually remains rather murky) reflecting the idea of the makers (similar criticism e.g. Henry and Odell 1994; Neeley and Barton 1994; Parkington 1986; Sieveking 1987). This assumption allows their grouping into discrete units that could be ordered into taxonomic classes and manipulated statistically. However, a close inspection of each stone piece convinced me that a number of them were not necessarily finished tools, or when finished many were modified for subsequent use.

Just as we now realize that the "Pompeii premise" is an overly simplified notion when considering a spatial configuration of a site, we should also recognize that a lithic artifact was not frozen in time upon its completion. Approaching artifacts as items that were a part of a production sequence, and that had use-lives, makes it clear that typological studies provide an overly rigid grid into which a large portion of recovered stone tools cannot fit.

The second problem with the typological approach that I see as a major obstacle lies on the interpretive level. Even if we were able to sort out all the artifacts into typological categories, this typology still remains only an organizing device that does not have much to say about the meaning or the use of the classes of tools. Yet these two levels — descriptive and interpretive — are regularly conflated (see Kelley and Hannen 1988: 183 for a discussion of classification in the context of explanation).

Once we embark on the path of typological variability over time and space, we are inevitably trapped in one of the three commonly evoked possibilities, or some permutations of them: an evolutionary model (based on temporal differences), functional differentiation, or ethnic differentiation (based on spatial variability). All of these scenarios accept the existence of tool types as a given, and only argue about the meaning of their numerical variability.

Dunnell (1971), in the most thorough discussion of classification to date (even if it relies heavily on ceramics as the prototypical variety of archaeological artifact), solves this problem by suggesting that prehistory should become "science". In his view explanation would be defined in terms of prediction and control, making all the observed phenomena ahistorical (Dunnell 1971: 20).

If we were to accept this premise, the only "explanation" that we could provide would deal with the physical or chemical properties of artifacts. However, this is a very narrow comprehension of "explanation" — only one step removed from description and barely within the realm of interpretation, all the more in that the techniques it would adopt for analysis are borrowed from other explanatory fields, while the knowledge it creates is not in turn exported to them. This approach also reflects an universal (not to mention masculine) understanding of science, where the ultimate goal is securing control over the subject under study, regardless of the research domain. In following Dunnel's vision of archaeology we thus run the risk of banality on the one hand, and over-simplification on the other. Such excesses of caution and parsimony only highlight the need for inquiries into the social and historical contingency of the practice of science.

Since Dunnell dismisses any contribution, or even connection, with socio-cultural anthropology, the why of the prehistoric phenomena is eliminated as a historical question that cannot be answered in a scientifically testable fashion. The conversion of classes based on artificial attributes into cultural classes is thus seen as unproblematic, as (Dunnell 1971: 193-194):

"...a pivotal operation in formal theory, the one upon which the discipline is based — creating features, usually modes, from artificial attributes and completing the articulation of the notion of artifact and that of culture. It is assumed that if a set of objects share the same feature, and that feature is artificial, then the objects share that feature because people responsible shared the same idea.... A certain amount of confusion is possible, and does indeed occur, with regard to this equation: a concern with "intent", contact, and genetic relationships (analogs/homologues). These post hoc queries are irrelevant so far as formal theory is concerned... The equation does not say how or why the sharing takes place. This latter concern is entirely inferential; it is an explanation of the distribution of class denotata, and it plays no role in their definition."

On one hand, Dunnell is absolutely correct in stating that explanation of the existence of classes of artifacts plays no role in their initial construction. On the other hand, he dismisses the concern with the reasons behind the existence of a pattern as inferential, historical, and consequently inaccessible in the science context, i.e. predictable and controllable[1]. If we were to follow his advice, the field of prehistory would be reduced to "artifact physics", and would have no reason to exist within the anthropology department, or for that matter to address issues of past societies beyond the most limited frame of their material remains. I am not certain that the physics department would appreciate such a contribution to its ranks as a major enrichment of the field.

Problems of measurement

In the previous chapters, I examined each typological category using traditional criteria that should be applicable to all Palaeolithic assemblages (Bordes 1961; Tixier 1974). The results of the analysis show that I found a much greater variation than the published reports would suggest (Felgenhauer 1959; Otte 1981; Svoboda 1994). Despite the fact that a number of artifact categories have been traditionally defined on the basis of measurements that are clearly replicable, and should not be viewed as subject to interpretations (e.g. a blade is any purposefully produced lithic that is at least twice as long as wide, or a microlith is any intentionally modified lithic measuring less than 5 cm), these criteria were observed with a great degree of liberty, in both the collections of Pavlov and of Willendorf II. A significant proportion of blades in the Pavlov sample (21.98%) did not meet the basic criteria (length was less than twice the width). The proportion of blades in the Willendorf II assemblages that were strictly speaking flakes, varied from 13.43% in W II/layer 5, to 4.95% in W II/layer 8.[2]

Several conclusions can be made on the basis of the first step: some researchers use more accurate measurements than others do. A definition of an artifact in terms of its metric size is no guarantee of accuracy or even adherence to the standard, as I discovered when judging each assemblage just in terms of its internal consistency. This was particularly the case with microliths that were all defined in terms of size, although the specific size varied in each assemblage, fluctuating between 4-5 cm (against the ideal French typological standard of 5 cm cf. Bordes 1971). Moreover, one of the distinct features of the Pavlovian was a claim about its significantly high proportion of microliths — 31% (Svoboda 1991, 1994). Yet when the measurement of microliths was applied consistently within each assemblage, the microlithic component changed quite significantly in each collection, and Pavlov had the lowest proportion of microliths compared to all the layers at Willendorf II. On the other hand, the variation in the microlith component in all the assemblages was not dramatic, to the contrary, it was remarkably similar ranging from 29% to 39%.

Table 8.1. Pavlov and Willendorf II: The microlithic component of all assemblages

cultural layer	*microliths*
Willendorf II/5	38%
Willendorf II/6	34%
Willendorf II/7	29.7%
Willendorf II/8	38.8%
Willendorf II/9	35.6%
Pavlov	29%

Consequently, I have to conclude that despite all the attempts for objective standards that would allow comparisons of any collections, the system of stone tool typology, and its subsequent statistical manipulation, should be viewed less as a hallmark of scientific archaeology, but rather as a temporary organizing device that needs to be tested through an external line of reasoning, and modified if need be. Moreover, we should recognize the interpretive spaces that exist even in the method itself, possible at every step. I would argue that the underlying assumptions of cultural uniqueness, or its lack, are a part of the whole analysis, and not only the final interpretation.

Measuring rationality

After measuring each individual lithic piece and calculating all the blade/flake/core counts in each layer, I determined the blade proportions in all the assemblages under study to have been between 70 to 80% . This information is generally used in typological approaches as an indicator of Upper Palaeolithic assemblages. However, with the dates available, the issue of temporal classification was not in doubt from the start, and we should consider other reasons why the proportions vary from each other over time, as well as why some of the indicators are strikingly similar despite the time difference.

Table 8.2. Willendorf II and Pavlov: Blade, flake and core proportions in cultural layers (%)

cultural layer	*blade*	*flake*	*core*
Willendorf II/5	73.67	24.0	2.33
Willendorf II/6	75.38	22.11	2.5
Willendorf II/7	82.33	16.25	1.42
Willendorf II/8	70.08	14.48	15.44
Willendorf II/9	69.16	27.74	3.1
Pavlov	76.7	19.7	3.6

The blade/flake ratio in the studied collections has changed over time but not in any simple progressive fashion. The flake percentage declined over time from 24% in W II/5 to 14.48% in W II/8, yet it increased to the highest level - 27.74% in W II/9. Similarly the percentage of cores changed over time from 2.33% in W II/5 to 15.44% in W II/8, only to drop to 3.1% in W II/9. While it may be tempting to consider cultural layer 9 as an exception to the pattern observed in the previous layers, it is important to note that the Pavlov collection, chronologically close to layers 8 and 9, had an almost identical proportion of cores (3.6%) as W II/9, and a somewhat lower but still quite significant percentage of flakes (19.7%).

Recently, the discussion of core/blade ratios has shifted from the traditional, and fairly sterile typological approach, to a consideration of prehistoric lifeways as the context in which the differences in assemblages should be considered (e.g. Jones and Madsen 1989; Kelly and Todd 1988; Kuhn 1992, 1994; Lurie 1990; Metcalfe and Barlow 1992; Nelson 1991; Shott 1986; Stiner and Kuhn 1992).

All these studies focus on mobile hunters and gatherers, and consider lithic tool kits from the perspective of transportation cost, and resource reliability. Some have dealt with specific archaeological cases (e.g. Jones and Madsen 1989; Kelly and Todd 1988) while others have constructed abstract models, so as to consider predictable outcomes that could be applied to the archaeological record (e.g. Kuhn 1994; Metcalfe and Barlow 1992). Welcome as these studies may be in terms of addressing prehistoric lifeways rather than just stone tools in isolation, they all share a striking hyper-rational perception of human behavior.

In the best tradition of New Archaeology, they all adopt the view that culture is "extrasomatic means of adaptation" (Binford 1972), and a successful adaptation at that. According to this literature, there has never been (and probably never will be) a more rational, calculating, predicting, future estimating, and past-experienced creature than a prehistoric hunter and/or gatherer.

Moreover, all these studies implicitly assume an egalitarian society, or even a homogeneous one, where each person is interchangeable, regardless of age, sex, or physical condition, has the same strength, will, ability, and chance to participate in all activities, decisions, and resulting consequences. Models for transportation cost of raw materials never specify who the carrier is, implying that the whole group consisted of equally able bodied individuals — or is the underlying assumption not spelled out, and the estimates are based on our modern day notions of gender and age roles? My hunch is that the latter is often the case.

Reading about these superheroes of the past, I was left with some questions: Since all these models are explicitly or implicitly based on ecological-evolutionary theories, how should we apply evolutionary theory to human rationality, considering the multiple displays of irrationality in every society known to us in the present, as well as in the recent past. Furthermore, even if we were to accept the possibility of some overarching rational spirit on the group level, the implicit assumption that all members of any group complied without resistance, discounts internal mechanisms of change, focusing on external influences.

On a practical level, as an archaeologist, I wonder how would one detect maladaptive behavior, or an erratic decision on the part of the prehistoric gatherers or hunters in the archaeological record. If all that we are observing from the prehistoric remains is successful adaptive behavior, how would we recognize the unsuccessful cases, those groups that miscalculated their risks, that ventured into inhospitable lands where they did not know how to cope, those groups that were seasonally facing hunger?

The traditional evolutionary ecology does not offer any models that would take such a grim view of the past. It is my sense that these models may reflect a particular social climate at a point in recent North American history, and express undue optimism about the ways that societies operate, rather than constituting ahistorical, scientifically testable models of the past . On the other hand, the complete silence about the possibility of irrational choices, mistakes that were not adaptive, risks that backfired, leaves me with an uneasy feeling that these have no place in the progressive agenda of evolutionary theories where only the rational and successful survive. The real debris of history remains in silence.

Thus while I do not dismiss predictive, normative models as entirely false or implausible, I do not find them particularly satisfying as an ultimate explanatory or interpretive form. I will therefore consider the differences and similarities in the assemblages on a microscale level, focusing on suggested prehistoric activities, and their possible seasonal differentiation.

It is my conviction that if we are to investigate prehistoric social contexts and to attempt to produce narratives of the past that involve potentially real human beings rather than bland copies of our norms, we have to start from the analysis of the recovered materials, and of the activities that could have produced them. The material evidence should be taken as the base line of the study of prehistory, not the endpoint reached by providing a detailed description.

Throughout this study I have assumed contemporaneity of the suggested activities in each layer, although it is plausible (or even very likely) that we are dealing with a palimpsest of separate occupational events. In that case each cultural layer at the location can be interpreted as having been visited for a particular seasonal activity. The evidence available at the moment supports the interpretation of seasonal occupations, whether these were singular of a longer duration, or repeated over a period of time. On the basis of artifacts recovered throughout the century, it is impossible to estimate the exact season, particularly since the faunal remains were unavailable for analysis, having been collected from Willendorf II as a whole, rather than by individual layers, and no botanical remains were collected at the time of excavation.

Factual artifacts: similarities and differences

W II/layers 5 and 6 were remarkably similar in their proportions of blades (73.7%, 75.4% respectively), flakes and cores, yet the observed traces of wear seem to suggest processed materials, as well as activities, that differ quite significantly in these two layers. The discrepancy cautions us that the same task may be carried out with different implements, that different activities may be accomplished with very similar tools, and that any proposed interpretation has to rely on several lines of evidence.

Typologically these two layers were described as being very different, since W II/layer 5 had a significantly higher proportion of microblades (11%) than W II/layer 6 (1%). However, a measurement of all lithics showed that the microlithic components in these particular layers were very similar — 38% in W II/layer 5 and 34% in W II/layer 6. Thus we are facing two chronologically separate assemblages consisting of very similar tool shapes and sizes, yet showing traces of different proposed prehistoric activities.

On the other hand, W II/layers 8 and 9, also chronologically distinct, were almost identical in the blade component (69-70%) but varied dramatically in the flake/core ratio. W II/layer 8 had the highest proportion of cores (15.4%) in all the Willendorf layers (and higher than Pavlov), a fact that needs to be accounted for in the interpretation of the collection.

Typologically, W II/layer 9 was dominated by endscrapers and burins (31% each), while W II/layer 8 although high in endscrapers (46%), had only a few burins (8%). As I have discussed in the previous chapters, I consider burins to be a rather ambiguous measure of distinction. Besides the number of possible ways how they could be produced, they also appear to have been frequently used for a wide range of activities, including the same function that an endscraper might have been used for. Thus the comparison of endscraper and burin frequencies may not necessarily be an indicator of any significant difference in life-style. However, if we consider the burin blow technique as a technological approach - rather than a typological issue - towards reducing cores, producing blades, or resharpening tools, it is noteworthy that some groups practiced it to a much higher degree than others.

The typological approach serves as a cautionary note, reminding us that change over time may not have been as straightforward as we imagine. The assumption that temporal and/or cultural difference is directly reflected in changing forms of material culture, or at least in changing proportions in the tool forms, appears a simplified notion that does not begin to describe the complexity of even the known remains of prehistoric lifeways. I would propose to circumvent the typological quandary by shifting focus to the observed use-wear traces on the lithic artifacts. I hope to address the interpretation of the prehistoric occupations from a more dynamic perspective focused on the activities that could have taken place, and to suggest further venues for such an approach that would strengthen the proposed interpretations.

Considering how unreliable the assignment of lithics into typological categories is, and the even more precarious nature of interpretation based on such a shaky ground, it may be worthwhile to pursue a different path in an attempt to give some meaning to the recovered assemblages. The typological approach was useful as a relative dating method prior to the advent of currently available absolute dating techniques (Parkington 1986, Renfrew 1979). The emergence, variability, and disappearance of different stone tool types allowed some chronological ordering, and a degree of spatial comparison, using a simplified concept of culture, now recognized as anthropologically problematic. But since the standard use of absolute dating methods, a number of chronologies based on lithic typology have had to be revised, and it has become clear that typological variability is not a simple time dependent variable.

As Renfrew put it: "Certainly in Europe the conventional framework for our prehistoric past is collapsing about our ears" (Renfrew 1979: 15). Yet the most general definitions of chronology based on stone tool typologies continue to be used, from the Palaeolithic, Mesolithic (questioned as it may be), Neolithic division, to the Aurignacian, Gravettian, Magdalenian industrial complexes. Questions are raised whenever a new excavated site does not fit the standard pattern. For example the recently recovered Austrian site Alberndorf dated to 20,500 BP was typologically assigned to the Aurignacian complex, attracting suggestions that an Epi-Aurignacian (or a prehistoric retro-style) may have to be created to explain the anomaly (Neugebauer-Maresch 1993, Trnka pers. comm.). Yet these cases are treated as exceptions to the rule, rather than a reason to question the rule, and our insistence on it. Having outlined the typological problem, let us try to examine the lithic artifacts from other angles, probing the possibility of revealing some traces of prehistoric social activity.

The main focus of the study presented here has been to investigate the possibility of circumventing the typological approach to prehistoric stone tool collections. I have chosen to examine tools in terms of the materials they probably processed, and to suggest activities that could have occurred at the sites. This type of study provides a better ground for claims of seasonal tasks, and allows a more accurate portrayal of differences and similarities over time and over the short geographic distance between a segment of the Willendorf II sequence (Layers 5-9) and the Dolní Vestonice/Pavlov site complex.

While I have noted some striking differences in the use-wear traces and their potential seasonal meanings, as well as in the use of tools for multiple tasks, I cannot state that any of this could be neatly arranged in a unidirectional line, and perceived as a clear case of development over time. Some similarities in processed materials and uses of the lithics appeared between the oldest Layer 5 and the most recent Layer 9 at Willendorf II, as well as the sample from Pavlov. Whether these similarities and differences can be translated into any "cultural" differences is rather questionable.

It is my firm belief that even if we were to attempt to use such a recent phenomenon as cultural or ethnic identity in the prehistoric past, we need much more detailed studies that address questions of social organization, everyday practices, production and consumption of material culture to make them meaningful. Only on the basis of such detailed, tedious work shall we ever be able to say that things "over there", or "back then", were very different or the same.

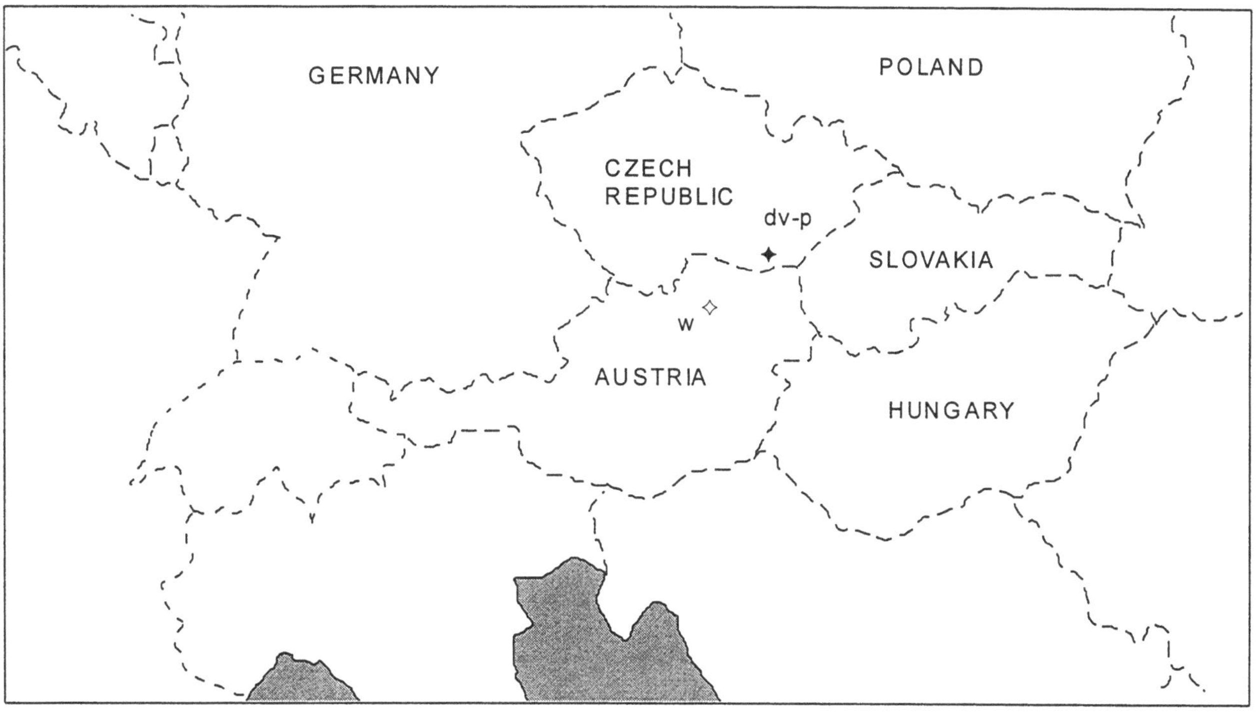

Map 8.1. Dolní Věstonice/Pavlov and Willendorf in Europe since 1992

Conclusions: writing on the margins and constructing the past

Throughout this text I have been conscious of the question of voice and the speaking subject. It is my conviction that the problem is even more confounded in archaeological writing than in general anthropology (even if not so widely discussed), since we are dealing with the "other " on several planes, not only geographically and culturally distant, but in addition chronologically removed. The increased consciousness of the past cultures not being "like us", has led to the quick solution of collapsing space and time, and treating the prehistoric[3] groups as if they were the equivalent of the anthropological "other", simply distant from us.

As anthropology faces the disconcerting fragmentation of the postmodern world, it also comes to recognize that the distant "other" is not necessarily mute, or even as distant. While individual societies are seen as much more fragmented, the world is more homogenized through the contacts made by modern technology, travel, trade, and migrations of masses of people. The former subjects of anthropological studies are frequently aware of their position in Western productions of knowledge, familiar with the writings, and more often than not participating in them, and even manipulating them. The subjects of studies are definitely not as passive as anthropology has presented them (Fabian 1983; Clifford and Marcus 1986).

Here, however, we have the crucial difference between the geographically distant "other" and the one temporally distant. Archaeologists study people who cannot be interviewed, or respond to the stories we tell. No matter how much leisure time the Western world creates, it is impossible to witness the archaeological equivalent of "Cannibal Tours" recognition of this difference should create a sense of responsibility, even caution in what one states, but also should lead to a more explicit admission of archaeology's special role in the process of Western knowledge production.[4]

For historical reasons stemming from the formation of the discipline, the implicit assumption in much anthropological discussion has been that one is dealing with cultures not exactly on par with one's own (whether this is expressed in terms of "traditional", "simple", or "developing"), and we had to find a way to write without expressing a judgment on other cultures' values, traditions. This issue, while still being debated, brings out of the shadows the notion of the "objective" anthropologist who is distant enough from the people that she is writing about, that she is capable of seeing the culture for "what it is", now she just has to learn not to be judgmental. Problematic as this notion may be, it does not address the problem of writing about one's own culture, tradition, group, or for that matter about one's own past.

The most obvious question that immediately jumps out is what is meant by the term "one's own". An automatic authority is not granted by speaking about one's fellow citizens, since no group is homogeneous, and the ability to speak sets one apart into a privileged position. Moreover in the context of central Europe I became much more conscious of the fact that it is often unclear where the alliances lie, what is the tradition that I can speak freely of as my own, be critical of, without having to answer for that criticism as a display of self-hate. Furthermore, archaeological work that deals with the Palaeolithic can by no means be treated in terms of one's own ancestry.

Geographical location of any Palaeolithic site, and modern day inhabitants of the region have an intimate connection forged purely by modern geography and its political context. Yet it is this tie to ancient places that has no obvious scientific validity but is based primarily on the results of scientific research that I am interested in.

When I started my field work in Czechoslovakia, I felt fortunate to be working in a country whose languages I spoke, whose general culture, traditions, and history I knew well. However, as I found out in the two years of working in Dolní Vestonice and Vienna, on one hand this close knowledge gives insights that are privileged, but the price for it is that it also abolishes the pretense of safe emotional distance that one can have from a context that is clearly conceived as "not one's own".

I had to take a stand on many issues of the country's past and present (as they are - in situ - inseparable). I felt that I could not play it safe and pretend no opinion, leaving it all behind once my archaeological research was done[5]. Any statements that I was making about the attitudes towards the past in Czechoslovakia and Austria, were partly my own, partly those perpetuated through historical traditions (not always neatly aligned, but neither at constant odds). Thus writing about national sentiments towards the land and its history and geography, resulted in generalizations that are self-conscious, a fact I have tried to convey rhetorically.

I used the pronoun "I" whenever I felt that I had to take an individual stand, be responsible for my actions, experiments, interpretations. On the other hand, most of my archaeological, anthropological, and historical knowledge is rooted in an academic tradition of which I am a part of, which I rely on, and making any statements resonating with this sense, I use the royal "we". This is a conscious attempt for an inclusive dialogue in the academic community, as well as among any readers that are not a part of it. Thus stating some of my assumptions in the form of - "if we accept that", I was giving a sense of journey along which some shared beliefs are necessary in order to find a common language, convey meanings, and accept as a possibility some of the presented interpretations.

I have also used the passive voice, even if I may have mixed feelings about it. The common practice in scientific writing is to distance oneself from the subject matter by using this form of speech. I have tried to evoke my presence in the text, so as to challenge the notion of the absent writing subject who only observes science as it reveals in front of him.[6] However, I do not feel that the simple answer is to replace all passive voice by an active one. The problem of the speaker is not merely a matter of personal involvement but also a matter of scientific tradition and cultural systems less clearly individuated into singular actors; not every part of an experimental or observational procedure is simply an action that one individual chooses to do.

I have been made much more aware of the dilemmas of action through my own position as a woman studying the most masculine of archaeologies — the Palaeolithic — and trying simultaneously to assert myself, and to subvert crude forms of evolutionism, based on a conception of the successful individual who overcomes all obstacles through struggle, defeating all those with whom he has to compete for survival. Thus we are facing the dilemma between generalized theories about "the people" or "the group", and the cacophony of individual statements that may not be representative of more than a small fragment of a society. Wherever one stands on this issue, the most productive approach may be a self-conscious representation of a segment of a group, and a reflection upon other groups that are enmeshed within a wider circle. Polarizing the question of interpretive representation into only two possibilities, general statements or personal confessions, leaves out a vast rhetorical space between these poles, a space that can resonate with stories reflecting both ends of the spectrum as well as gradations between them.

I have tried to solve this problem by including parts of the diary that I kept throughout my fieldwork, recording events, dialogues, comments other people made, and stories they told me. These stories are included in the text partly to evoke a sense of the context in which I was working, and to diffuse the notion that science, even when conducted with specialized expensive instruments, operates in isolation from the outside worlds. In addition, some of the stories were written down to evoke other themes, to remind the reader of more distant connections, and weave in threads that are not necessarily immediately obvious.

On discoveries

> It is homesickness that gave rise to the adventures through which subjectivity (whose fundamental history is presented in the Odyssey) escapes from the prehistoric world. The quintessential paradox of the epic resides in the fact that the notion of homeland is opposed to myth — which the fascist would falsely present as homeland. Here there is a reminiscence of history, in which the domiciled and settled life, the prerequisite for any homeland, followed the nomadic age. If it is the fixed order of property dependent on the settled life that grounds the human alienation in which originates all homesickness and all longing for the lost primal state of man, it is nevertheless the settled life and fixed property (only in which a notion of a homeland can appear) to which all longing and homesickness are directed. Novalis' definition, according to which all philosophy is homesickness, holds true only if this longing is not dissolved into the phantasm of a lost remote antiquity, but represents the homeland, nature itself, as wrested form myth. Homeland is the state of having escaped. ...The translation of myths into the novel, as in the adventure story, does not so much falsify the myths as sweep myth into time, concealing the abyss that separates it from homeland and expiation. *The vengeance wrecked by civilization on the prehistoric world is a fearful one...* (Horkheimer and Adorno. 1987 [1944], 78; emphasis added).

Like most longer essays, this text has opened a number of complex issues, none easily closed or solved. Although I have sought to offer an opinion on the topics examined, I by no means pretend to have the final answer, or ability to resolve them all into a single line. Yet in the long process that led to writing this text, I have learned a great deal, and have come to the recognition that I find the endeavor itself at least as fulfilling as the results produced. A male cultural anthropologist (a member of the National Academy of Sciences) once told me in an attempt to defend "science", that scientific discovery is like an orgasm; it gives you a thrill, and the recognition that all the trouble was worthwhile. Many years after this conversation, I have come to the conclusion that I do not share his opinion, and that what matters to me in research is the trouble that one goes through, the many small detours along the way that lead into unexpected niches and illuminate one's life. The top of any mountain does not always offer the most rewarding view. Similarly, reaching the climatic "truth" is neither always possible nor even crucial to one's endeavor, for the process of research may still foster understanding, and contribute to the academic fields of which we are a part. My hope is that this text constitutes such a contribution, taking the reader on detours beyond the linear, offering greater possibility of new prospects.

Notes:

[1] One might add that the fundamental uncertainty of prehistory — the lack of control over the preservation of elements of the past — is hardly solved by artifact physics, only locked in the closet..

[2] This excludes all the broken pieces that were not included in the calculation.

[3] Reference to ethnographic film of that title by Dennis O'Rourke, which chronicles a tourist expedition upriver in New Guinea. Theme parks, museum exhibits, or creations like Jurassic Park are not included in this context.

[4] I stress the term "Western", as one of the points that I hope to make is that the practice of archaeology, and a cataloguing interest in the past is not universal, but rather a fairly recent historical invention, peculiar to Euro-American cultures, intensively as it may be exported to other parts of the world (Trigger 1991). I might add that in a work focused on Europe's eastern boundary, the term "Western" echoes in rich and misleading ways.

[5] As fate would have it, Czechoslovakia broke up during the first year of my fieldwork. I have rarely felt as much a victim of history as on October 28, 1992 (the date of the creation of the former Czechoslovakia in 1918), when, together with my mother and my sister, I stood in the auditorium of a concert hall, trying to sing the Czechoslovak national anthem for the last time, as a part of a protest against the upcoming split between Slovakia and the Czech Republic.

[6] I purposefully use the pronoun "he" in this context, since the assumed gender in most science writing is male.

Epilogue: The border

It was still dark outside. Morning cold air enveloped all of us waiting for the bus. Most people were smoking, a few talked to each other between the long pulls from a cigarette. At least something kept them warm. We all knew each other from sight, taking the same bus every morning for an hour on the way to work. Four, sometimes five buses crossed the border between Slovakia and Austria every morning at seven. All these people worked in Vienna, most of them illegally, for low wages that supported their families on the other side of the border. Nearly all of them were women who now brought in the main income for the household, even if it meant washing floors in a hospital, taking care of other women's small children, or, if they were lucky, a clerical job. They spoke very little German, yet had been doing it for a year or more. I crossed the border with them every morning and night, all winter long that year, sitting in the darkness of the bus, listening to stories they told in tired voices, dozing off, as we drove through orderly Austrian villages, passing houses with bright window lights.

Resigned, the women complained about the double burden. They had to do all the housework, shopping, child care, even when bringing twice or three times as much money than their husbands. They were all married. Being single and working in Vienna was considered a sign of personal greed. Married women had to do it for the family, to help their husbands in "keeping the standard". I wore a wedding band all year long to avoid questions, unpleasant suspicions, and unwelcome advances. This way we were all equal, no one asked about work conditions, pay, or work permit, knowing the answer, even if it were not always accurate.

The physical removal of the old border, with its barbed wire, electricity, and trained dogs, had had very little immediate effect on the way people thought, or behaved towards each other. It also hardly changed gender relations at all. To the contrary, as if in defense, women had become a conservative element, vigorously upholding "tradition". Reversed economic power did not affect cultural practices in the male dominated society of Slovakia (or the Czech Republic), at all. What mattered was that the family as a whole benefited financially. Consumer goods were regularly passed along the aisle of the bus for everyone to admire. Years of shortage of any colorful items made them seem more desirable than sleep, pleasure, and walks in the afternoon, more desirable, in fact, than anything else. Barbie dolls, make-up kits for little girls, "sexy" skimpy clothing were all purchased and imported across the border. To all these women life in Austria seemed so colorful, easy, pleasant, so very different from the monotonous tiring routine back home. Yet whenever they talked about women in Vienna, it was in a condescending tone, condemning them for not caring enough for their households, their husbands, their cooking, for being too frivolous, selfish, and career minded. While the border may have disappeared as a physical barrier, it continued to separate the two worlds, the categories "us" and "them" were firmly in place. I stayed silent, knowing how much I was being judged, how much I had become a foreigner in my own land. "Things are very different over there," a woman told me one night, as we shared a piece of a poppy seed cake on the way back home.

(somewhere along the way between Vienna - Wolfsthal - Bratislava, December 1993).

Appendices

A. Known Palaeolithic sites in the region .

B. Pavlov 1952/53, site plan.

C. Pavlov 1952/3, location of hearths.

D. Pavlov 1952/3, examples of examined stone tools. Used and unused stone pieces included. Uninterrupted line = traces of use-wear; interrupted line = traces of hafting.

E. Pavlov 1952/3, examples of bone implements with clear marks of modification.

F. Willendorf II, profile and site location.

G. Willendorf II, stratigraphy with dates.

H. Willendorf II, examples of examined stone tools from layers 5-9. Traces of use-wear and traces of hafting noted. Some selections exhibited patination or indeterminate marks and a few could not be located in the current museum collection.

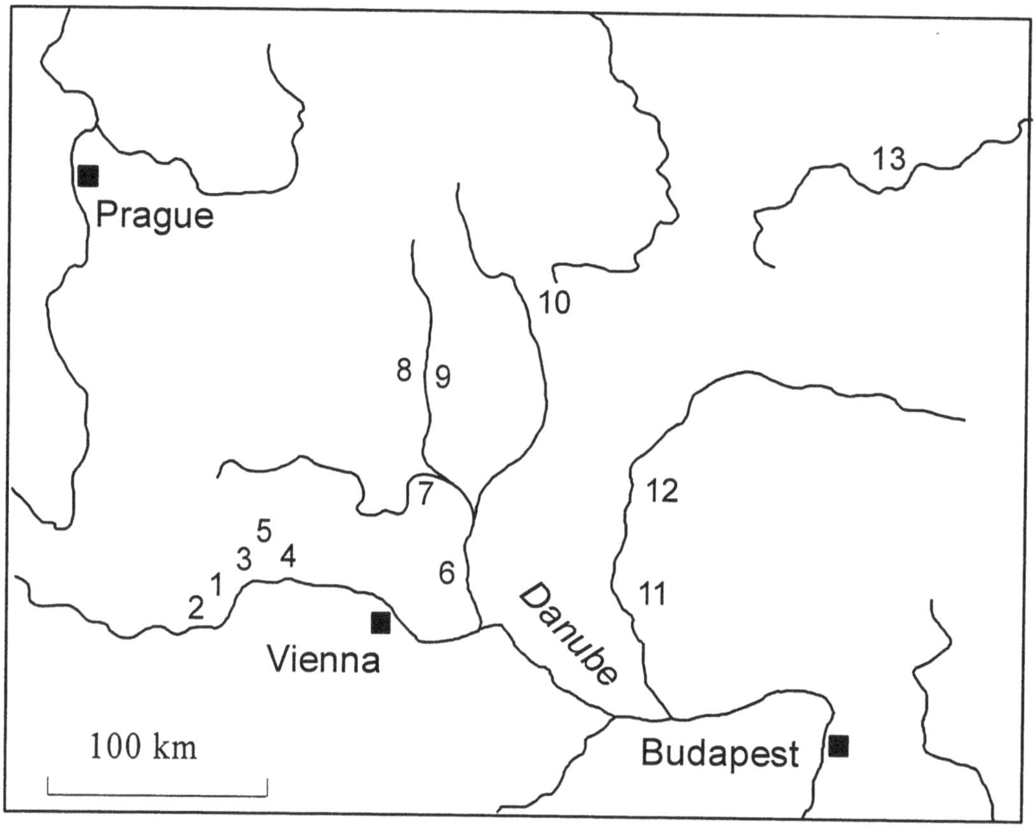

Appendix A: Known Palaeolithic sites in the region
(after Haesaerts 1990)

1. Willendorf and Schwallenbach
2. Aggsbach
3. Krems
4. Langenlois
5. Stratzing
6. Stillfried
7. Dolní Věstonice and Pavlov
8. Bohunice
9. Stránská Skála
10. Předmostí
11. Nitra Čermán
12. Moravany
13. Spadzista

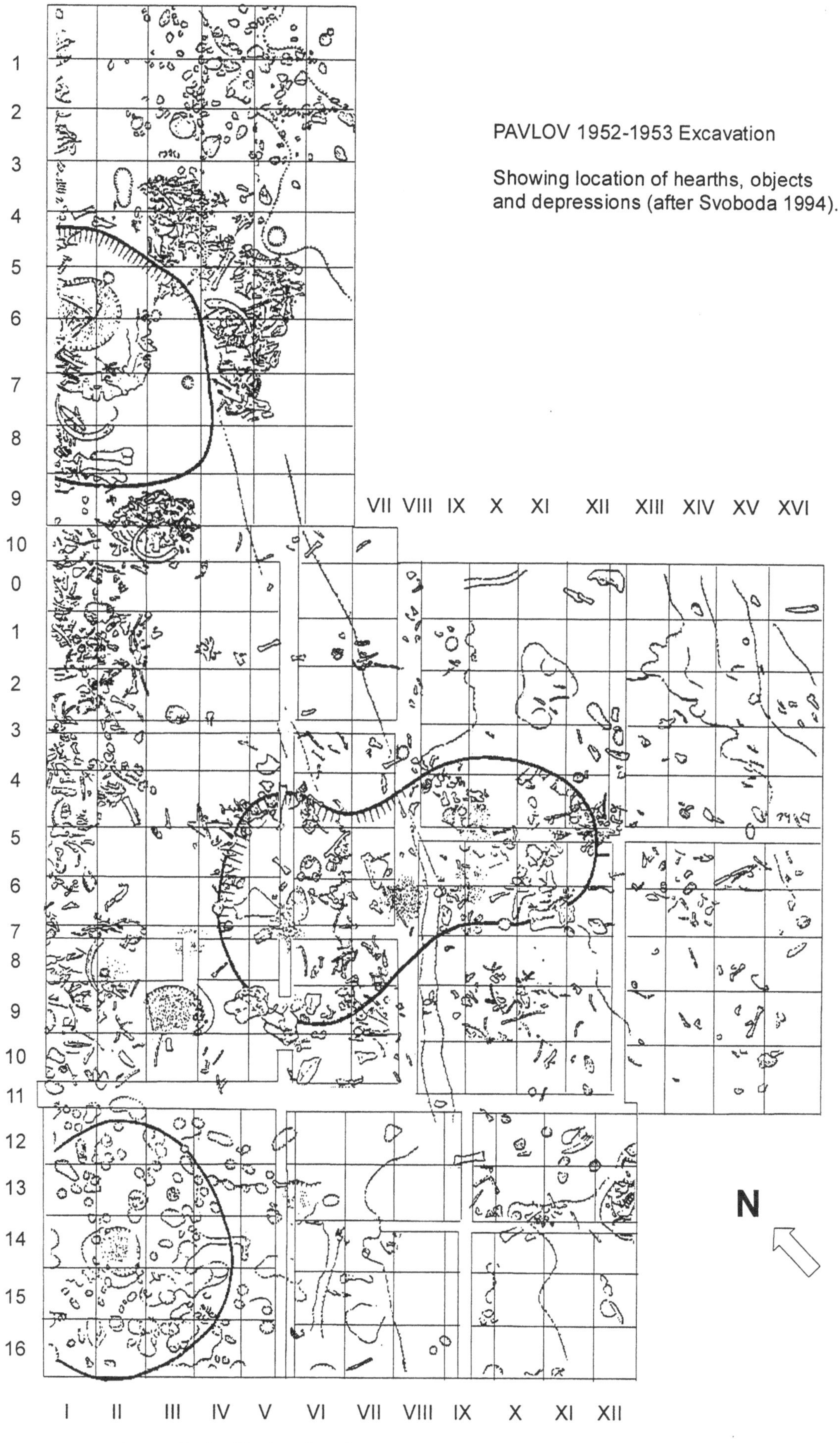

Appendix B: Pavlov site plan

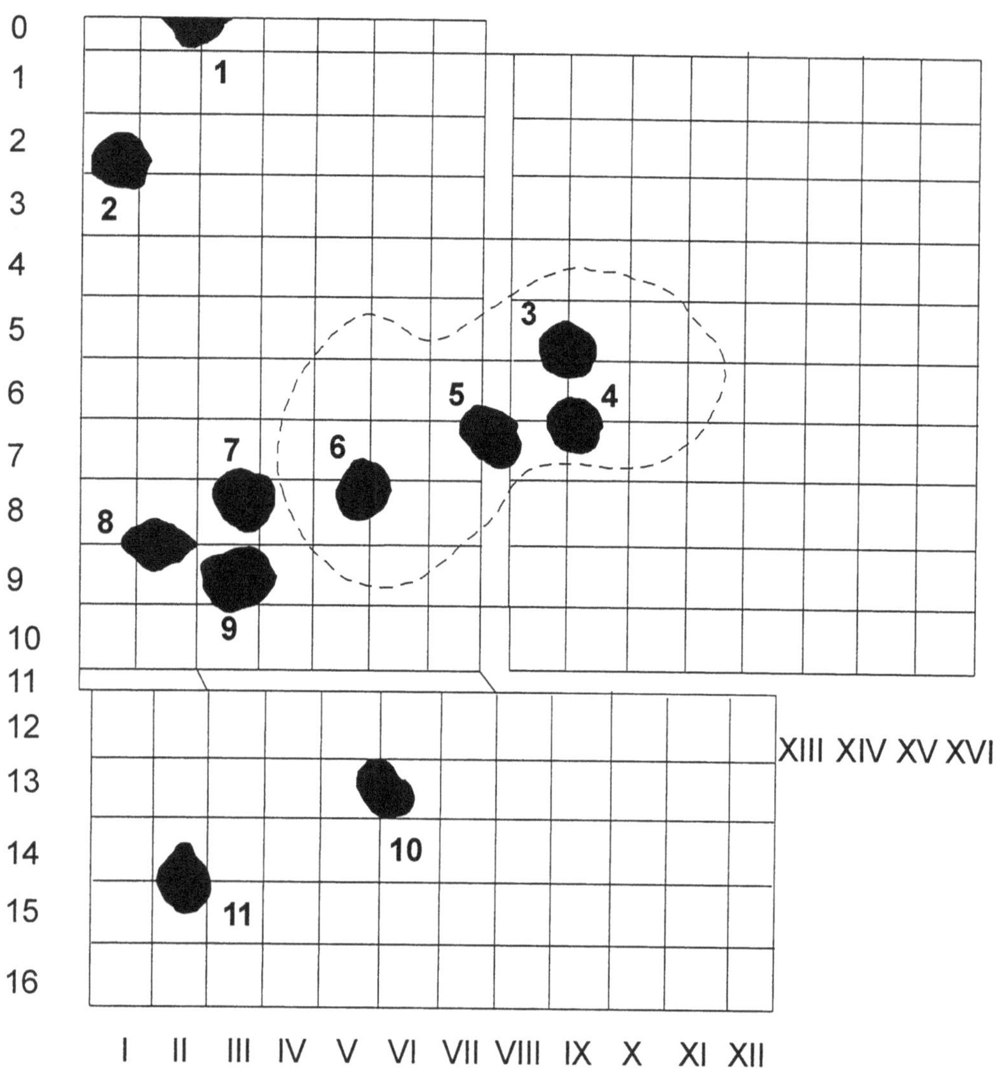

Appendix C: Pavlov 1952/3, location of hearths
(after Svoboda 1994)

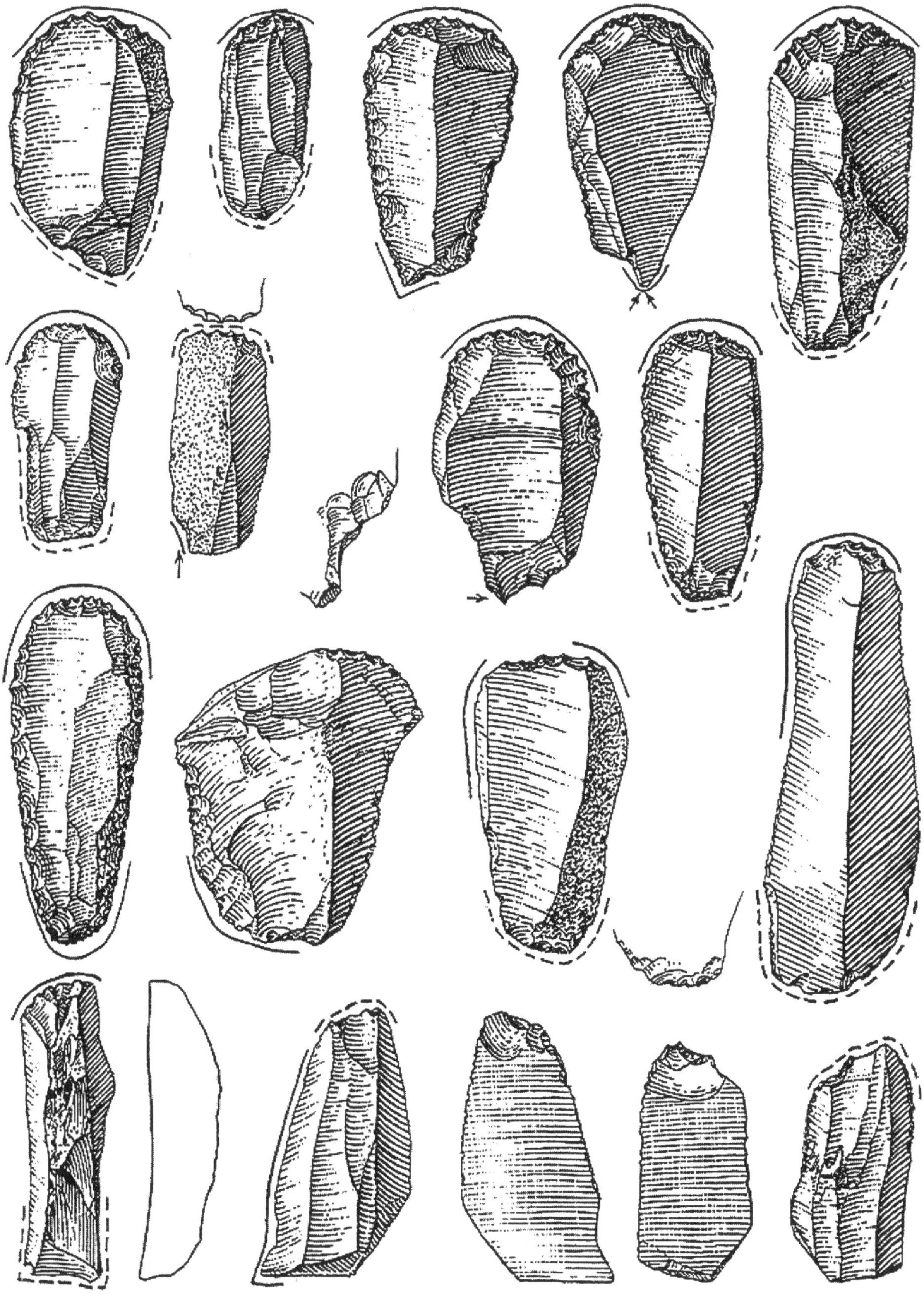

Appendix D1: Pavlov 1952/3, examples of examined stone tools. Used and unused stone pieces included Uninterrupted line = traces of use-wear, interrupted line = traces of hafting. To scale (after Svoboda 1994: 79).

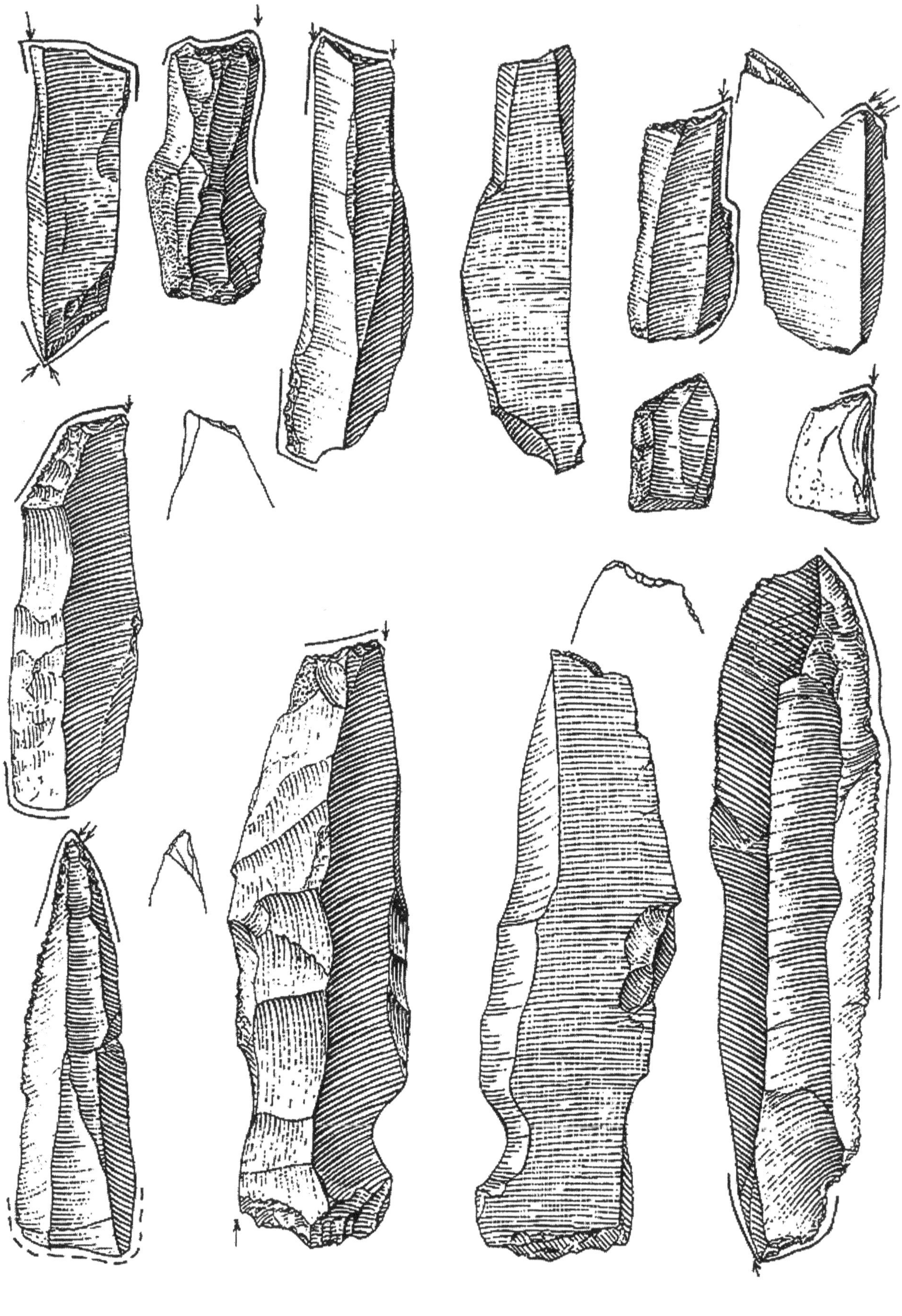

Appendix D2: Pavlov 1952/3, examples of examined stone tools. Used and unused stone pieces included Uninterrupted line = traces of use-wear, interrupted line = traces of hafting. To scale (after Svoboda 1994: 80).

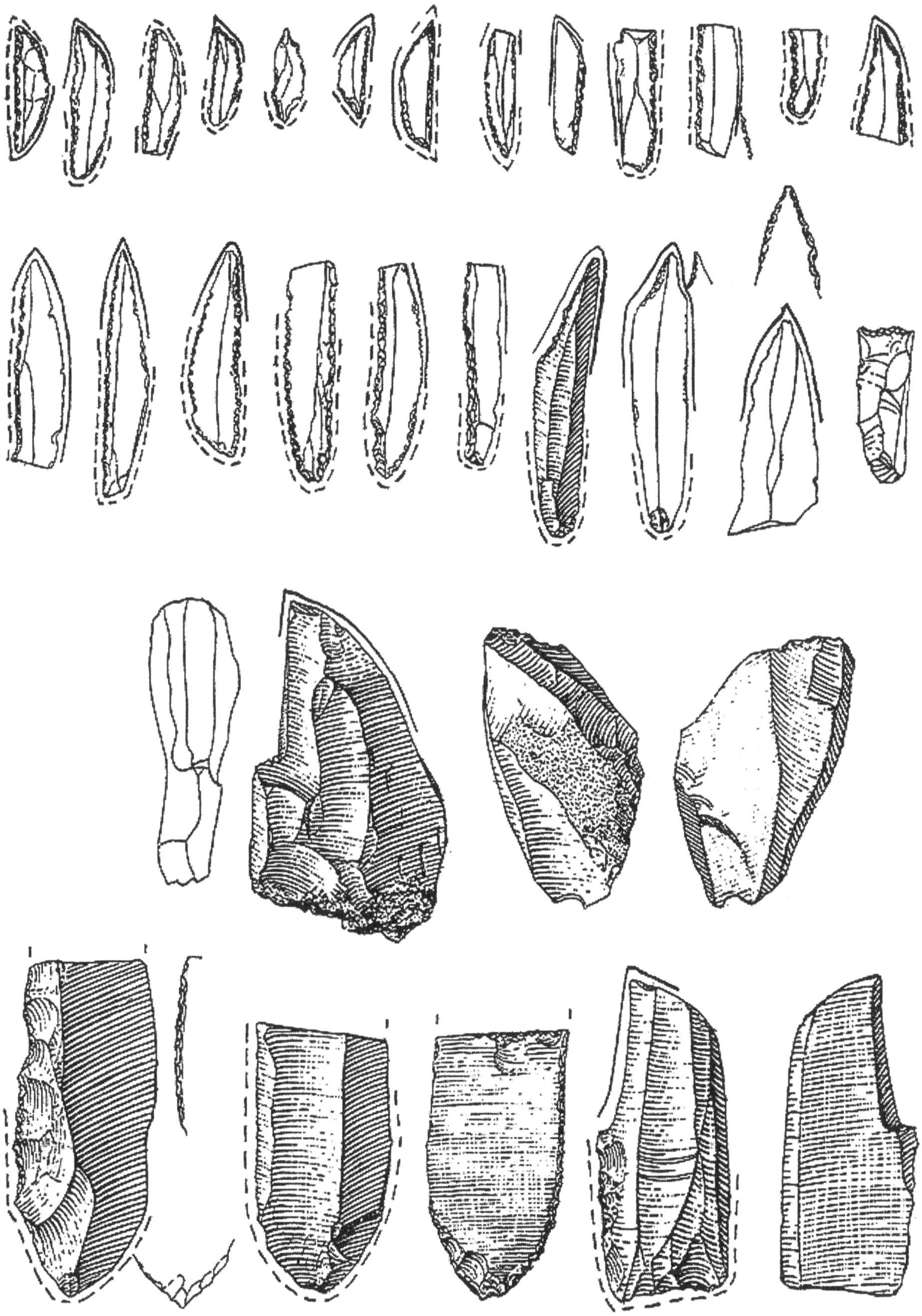

Appendix D3: Pavlov 1952/3, examples of examined stone tools. Used and unused stone pieces included Uninterrupted line = traces of use-wear, interrupted line = traces of hafting. To scale (after Svoboda 1994: 81).

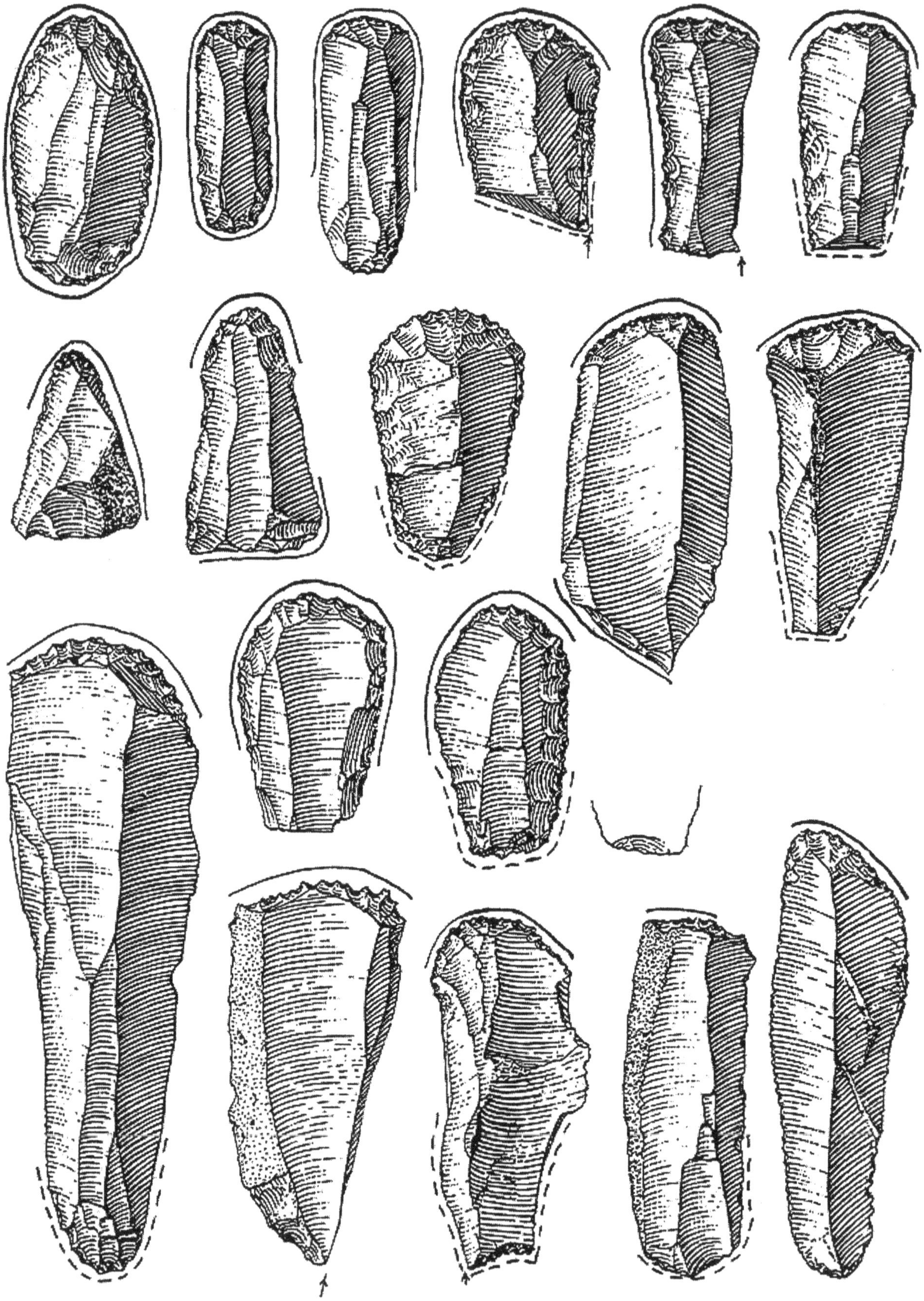

Appendix D4: Pavlov 1952/3, examples of examined stone tools. Used and unused stone pieces included Uninterrupted line = traces of use-wear, interrupted line = traces of hafting. To scale (after Svoboda 1994: 83).

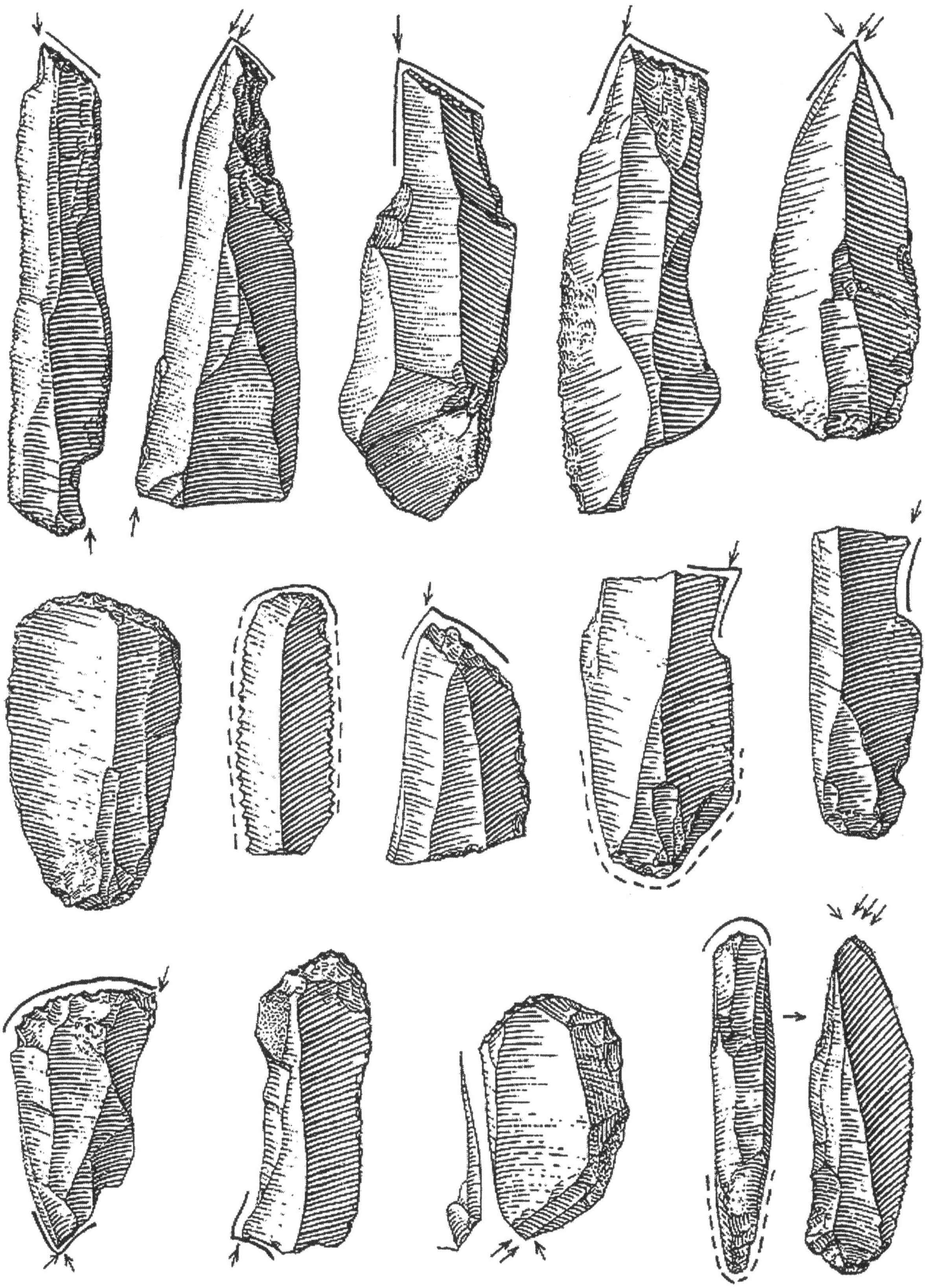

Appendix D5: Pavlov 1952/3, examples of examined stone tools. Used and unused stone pieces included Uninterrupted line = traces of use-wear, interrupted line = traces of hafting. To scale (after Svoboda 1994: 87).

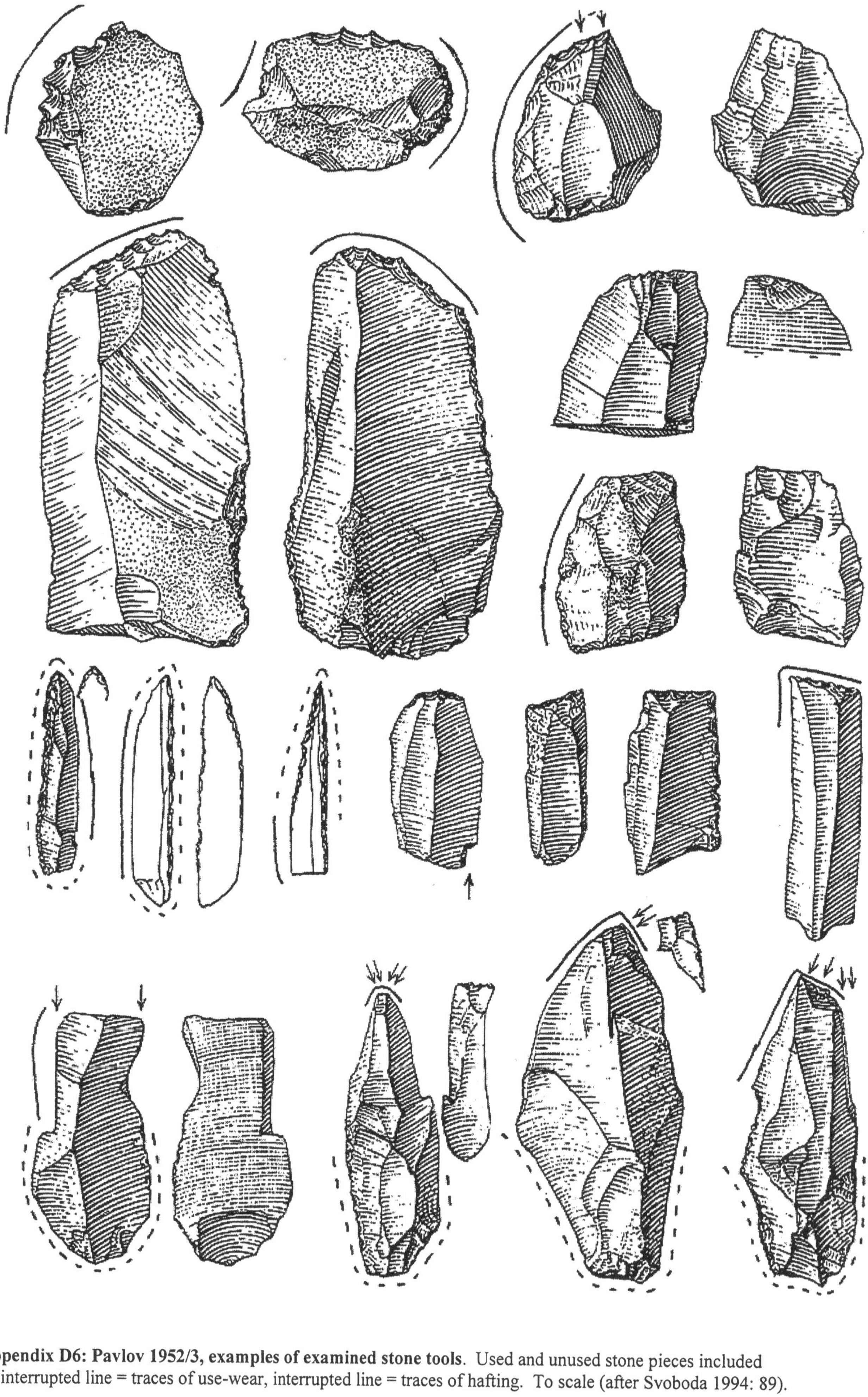

Appendix D6: Pavlov 1952/3, examples of examined stone tools. Used and unused stone pieces included Uninterrupted line = traces of use-wear, interrupted line = traces of hafting. To scale (after Svoboda 1994: 89).

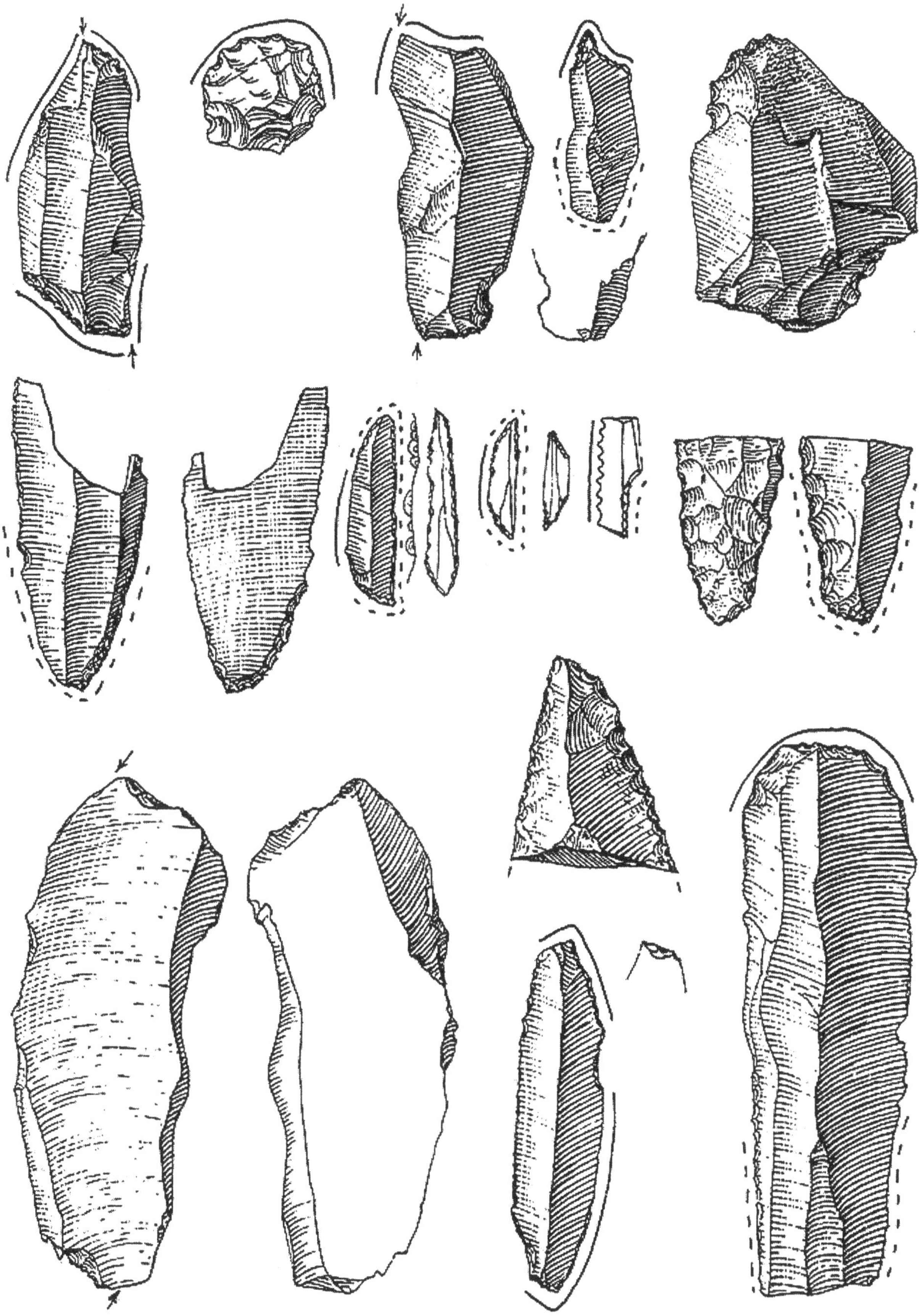

Appendix D7: Pavlov 1952/3, examples of examined stone tools. Used and unused stone pieces included Uninterrupted line = traces of use-wear, interrupted line = traces of hafting. To scale (after Svoboda 1994: 92).

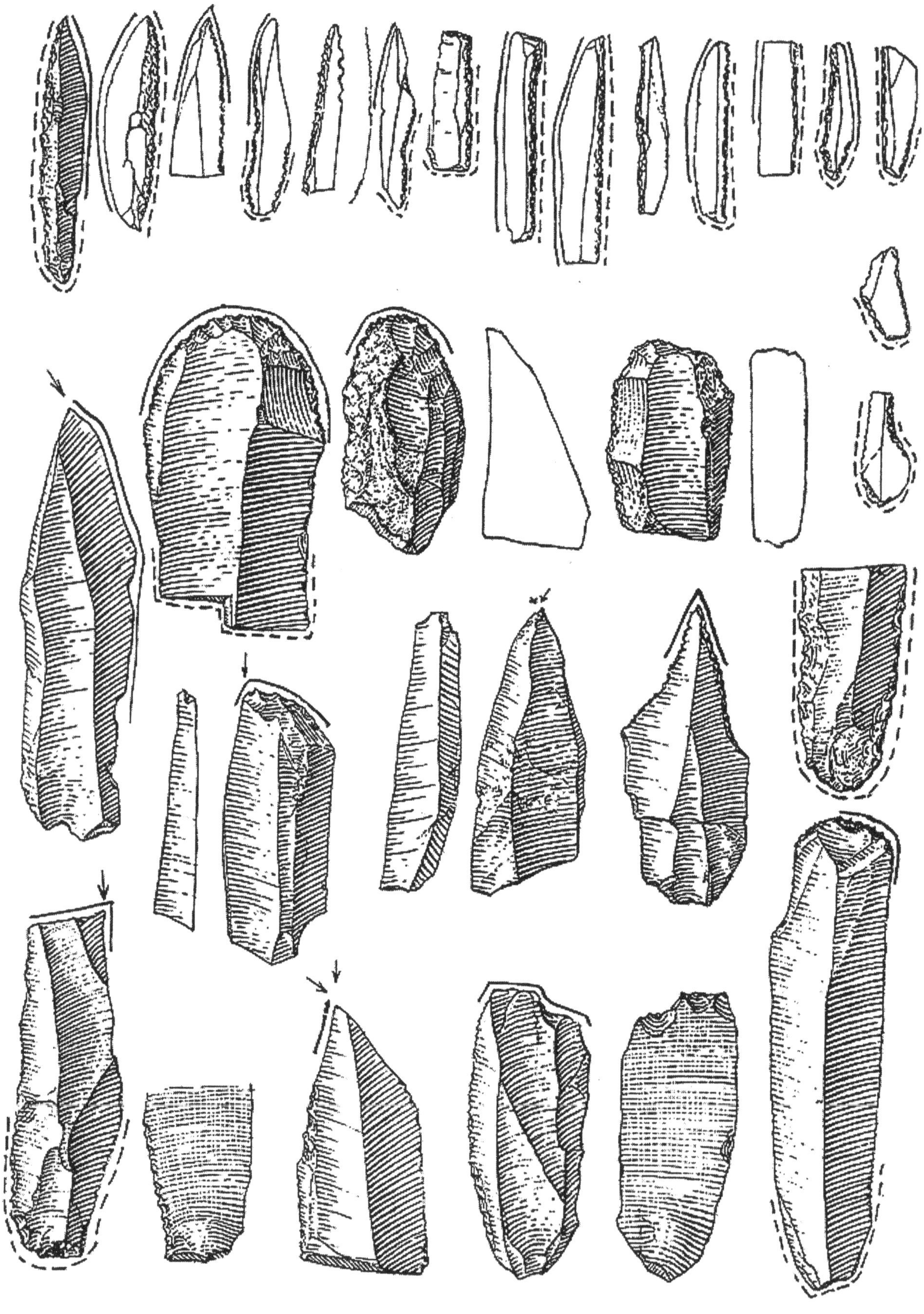

Appendix D8: Pavlov 1952/3, examples of examined stone tools. Used and unused stone pieces included Uninterrupted line = traces of use-wear, interrupted line = traces of hafting. To scale (after Svoboda 1994: 82).

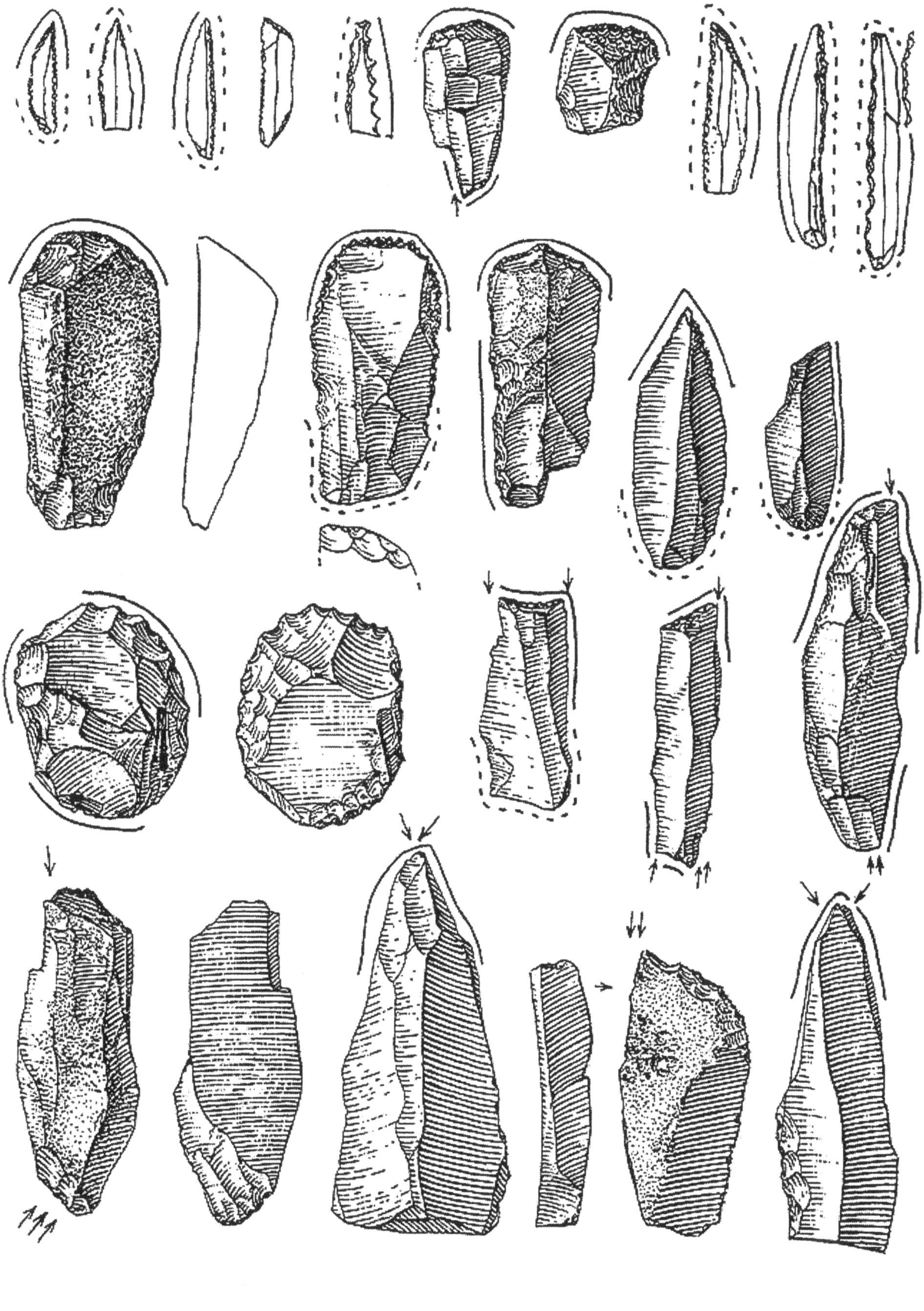

Appendix D9: Pavlov 1952/3, examples of examined stone tools. Used and unused stone pieces included Uninterrupted line = traces of use-wear, interrupted line = traces of hafting. To scale (after Svoboda 1994: 90).

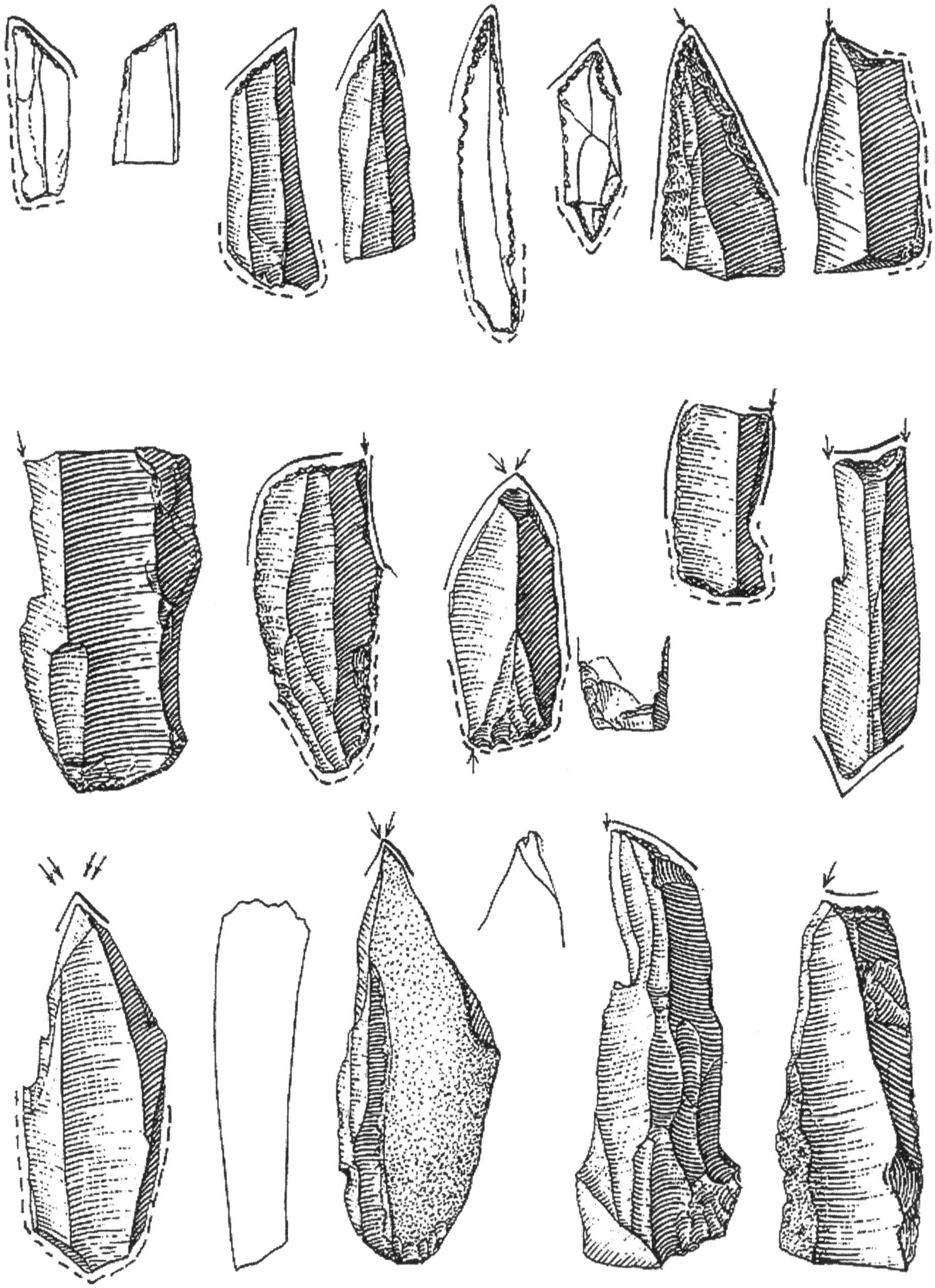

Appendix D10: Pavlov 1952/3, examples of examined stone tools. Used and unused stone pieces included Uninterrupted line = traces of use-wear, interrupted line = traces of hafting. To scale (after Svoboda 1994: 84).

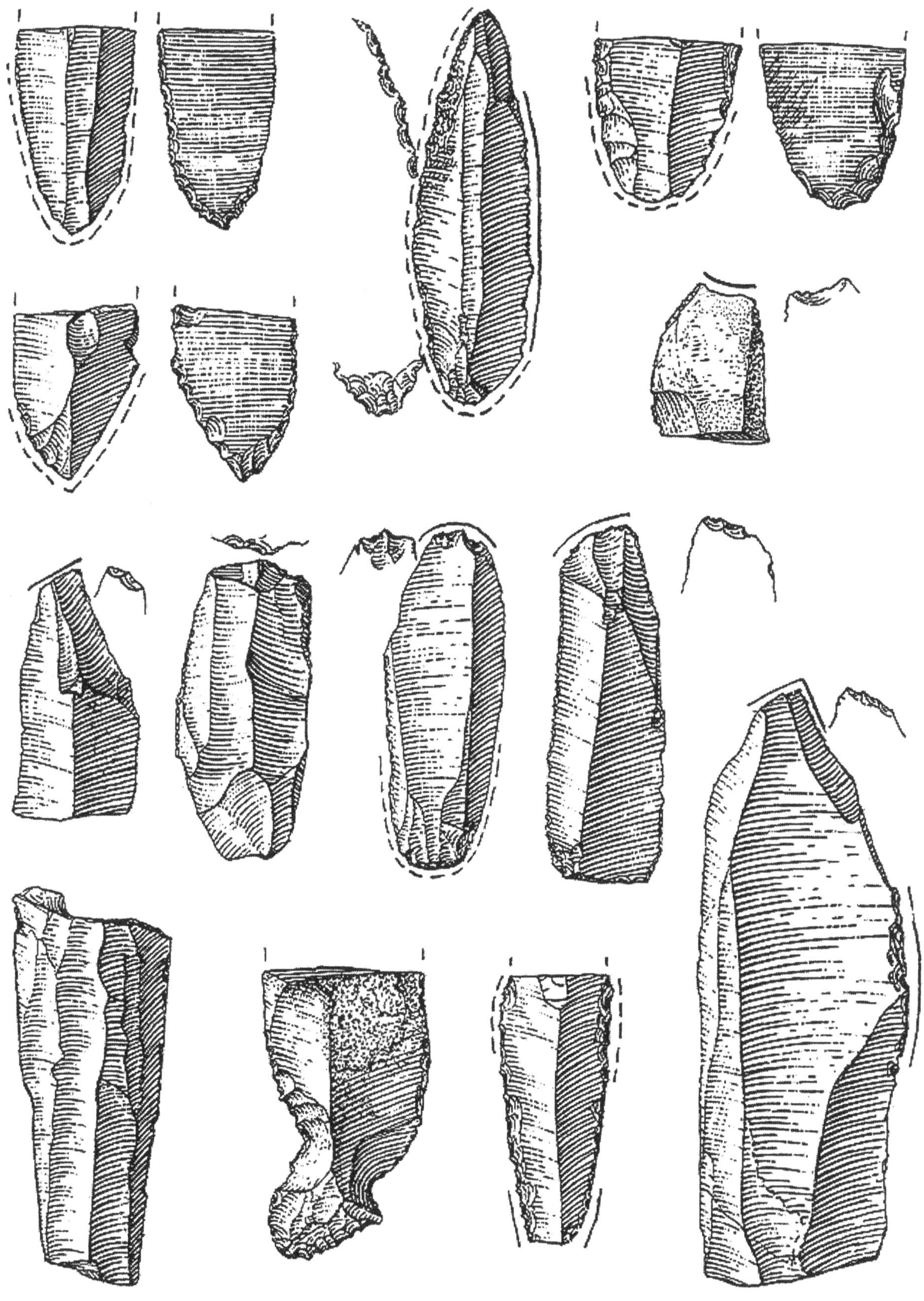

Appendix D11: Pavlov 1952/3, examples of examined stone tools. Used and unused stone pieces included Uninterrupted line = traces of use-wear, interrupted line = traces of hafting. To scale (after Svoboda 1994: 86).

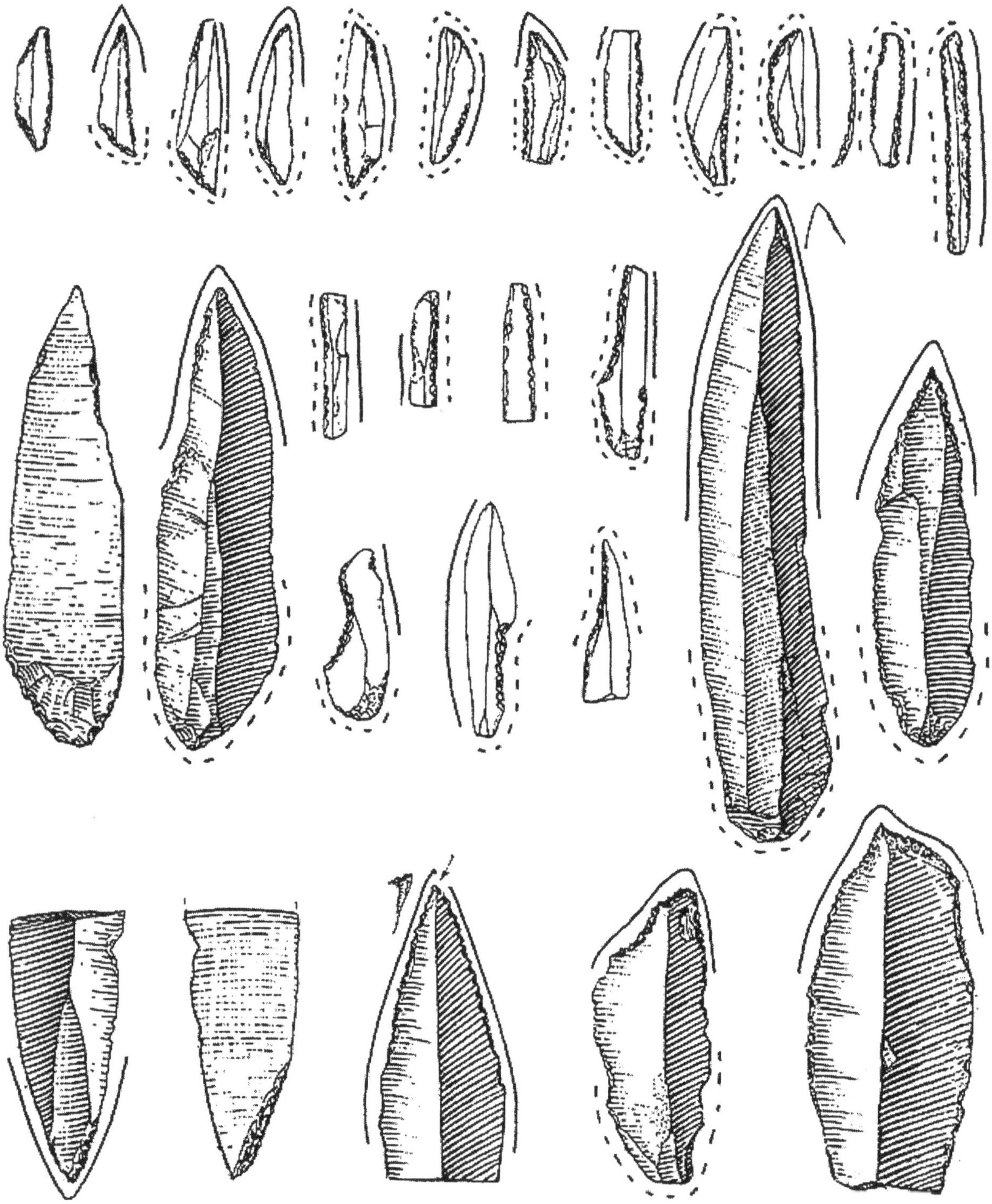

Appendix D12: Pavlov 1952/3, examples of examined stone tools. Used and unused stone pieces included Uninterrupted line = traces of use-wear, interrupted line = traces of hafting. To scale (after Svoboda 1994: 88).

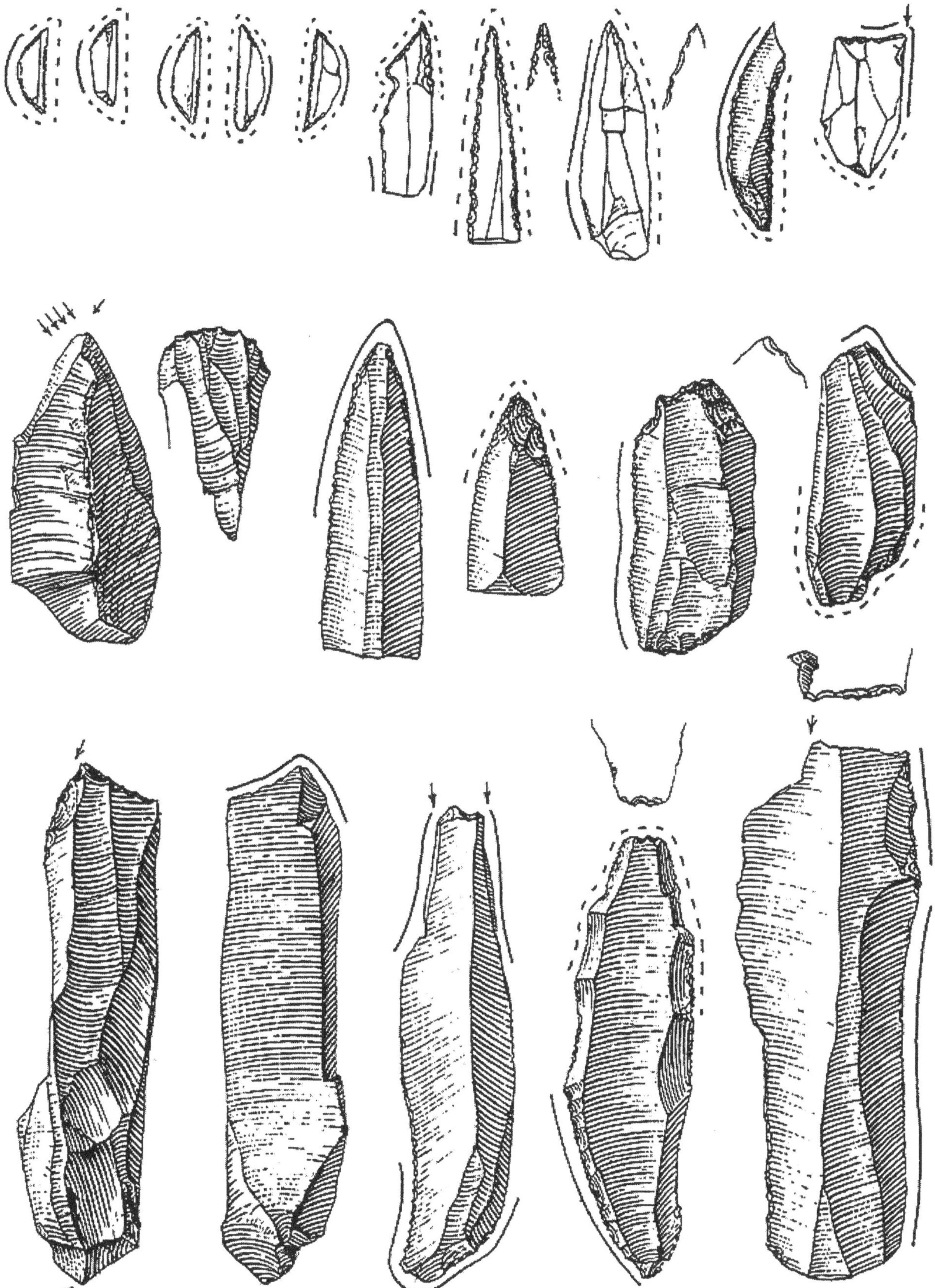

Appendix D13: Pavlov 1952/3, examples of examined stone tools. Used and unused stone pieces included Uninterrupted line = traces of use-wear, interrupted line = traces of hafting. To scale (after Svoboda 1994: 91).

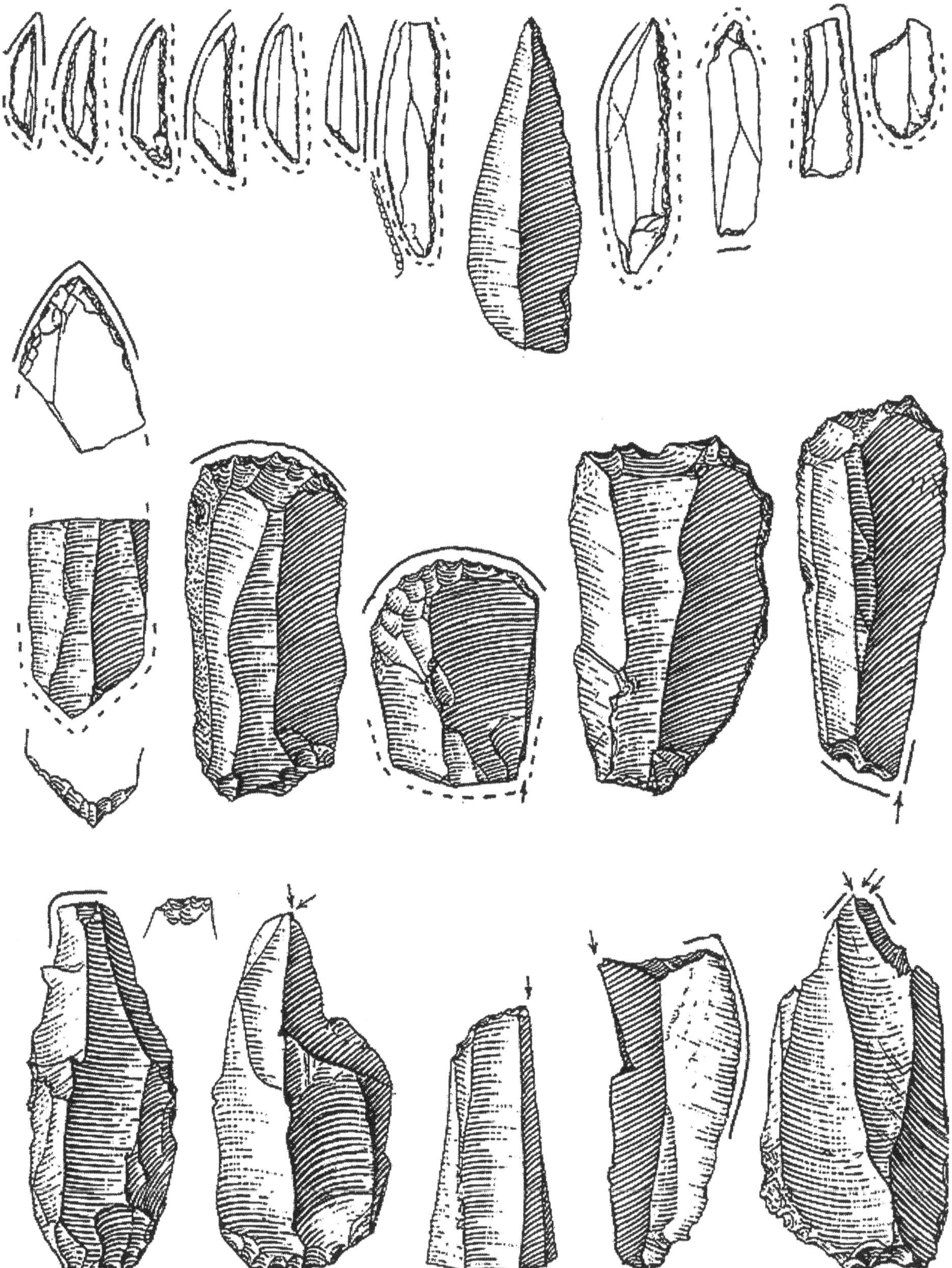

Appendix D14: Pavlov 1952/3, examples of examined stone tools. Used and unused stone pieces included Uninterrupted line = traces of use-wear, interrupted line = traces of hafting. To scale (after Svoboda 1994: 91).

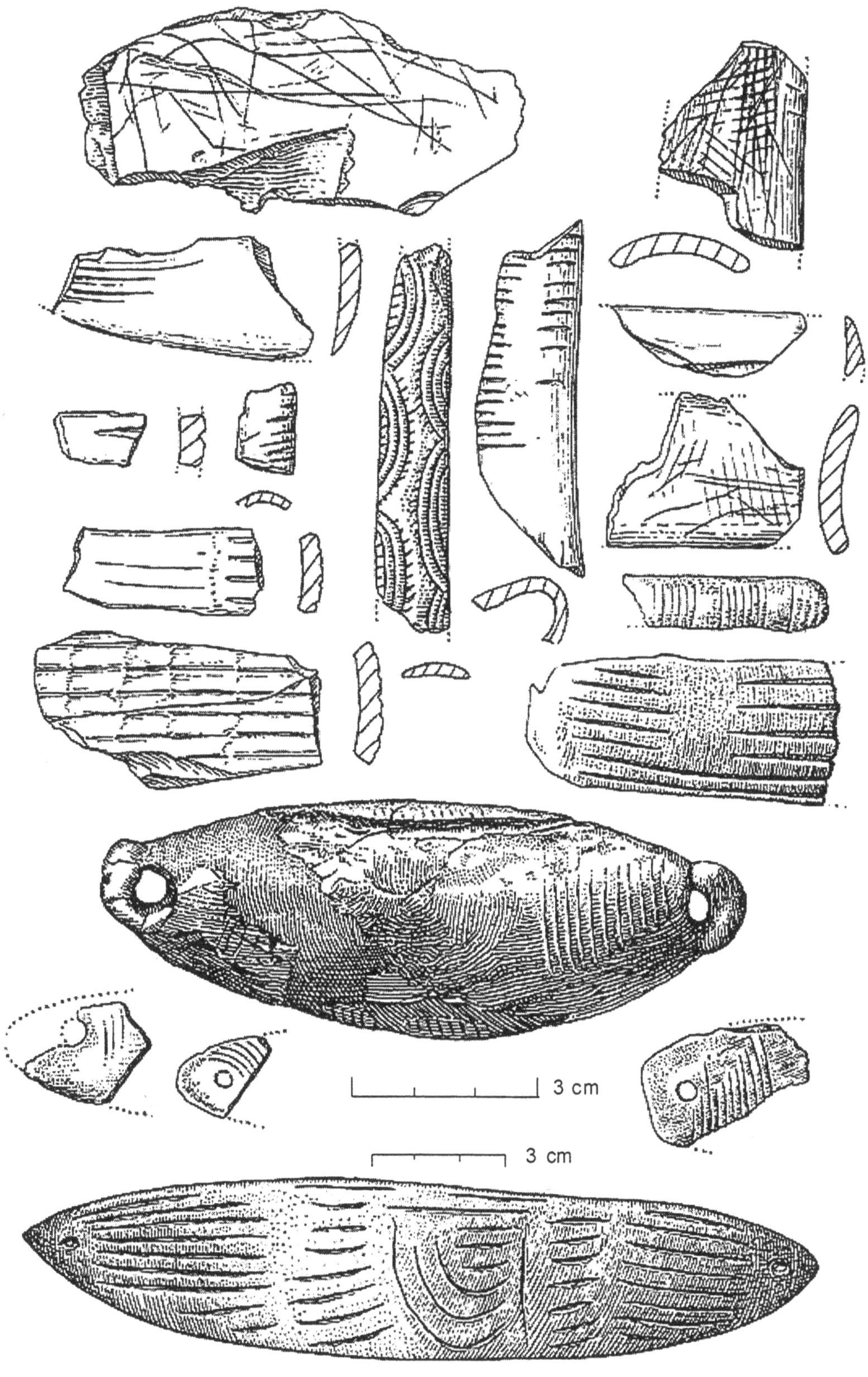

Appendix E1: Pavlov 1952/3, examples of bone implements with clear marks of modification. (after Svoboda 1994: 156).

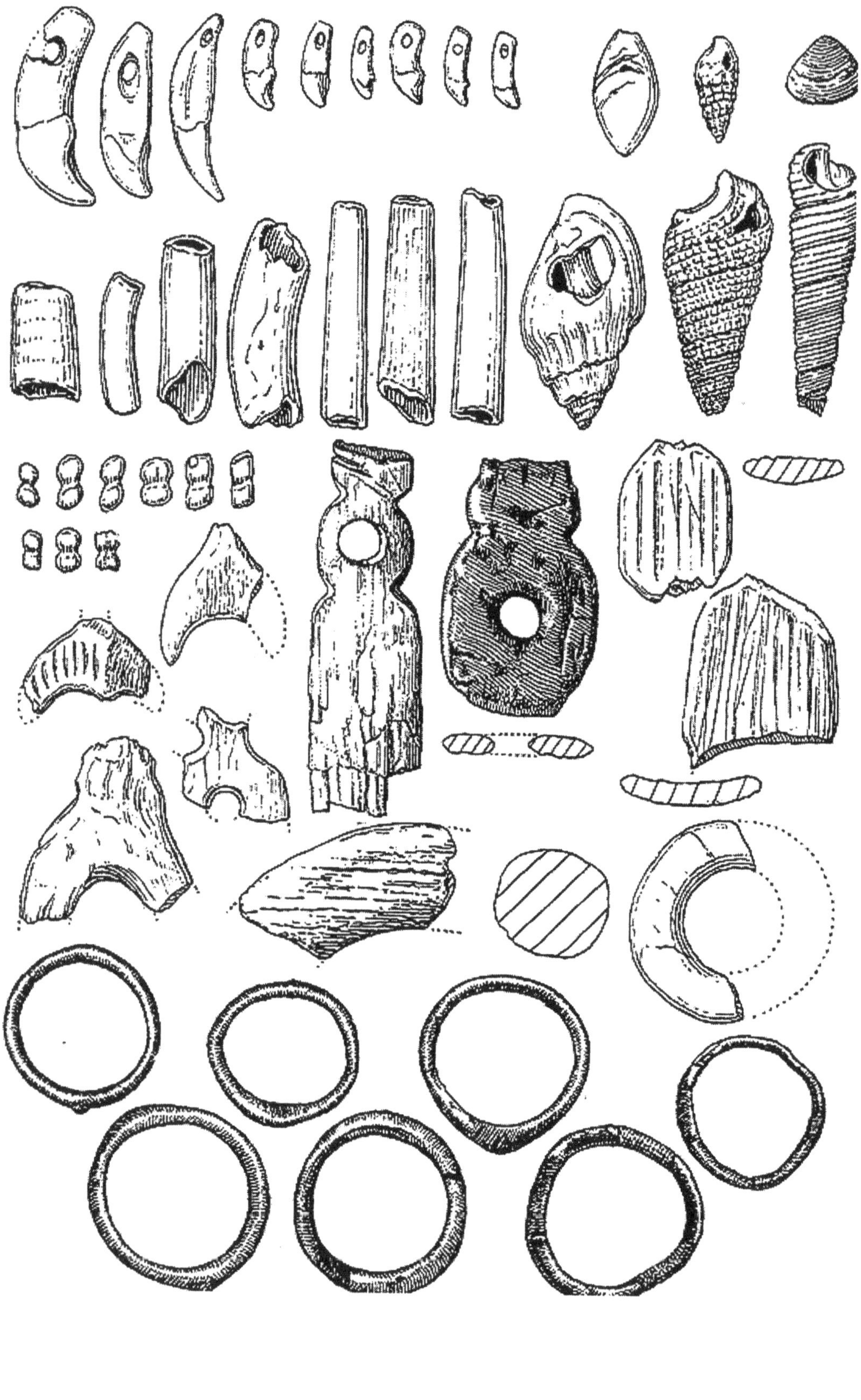

Appendix E2: Pavlov 1952/3, examples of bone implements with clear marks of modification. (after Svoboda 1994: 157).

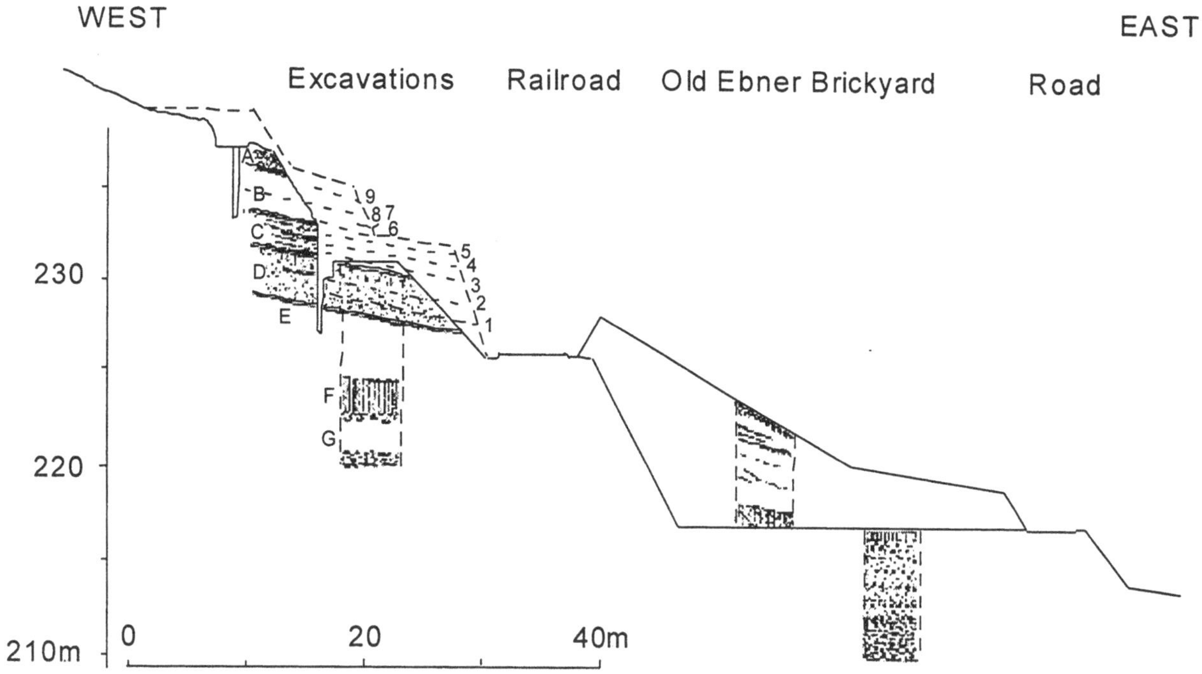

Appendix F: Willendorf II, profile and site location
(After Haesaerts et. al., 1996, from Brandtner 1956-9)

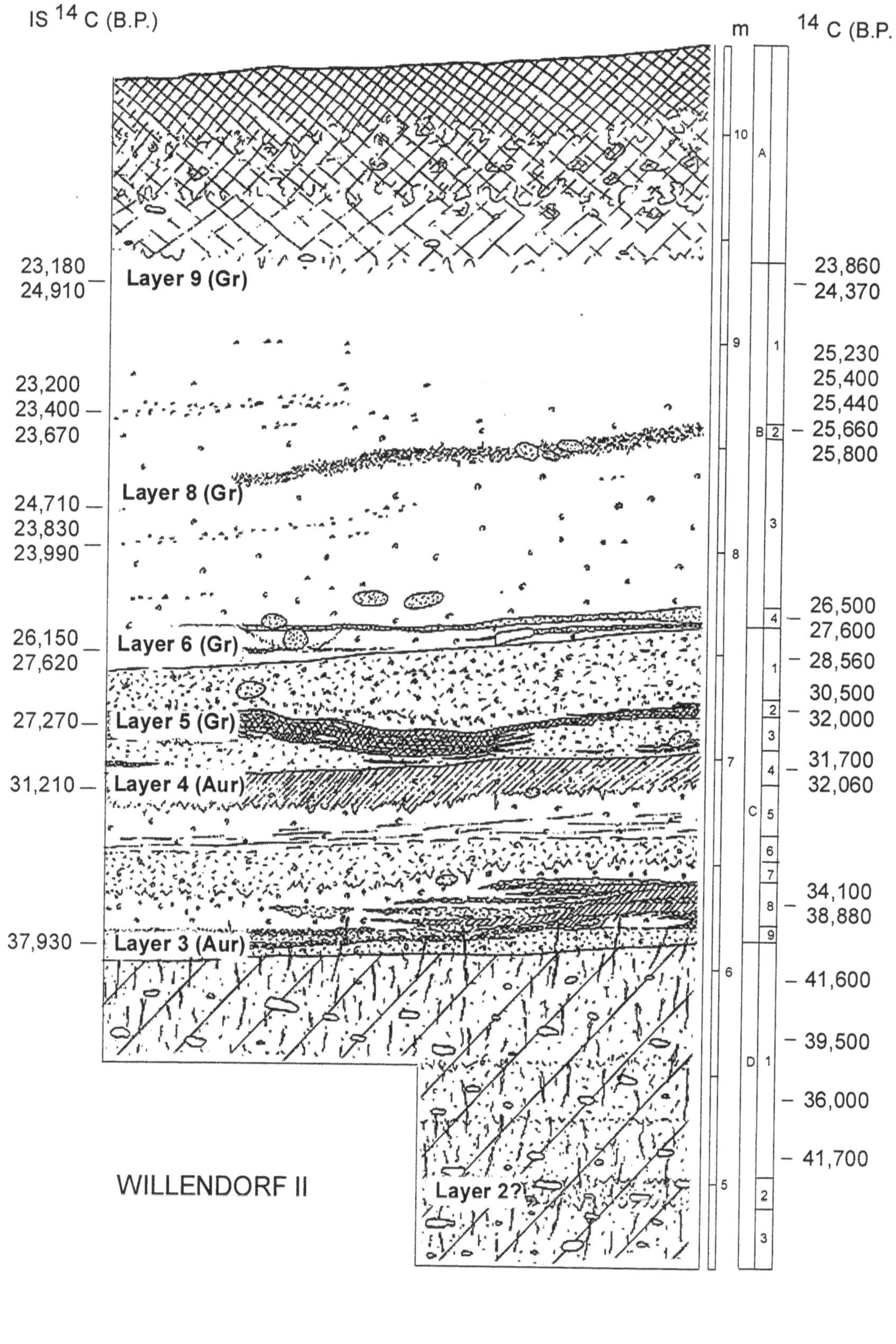

Appendix G: Willendorf II, stratigraphy with dates
(after Haesaerts et. al., 1996)

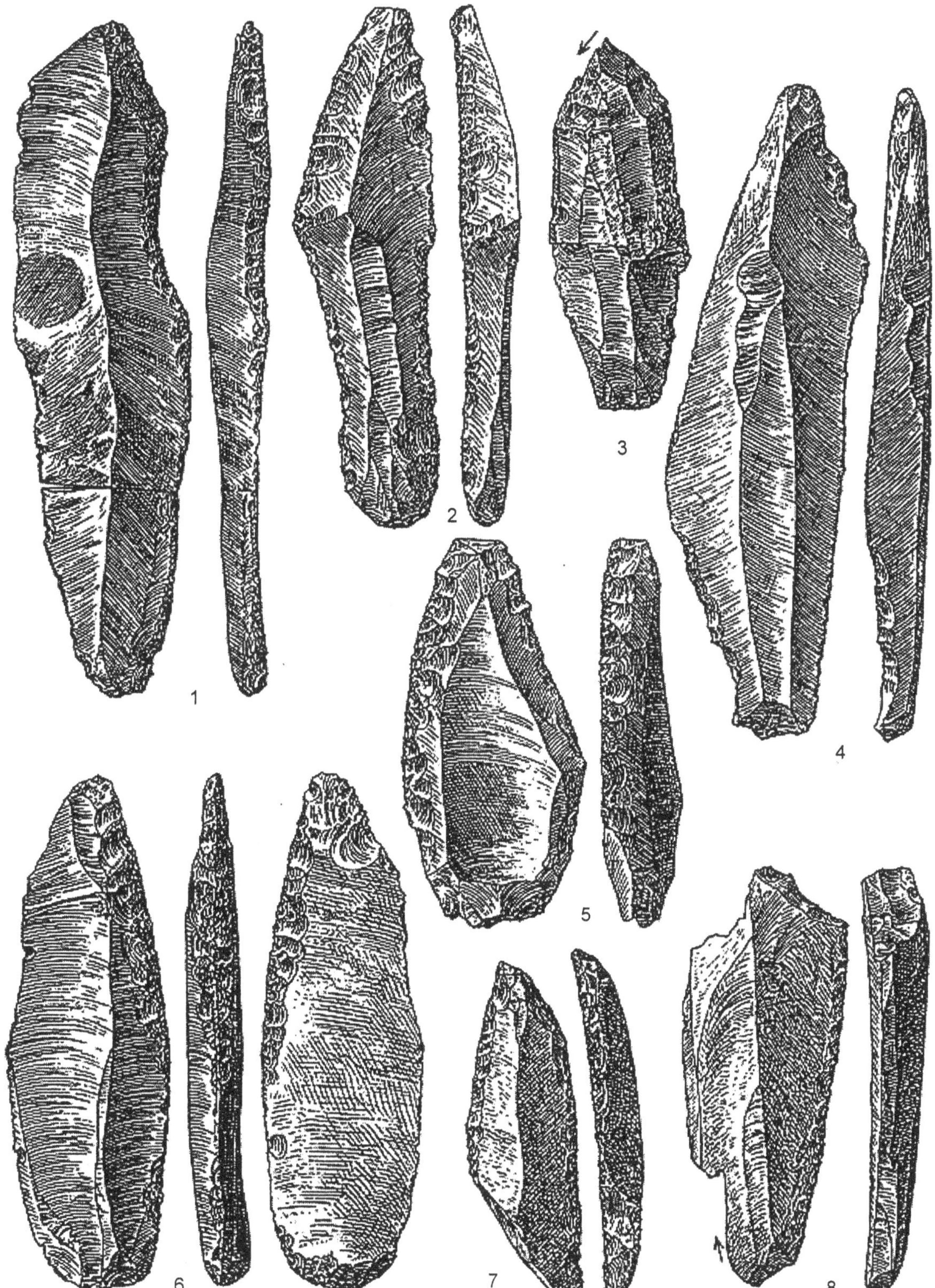

Appendix H1: Willendorf II, examples of stone tools from layer 5.
(After Felgenhauer 1959: Inventory Appendix 27)
Traces of Hafting: 2, 3, 4, 6
Traces of Use: 2, 3, 4, 5, 6

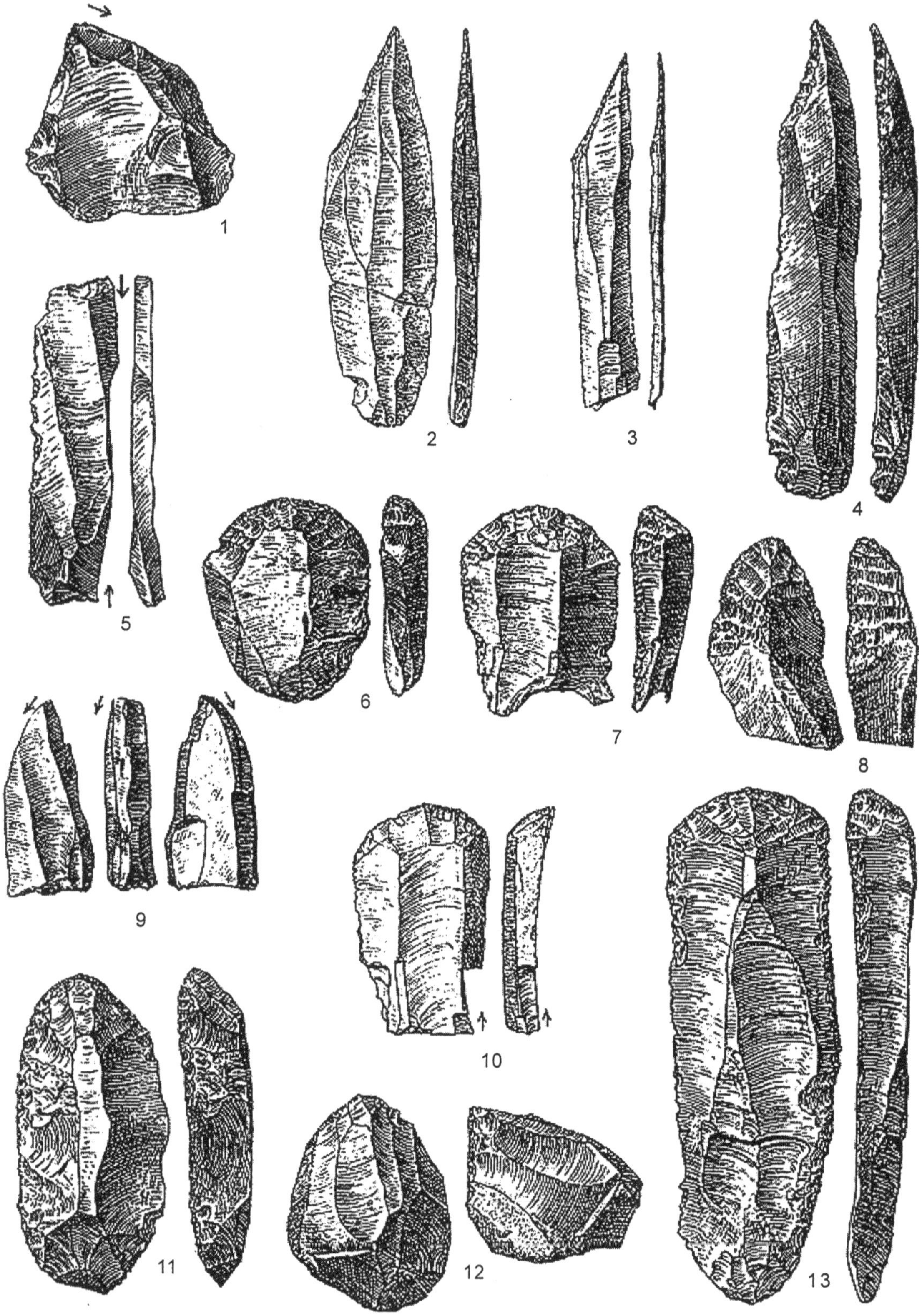

Appendix H2: Willendorf II, examples of stone tools from layer 5.
(After Felgenhauer 1959: Inventory Appendix 28)

Traces of Hafting:	2, 3, 4, 5, 6, 7, 8, 10, 11, 13
Traces of Use:	2, 3, 4, 5, 6, 7, 8, 9, 10, 11, 13

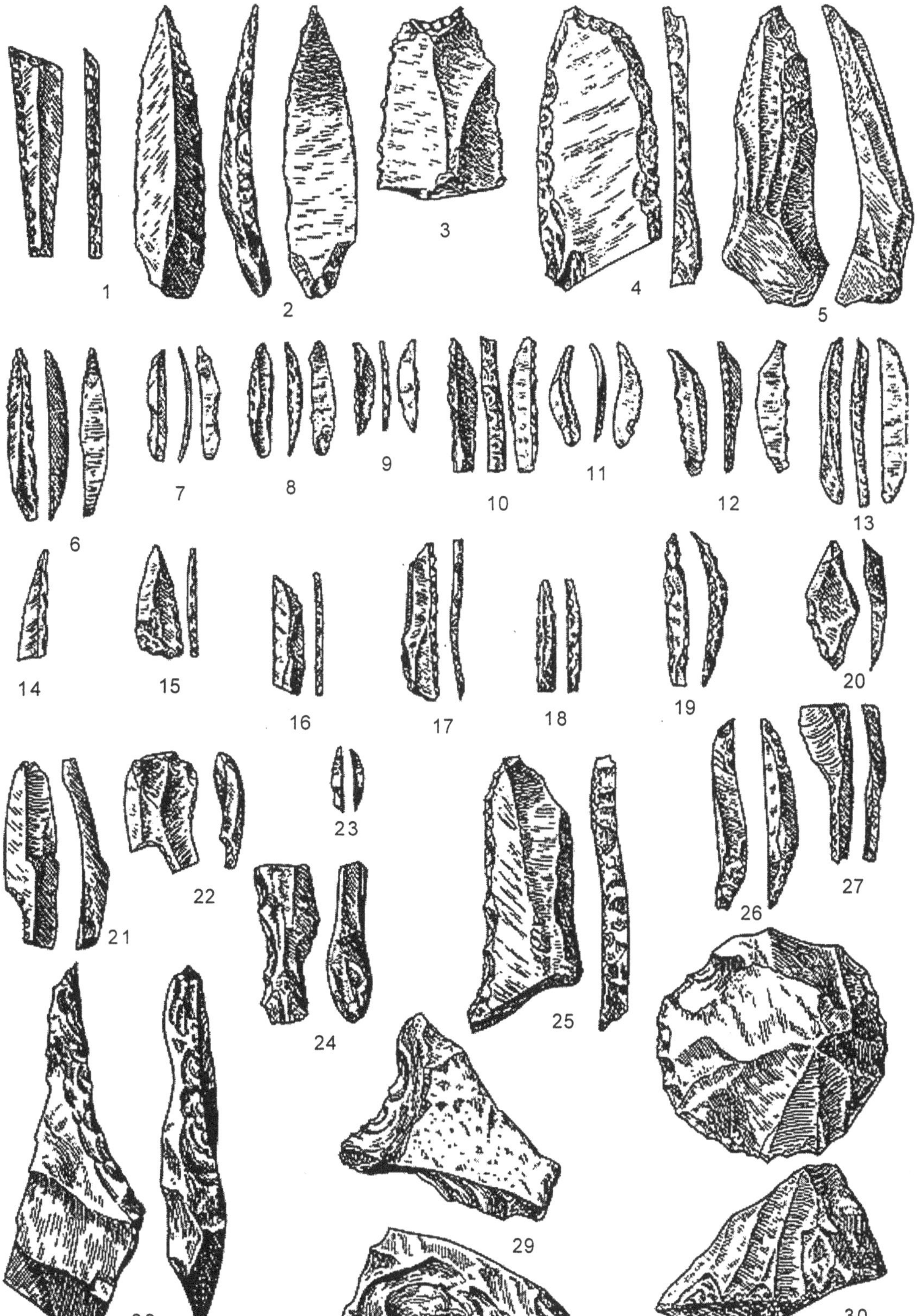

Appendix H3: Willendorf II, examples of stone tools from layer 5.
(After Felgenhauer 1959: Inventory Appendix 29)
Traces of Hafting: 3, 4, 6, 8, 11, 13, 14, 15, 16, 18, 19, 21, 22, 25
Traces of Use: 3, 6, 8, 9, 11, 12, 13, 14, 15, 16, 18, 19, 21, 25, 26, 28

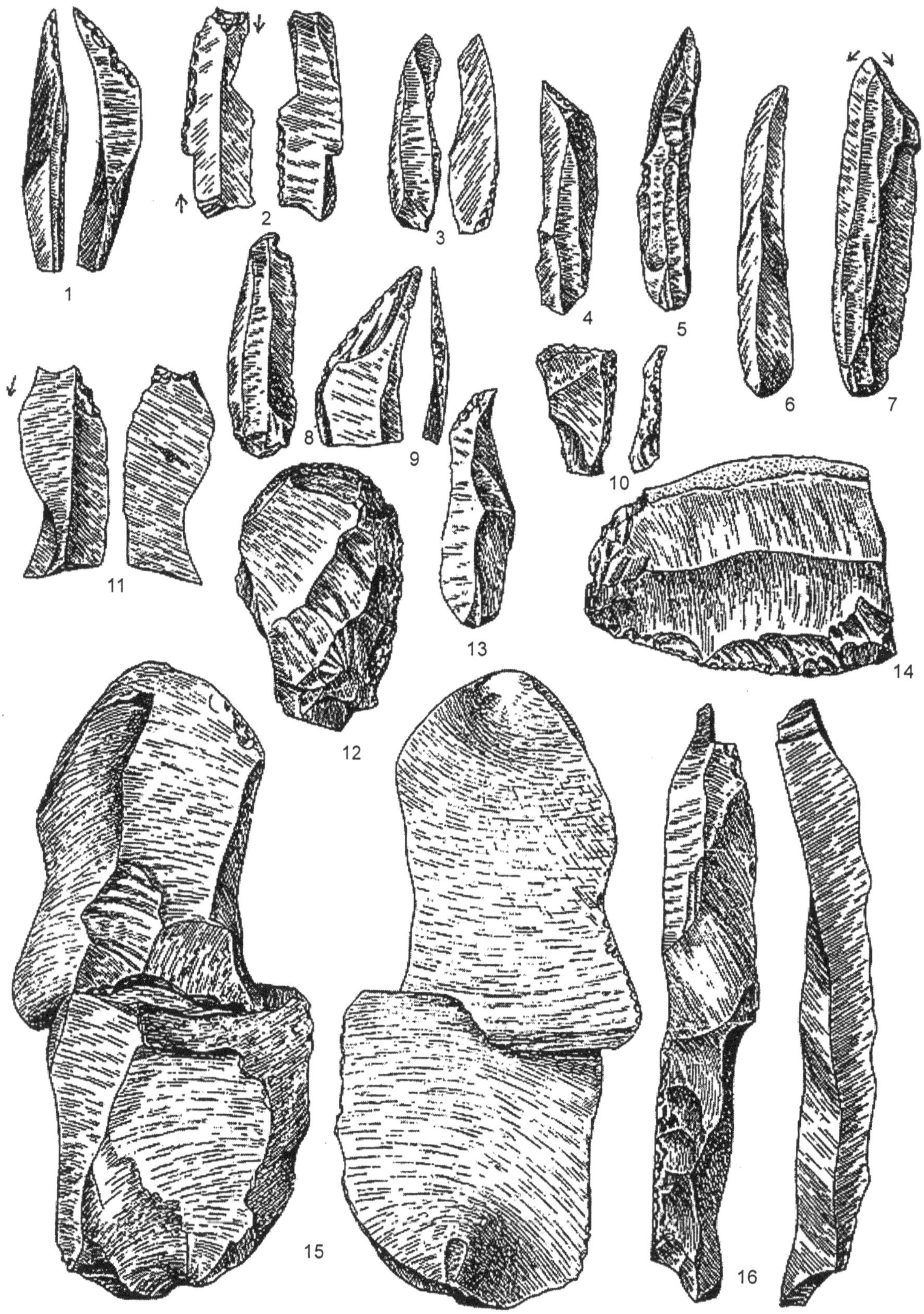

Appendix H4: Willendorf II, examples of stone tools from layer 5.
(After Felgenhauer 1959: Inventory Appendix 30)

Traces of Hafting:	4, 8, 11, 14
Traces of Use:	1, 4, 8, 9, 11

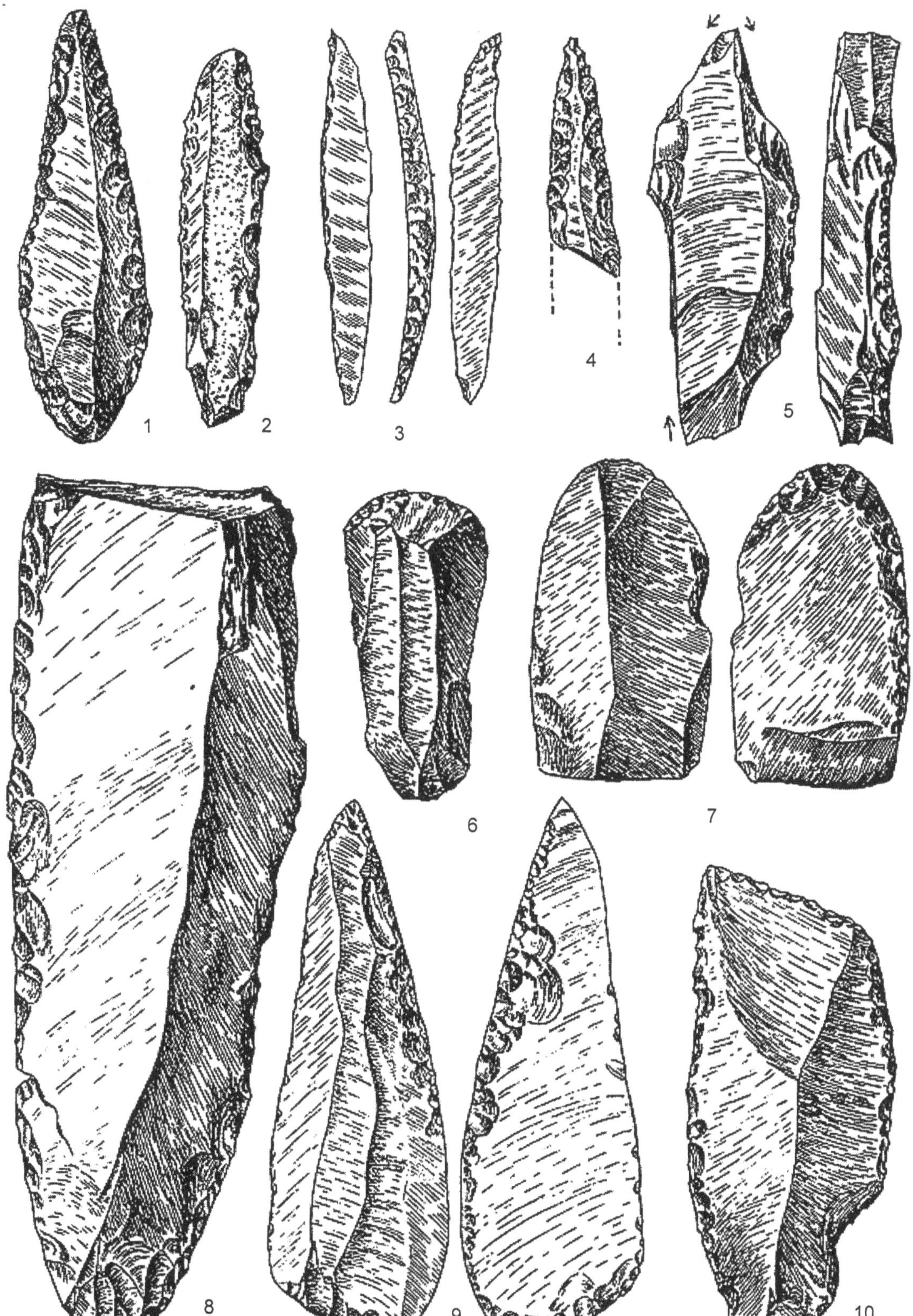

Appendix H5: Willendorf II, examples of stone tools from layer 6.
(After Felgenhauer 1959: Inventory Appendix 31. Large tool at left from layer 5)
Traces of Hafting: 3, 4, 6, 7
Traces of Use: 1, 2, 3, 4, 5, 6, 8, 9

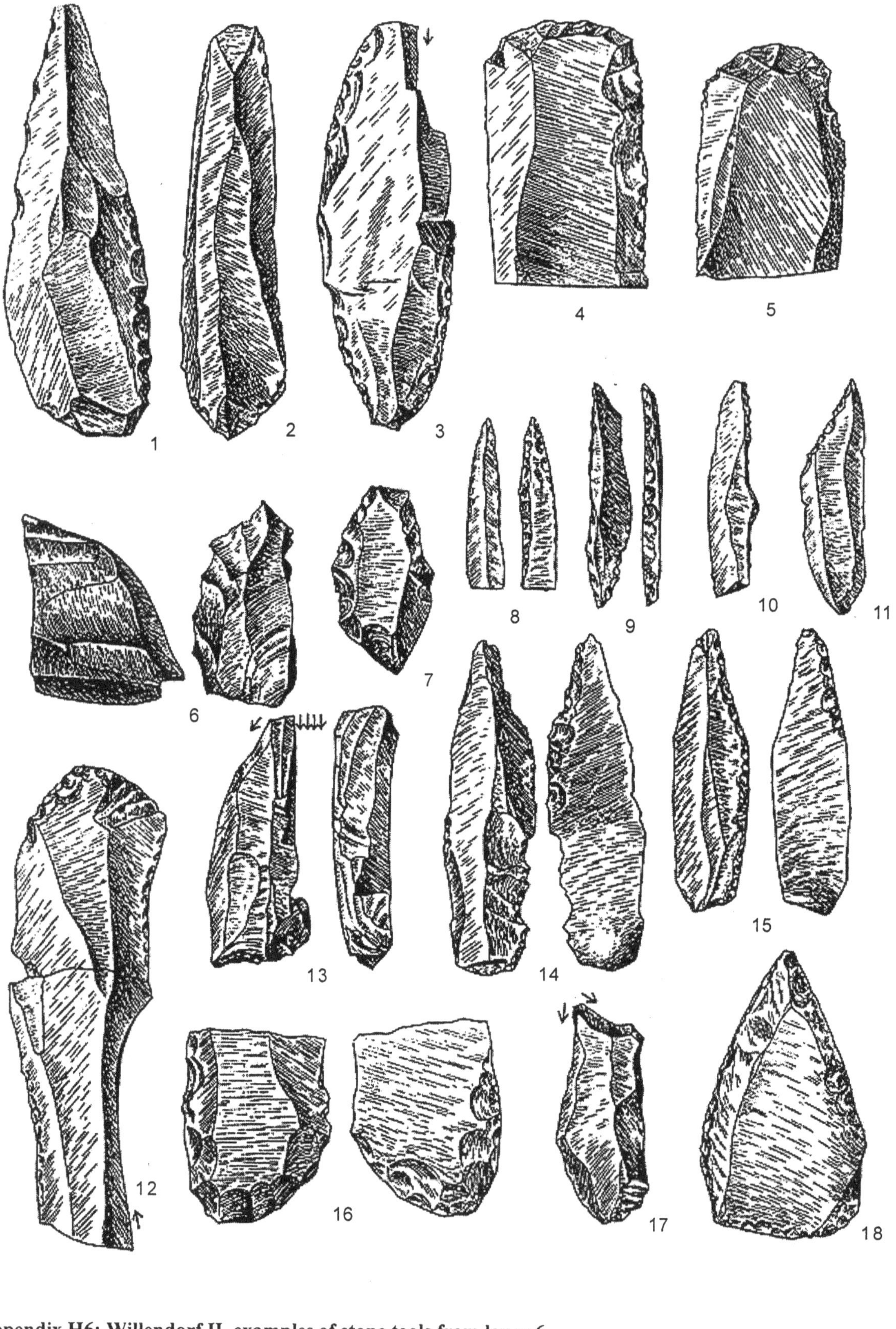

Appendix H6: Willendorf II, examples of stone tools from layer 6.
(After Felgenhauer 1959: Inventory Appendix 32)
Traces of Hafting: 1, 2, 3, 5, 7, 8, 11, 15, 17, 18
Traces of Use: 1, 2, 3, 5, 7, 8, 11, 13, 15, 17, 18

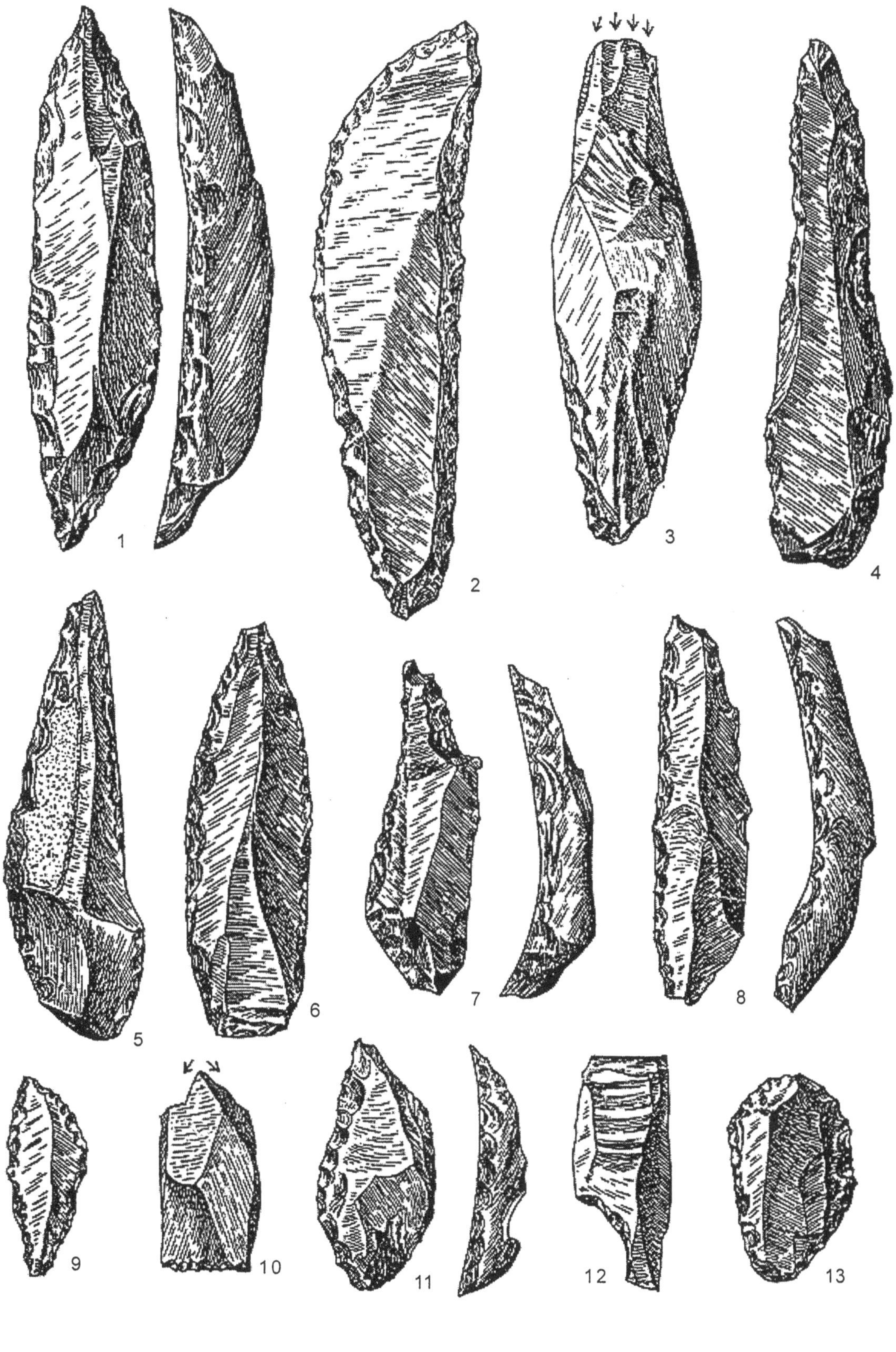

Appendix H7: Willendorf II, examples of stone tools from layer 7.
(After Felgenhauer 1959: Inventory Appendix 33)
Traces of Hafting: 4, 5, 6, 7, 9, 10, 12, 13
Traces of Use: 1, 2, 3, 4, 5, 6, 7, 9, 10, 12, 13

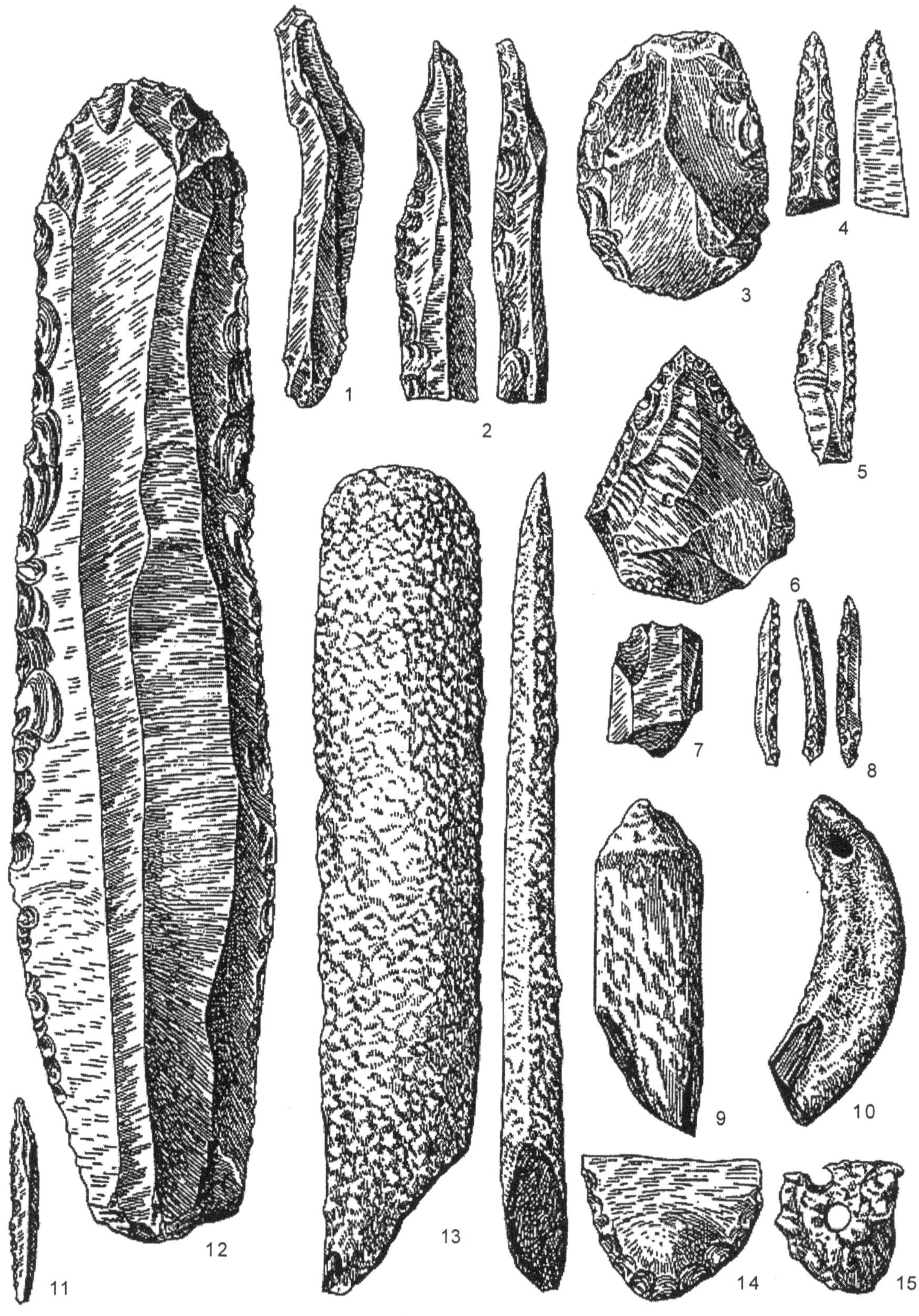

Appendix H8: Willendorf II, examples of stone tools from layer 7.
(After Felgenhauer 1959: Inventory Appendix 34)

Traces of Hafting:	1, 2, 5, 6, 8, 11
Traces of Use:	1, 2, 3, 4, 5, 6, 8, 11, 12, 14

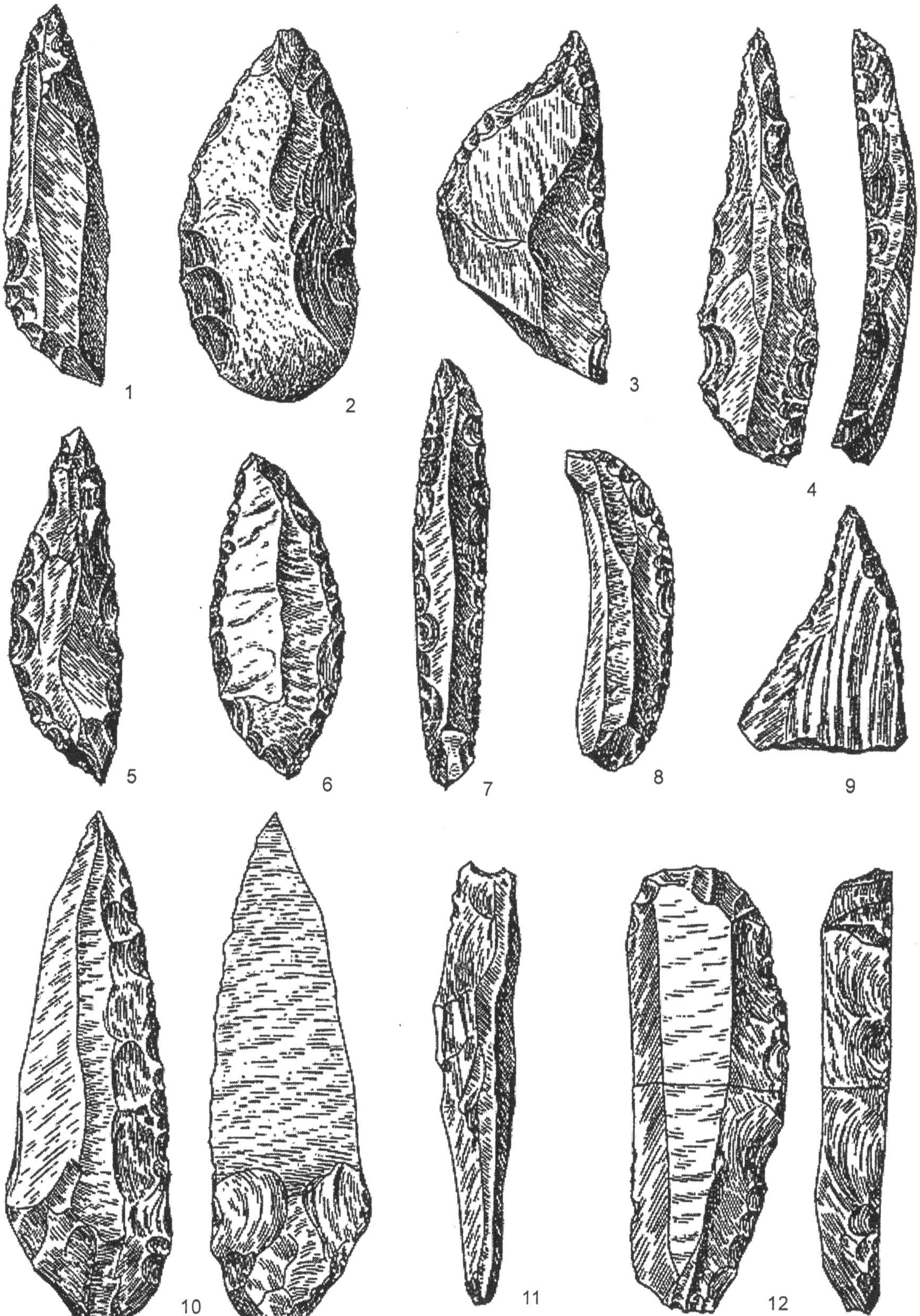

Appendix H9: Willendorf II, examples of stone tools from layer 8.
(After Felgenhauer 1959: Inventory Appendix 35)
Traces of Hafting: 1, 2, 4, 5, 6, 7, 8, 9, 10
Traces of Use: 1, 2, 4, 5, 6, 7, 8, 9, 10, 12

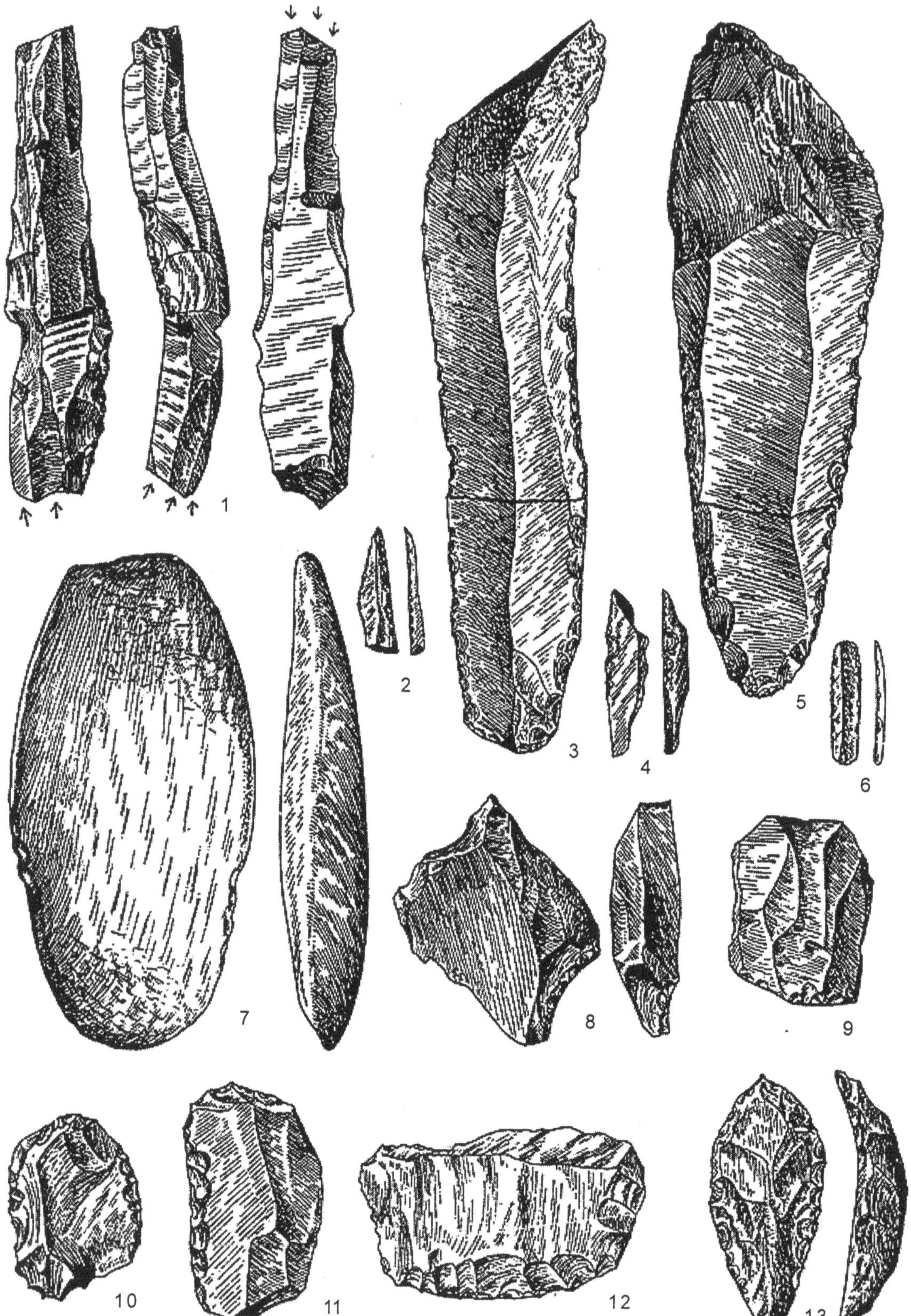

Appendix H10: Willendorf II, examples of stone tools from layer 8.
(After Felgenhauer 1959: Inventory Appendix 36)

Traces of Hafting:	2, 3, 4, 10, 11
Traces of Use:	2, 3, 4, 10, 11

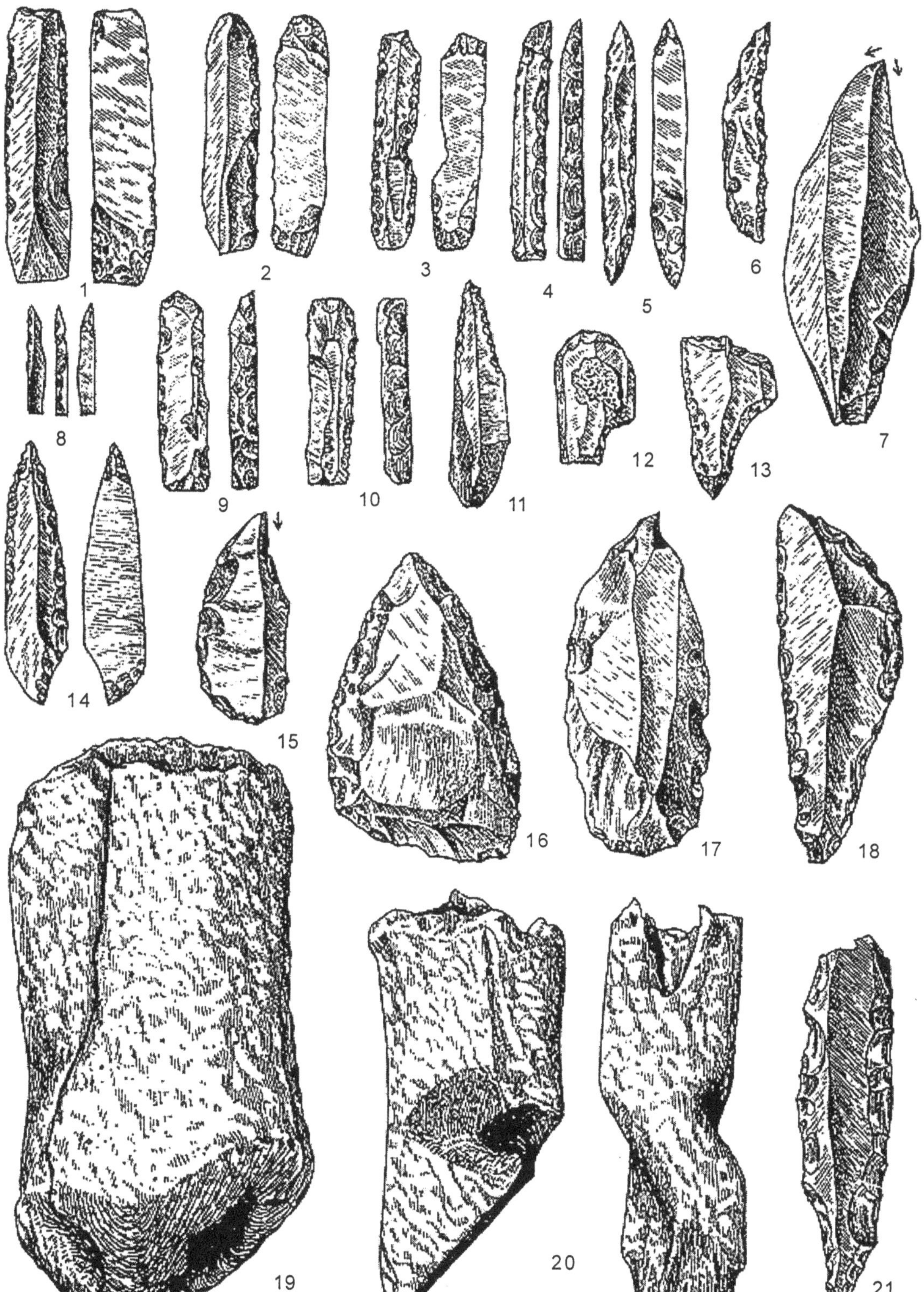

Appendix H11: Willendorf II, examples of stone tools from layer 8.
(After Felgenhauer 1959: Inventory Appendix 37)
Traces of Hafting: 2, 4, 5, 6, 9, 12, 14, 17, 18
Traces of Use: 2, 4, 5, 6, 9, 17, 18

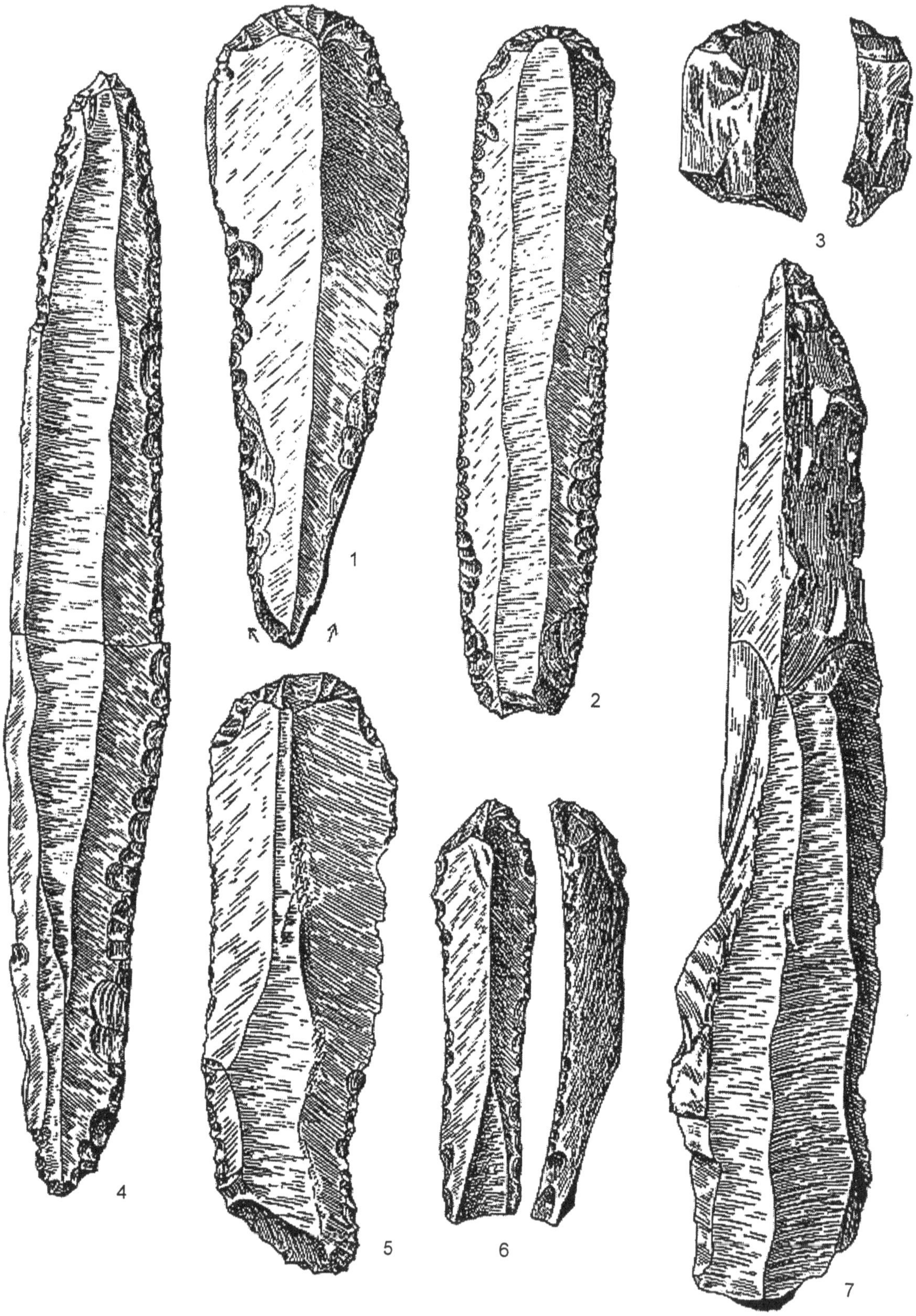

Appendix H12: Willendorf II, examples of stone tools from layer 9.
(After Felgenhauer 1959: Inventory Appendix 38)
Traces of Hafting: 2, 4, 5
Traces of Use: 1, 2, 4, 5

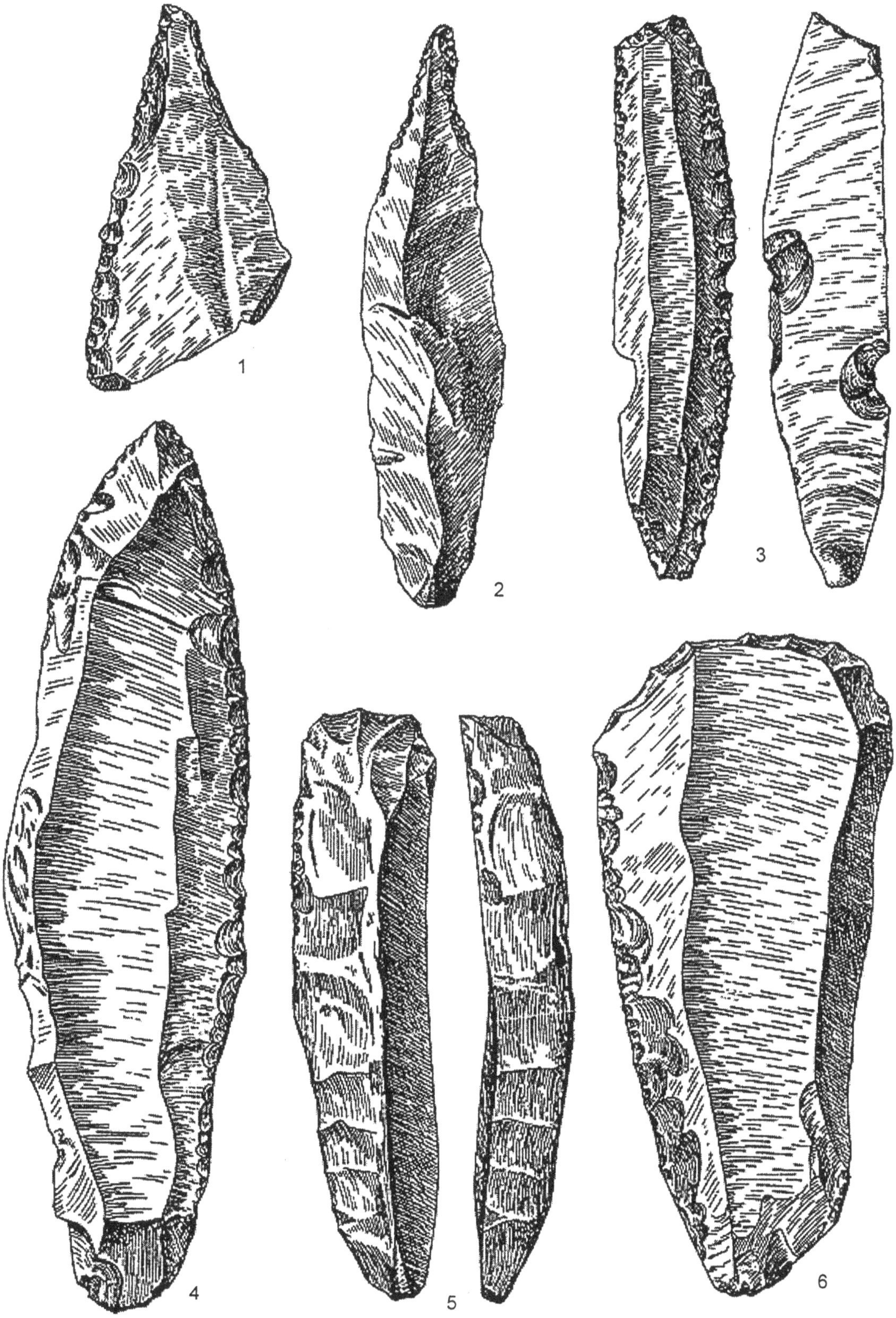

Appendix H13: Willendorf II, examples of stone tools from layer 9.
(After Felgenhauer 1959: Inventory Appendix 39)
Traces of Hafting: 2, 3, 4, 6
Traces of Use: 1, 2, 3, 4, 6

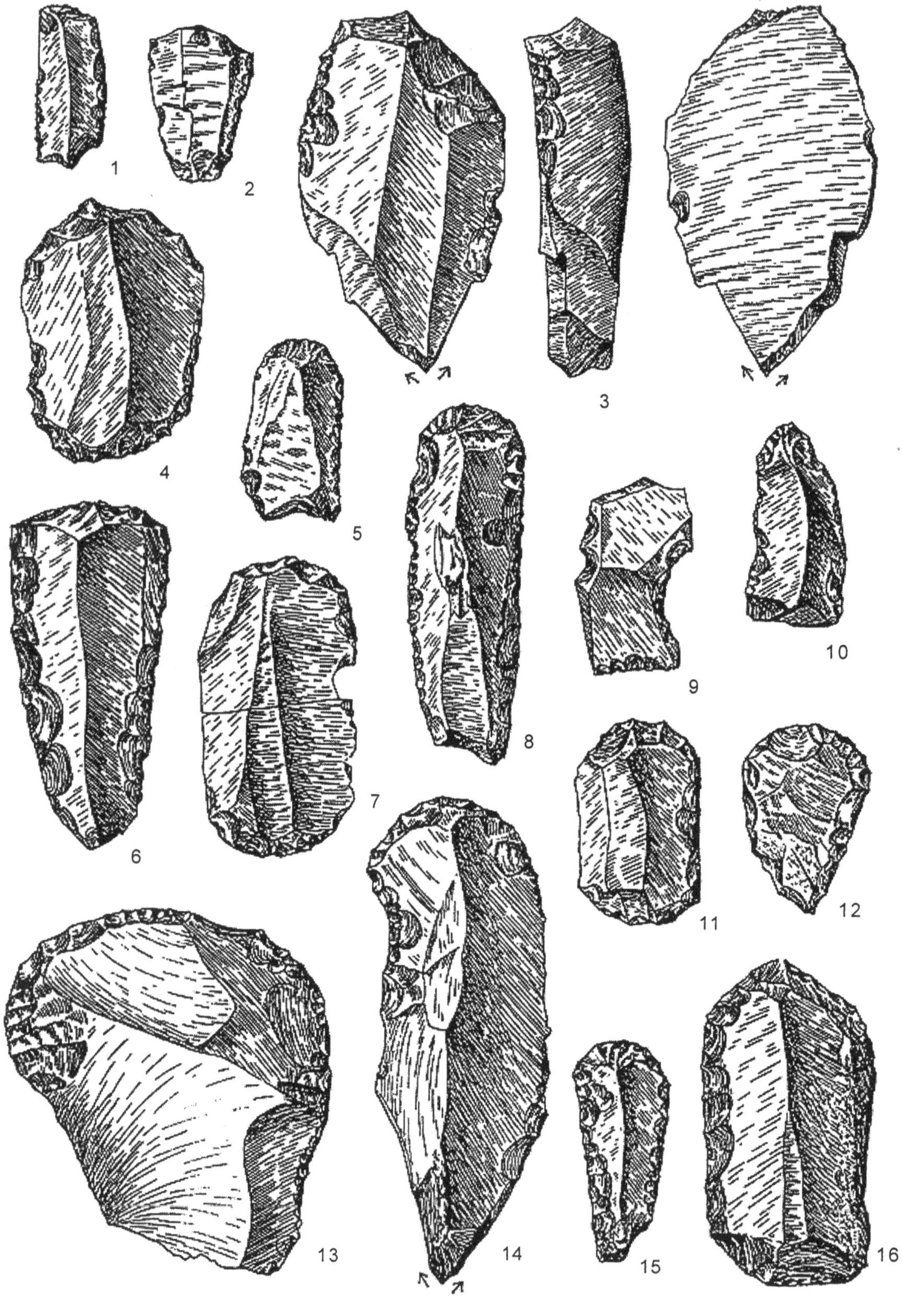

Appendix H14: Willendorf II, examples of stone tools from layer 9.
(After Felgenhauer 1959: Inventory Appendix 40)
Traces of Hafting: 2, 3, 5, 6, 8, 12, 14, 15
Traces of Use: 2, 3, 4, 5, 6, 8, 11, 12, 13, 14, 15, 16

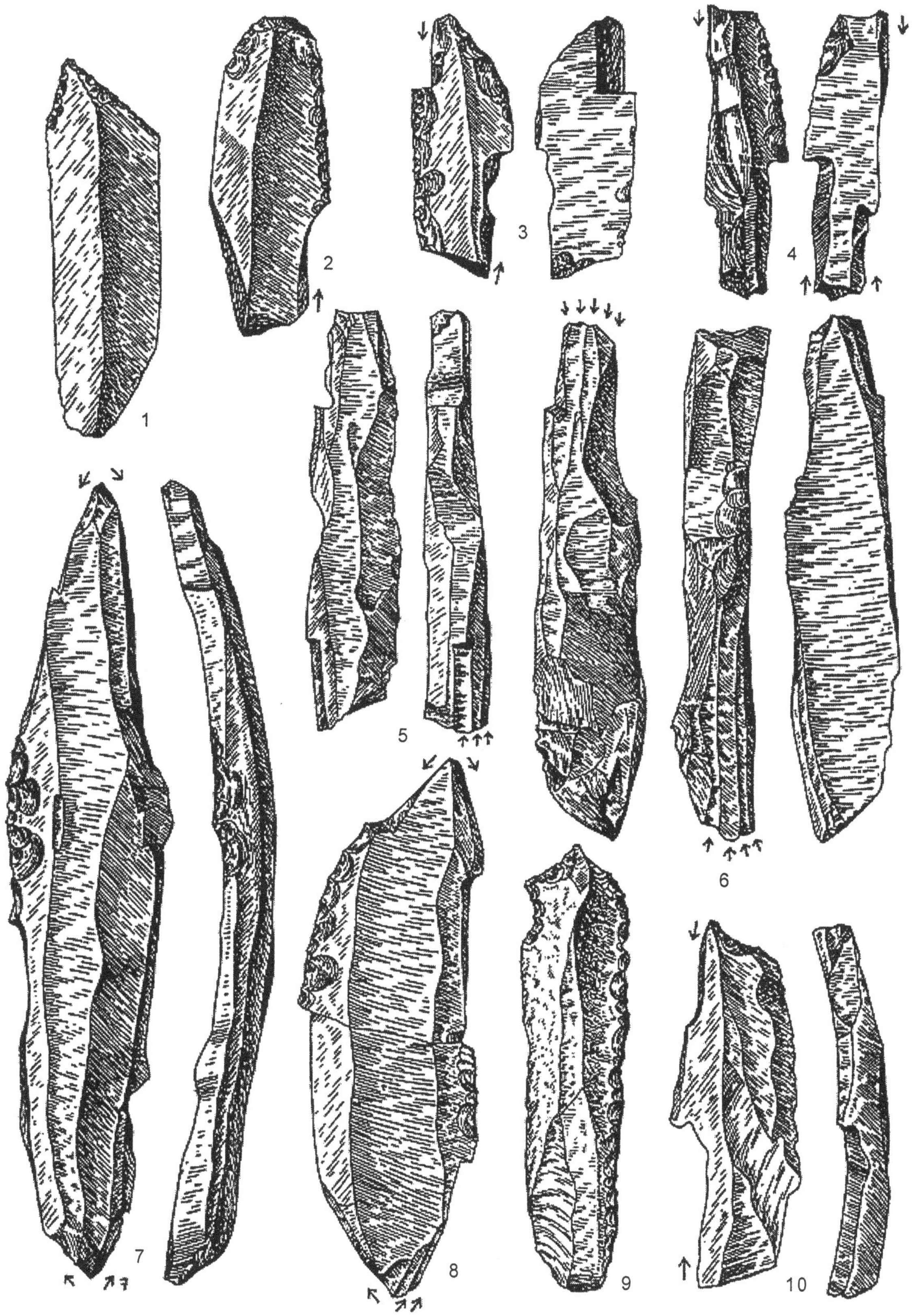

Appendix H15: Willendorf II, examples of stone tools from layer 9.
(After Felgenhauer 1959: Inventory Appendix 41)
Traces of Hafting: 2, 4, 7, 9
Traces of Use: 1, 2, 3, 4, 7, 8, 9

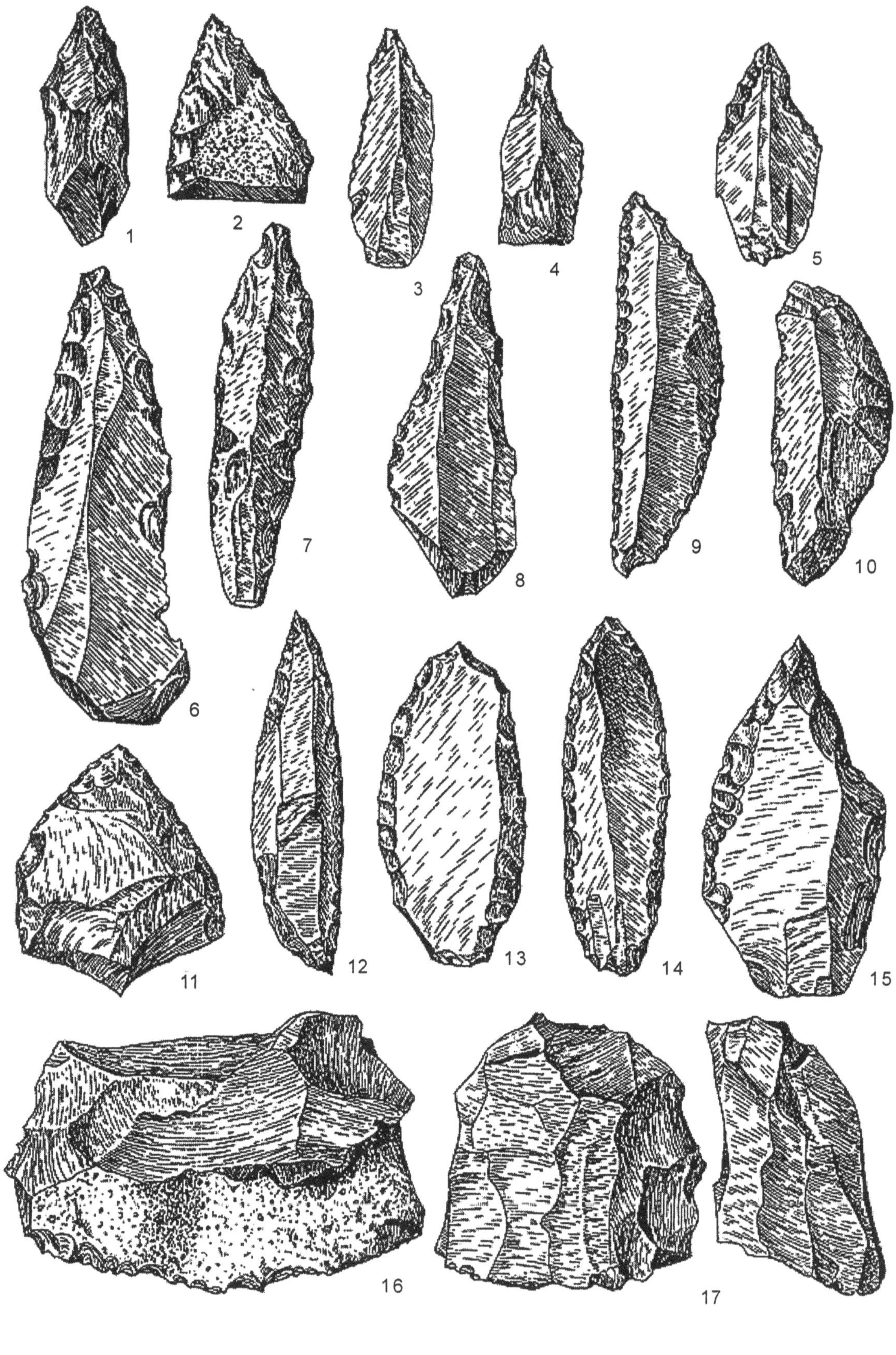

Appendix H16: Willendorf II, examples of stone tools from layer 9.
(After Felgenhauer 1959: Inventory Appendix 42)
Traces of Hafting: 1, 2, 3, 5, 12, 13, 14
Traces of Use: 1, 2, 3, 4, 5, 9, 11, 12, 13, 14, 15

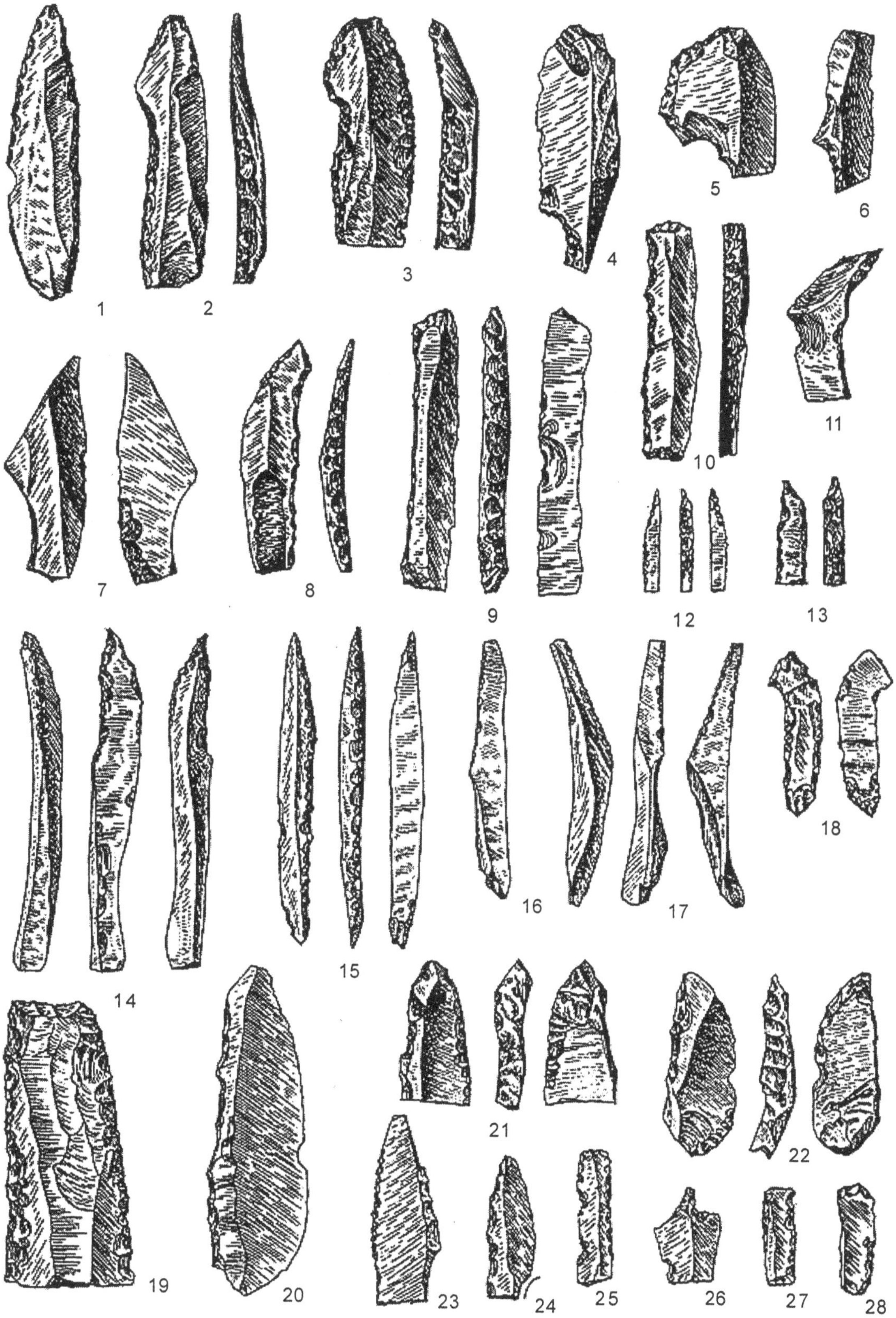

Appendix H17: Willendorf II, examples of stone tools from layer 9.
(After Felgenhauer 1959: Inventory Appendix 43)

Traces of Hafting:	3, 5, 7, 8, 12, 13, 18, 23, 24, 25, 28
Traces of Use:	3, 7, 8, 12, 13, 18, 19, 23, 24, 25, 28

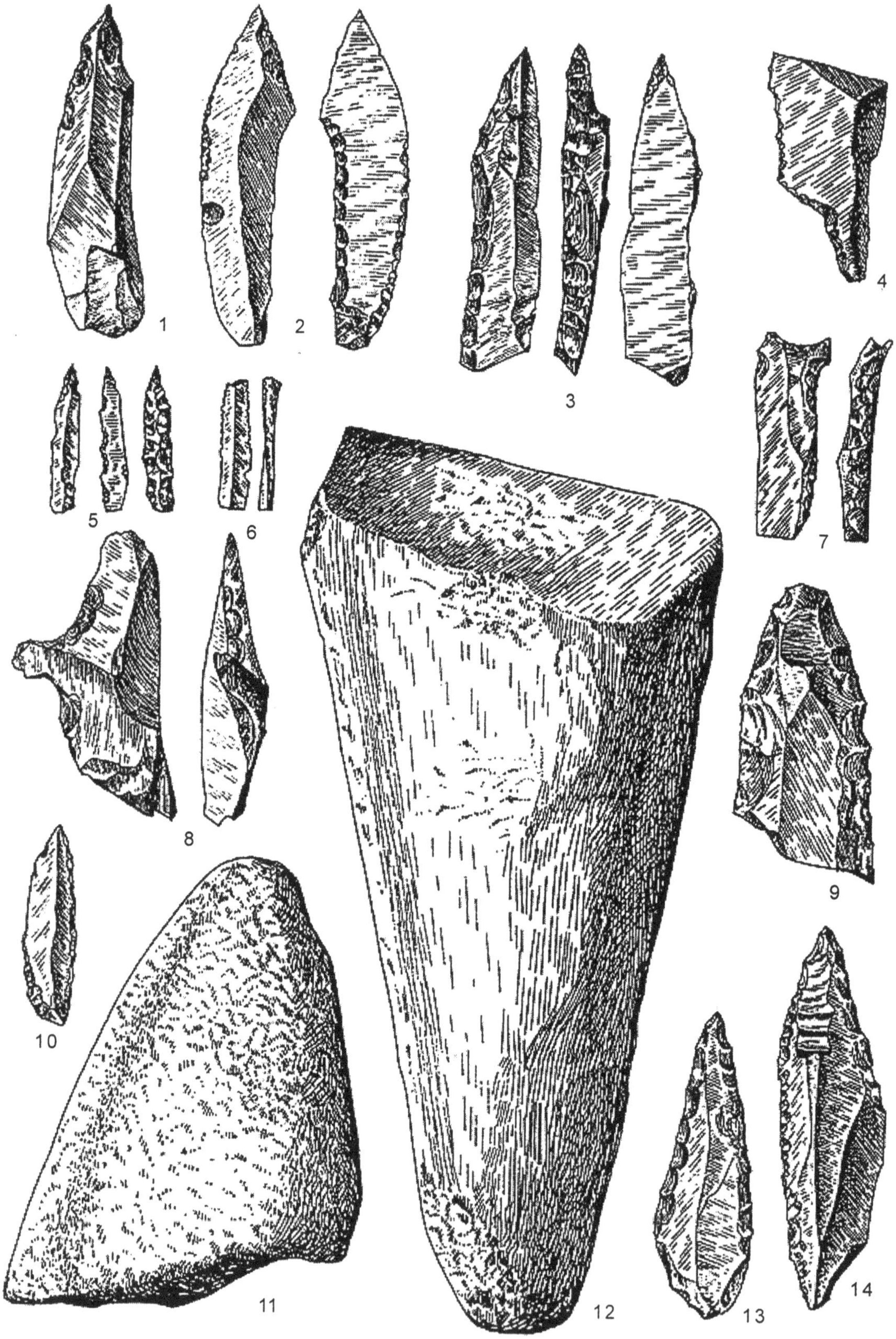

Appendix H18: Willendorf II, examples of stone tools from layer 9.
(After Felgenhauer 1959: Inventory Appendix 44)

Traces of Hafting:	3, 7, 10, 14
Traces of Use:	3, 5, 7, 9, 10, 12, 14

Bibliography

Absolon, K. 1925. 'A discovery as wonderful as that of Tutenkhamens Tomb. Moravia over 20,000 years ago', *The Illustrated London News* October 31, 1925.

Absolon, K. 1926. 'Bericht über die palaeolitische Abteilung am mährischen Landes-Museum und die palaeolithische Forschung in Mähren', *Časopis Moravského Zemského Muzea* 24/11.

Absolon, K. 1927. 'Une nouvelle et importante station Aurignacienne en Moravie', *Revue Anthropologique* v. 27: 73-88.

Absolon, K. 1928. 'Über die große Aurignac-Station bei Unter-Wisternitz in Mähren', *Tagungsberichte der deutschen Anthropologischen Gesellschaft,* Bericht über die 49. Versammlung in Köln, 57-61, Leipzig.

Absolon, K. 1929. 'An amazing Paleolithic Pompeii in Moravia - Part I, II, III', *The Illustrated London News* November 23, 1929; November 30, 1929; December 14, 1929.

Absolon, K. 1935. 'Ein Anhängsel aus einem fossilen Menschenzahn', *Zeitschrift für Rassenkunde*, p. 317

Absolon, K. 1936. 'A vast prehistoric Pompeii revisited', *The Illustrated London News* March 21, 1936; March 28, 1936.

Absolon, K. 1937. 'The worlds earliest portrait - 30 thousand years', *The Illustrated London News* October 2, 1937.

Absolon, K. 1938a. *Výzkum diluvialní stanice lovců mamutů v Dolnich Vestonicích na Pavlovských kopcích na Moravě.* Pracovní zpráva za první rok 1924. Palaeoethnologická Serie c. 4. Brno.

Absolon, K. 1938b. *Die Erforschung der diluvialen Mammutjägerstation von Unter-Wisternitz an den Pollauer Bergen in Mähren.* Arbeitsbericht über das zweite Jahr 1925. Brno.

Absolon, K. 1945. *Výzkum diluvialní stanice lovců mamutů v Dolnich Vestonicích na Pavlovských kopcích na Moravě.* Pracovní zpráva za tretí rok 1926. Brno.

Absolon, K. and B. Klíma 1977. *Předmostí, ein Mammutjägerplatz in Mähren.* Praha: Academia.

Adamenko, O. M. and V. N. Gladilin 1989. 'Korolevo: un des plus anciens habitats Acheuléens et Moustériens de Transcarpatie Soviétique', *L'Anthropologie* (Paris) 93/4: 689-712.

Adams, W. Y. and E. W. Adams 1991. *Archaeological typology and practical reality.* Cambridge: Cambridge University Press.

Adovasio J., O. Soffer and B. Klíma 1996. 'Upper Paleolithic fiber technology: interlaced woven finds from Pavlov I, Czech Republic, c. 26,000 years ago', *Antiquity* 70/269: 526-534.

Albrecht, G. and H. Berke, 1988. 'The Bruder Valley near Engen/Hegau: Varying land use in the Magdalenian', in M. Otte (ed.) *De la Loire á l'Oder.* Actes du Colloque de Liége, Décembre 1985, Volume ii. BAR International Series 444(ii).

Allsworth-Jones, P. 1986. *The Szeletian and the Transition from Middle to Upper Paleolithic in Central Europe.* Oxford: Oxford University Press.

Allsworth-Jones, P. 1990. 'The Szeletian and the stratigraphic succession in central Europe and adjacent areas: main trends, recent results, and problems for resolution', in P. Mellars *The Emergence of Modern Humans.* Ithaca, N.Y.: Cornell University Press.

Ames, K. M. 1985. 'Hierarchies, stress, and logistical strategies among hunter-gatherers in northwestern North America', in T. D. Price and Brown, J. A. (eds.) *Prehistoric Hunter-Gatherers.* New York: Academic Press.

Amick, D. S. and R. P. Mauldin, (eds.) 1989. *Experiments in Lithic Technology.* BAR International Series 528.

Andel, T. H. van 1990. 'Living in the last high glacial - an interdisciplinary challenge', in O. Soffer and Gamble C. (eds.) *The World at 18,000 BP, vol.1, High Latitudes.* London: Unwin Hyman.

Anderson, B. 1983. *Imagined Communities: Reflections on the Origin and Spread of Nationalism.* London: Verso Books

Andrews, J. T. 1982. 'On the reconstruction of Pleistocene ice sheets: a review', *Quaternary Science Reviews* 1: 1-30.

Andriolo, K. R. 1979. 'Kulturkreislehre and the Austrian mind', *Man* 14: 133-44.

Arnold, B. 1990. 'The past as propaganda: totalitarian archaeology in Nazi Germany', *Antiquity* 64: 464-478.

Arnold, D. 1988. *Famine.* Oxford: Basil Blackwell.

Arutiunov, S. A. and A. M. Khazanov 1981. 'Das Problem der archaeologischen Kriterien mit ethnischer Spezifik', *Ethnographisch Archaeologische Zeitschrift* 22: 669-685.

Bahn, P. 1982. 'Intersite and interregional links during the Upper Paleolithic: the Pyrenean evidence', *Oxford Journal of Archaeology* 1: 247-268.

Bailey, G. N. and P. Callow (eds.) 1986. *Stone Age Prehistory.* Cambridge: Cambridge University Press.

Bailey, G. N. (ed.) 1983. *Hunter-Gatherer Economy in Prehistory: A European Perspective*. Cambridge: Cambridge University Press.

Bailey, R. C. and R. Aunger. 1989. 'Net hunters versus archers - variation in women's subsistence strategies in the Ituri forest', *Human Ecology* 17 (3): 273-297.

Bamforth, D. B. 1988. 'Investigating microwear polishes with blind tests: the institute results in context', *Journal of Archaeological Science* 15: 11-23.

Bamforth, D. B., G. Burns, and C. Woodman. 1990. 'Ambiguous use traces and blind test results: new data', *Journal of Archaeological Science* 17/4: 413-430.

Bamforth, D. B., and P. Bleed 1997. 'Technology, flaked stone technology, and risk', in C. M. Barton and G. Clark (eds.) *Rediscovering Darwin: Evolutionary Theory and Archaeological Explanation*. Archaeological Papers of the American Anthropological Association, No. 7

Bánesz, L. 1976. 'Prírodné prostredie, hospodárska základňa a materiálna kultúra Aurignacienu strednej Európy', *Slovenská Archaeológia*. 24: 5-82.

Bárta, J. 1965. *Slovensko v Staršej a Strednej Dobe Kamennej*. Bratislava: Vydavatelstvo Slovenskej Akadémie Vied.

Bárta, J. 1980. *Významné paleolitické lokality na strednom a západnom Slovensku*. Nitra: Archeologický Ústav ČSAV.

Bárta, J. 1987. 'Problems of dwelling structures in the Paleolithic of Western Slovakia', *Anthropologie* 25: 105-110.

Bárta, J. 1989. 'Trenčianske Bohuslavice, a Gravettian settlement in western Slovakia', *Anthropologie* 93: 173-182.

Barton, M. C. 1988. *Lithic Variability and Middle Paleolithic Behavior: New Evidence from the Iberian Peninsula*. BAR International Series S-408.

Barton, M. C. 1989. 'Beyond Style and Function: A View from the Middle Paleolithic', *American Anthropologist* 92/1: 57-72

Barton, M. C. 1991. 'Retouched tools, fact or fiction? Paradigms for interpreting Paleolithic chipped stone', in G. A. Clark (ed.) *Perspectives on the Past*. Philadelphia: University of Pennsylvania Press.

Bassin, M. 1991. 'Russia between Europe and Asia: the ideological construction of geographical space', *Slavic Review* 50/1: 1-16.

Baumann, W., Mania, D., Toepfer, V. and Eissmann, L. 1983. *Markkleeberg*. Berlin: Veröffentlichungen des Landesmuseums für Vorgeschichte Dresden, Band 16.

Bayer, J. 1924. 'Eine Mammutjägerstation im Löß bei Pollau in Südmahren', in *Die Eiszeit* 1/2: 81-88.

Beglar, E.B. and Keatinge, R.W. 1979. 'Theoretical goal and methodological realities: problems in the reconstruction of subsistence economies', *World Archaeology* 11: 208-26.

Behm-Blancke, G. 1967. 'Zur Datierung der altsteinzeitlichen Artefakte von Ehringsdorf bei Weimar', *Ausgrabungen und Funde*. 12: 247-51.

Behrens, H. 1973. *Die Jungsteinzeit im Mittelelbe-Saale-Gebiet.* Berlin: Deutscher Verlag der Wissenschaften,

Ben-David, J. 1984. *The Scientist's Role in a Society: A Comparative Study.* Chicago: University of Chicago Press.

Bender, B. and B. Morris. 1988. 'Twenty years of history, evolution and social change in gatherer-hunter studies', in Ingold, T., Riches, D. and Woodburn, J. (eds.) *Hunters and Gatherers 1: History, evolution and social change*. Oxford: Berg.

Berke, H., J. Hahn and C. -J. Kind (eds.) 1984. 'Jungpalaolitische Siedlungstrukturen in Europa. Koloquium 1983 Reisenburg Gunzburg', *Urgeschichtliche Materialhefte* 6.

Bettinger, R.L. 1980. 'Explanatory predictive models of hunter-gatherer adaptation', in M. B. Schiffer *Advances in Archaeological Method and Theory* 3. New York: Academic Press.

Beug, H. -J. 1982. 'Vegetation History and Climatic Changes in Central and Southern Europe', in A. F. Harding (ed.) *Climatic Change in Prehistory*. Edinburgh: Edinburgh University Press.

Binford, L. 1965. 'Archaeological systematics and the study of culture process', *American Antiquity* 31. 203-210.

Binford, L. 1973. 'Interassemblage variability: the Mousterian and the 'functional' argument', in C. Renfrew (ed.) *The Explanation of Culture Change*. London: Gerald Duckworth.

Binford, L. 1979. 'Organization and formation processes: looking at curation technologies', *Journal of Anthropological Research* 35: 255-273.

Binford, L. 1980. 'Willow smoke and dogs' tails: Hunter-gatherer settlement systems and archaeological site formation', *American Antiquity* 45: 4-20.

Binford, L. 1982a. 'Meaning, inference and the material record', in C. Renfrew and S. Shennan (eds.) *Ranking, Resource, and Exchange.* Cambridge: Cambridge University Press.

Binford, L. 1982b. 'The Archaeology of Place', *Journal of Anthropological Archaeology* 1: 5-31.

Binford, L., and J. Sabloff. 1982. 'Paradigms, systematics, and archaeology', *Journal of Anthropological Research* 38: 137-153.

Bintliff, J. L., Davidson, D. A. and E. G. Grant, (eds.) 1988. *Conceptual Issues in Environmental Archaeology.* Edinburgh: Edinburgh University Press.

Birks, H. J. B. and H. H. Birks. 1980. *Quaternary Palaeoecology.* London: Edward Arnold.

Boas, F. 1948. *Race, Language and Culture.* New York: Macmillan.

Bohmers, A. 1940. 'Das eiszeitliche Fauenkopfchen von Unter-Wisternitz', in *Kleine Kostbarkeiten aus Kunst und Geschichte.* Ahnenerbe Stiftung-Verlag, Berlin-Dahlem.

Bohmers, A. 1941a. 'Die Ausgrabungen bei Unter-Wisternitz', *Forschungen und Fortschritte* 17/3.

Bohmers, A. 1941b 'Reiche Funde eiszeitlicher Bildkunst, die Ausgrabungen bei Unter-Wisternitz', *Germanien* 2: 45-57

Bohmers, A. 1943. 'Die Erforschung von Unter-Wisternitz Bericht über die Ausgrabungen von 1939-1942', *Anthropos Forschungen* Bd. 1

Bohmers, A. 1951. 'Die Höhlen von Mauern', *Palaeohistoria* 1: 1-107.

Bohmers, A. 1960. 'Statistiques et graphiques dans l'étude des industries lithiques préhistoriques. Considérations générales au sujet du Hambourgien, du Tjongerien, du Magdalénien et de l'Azilien', *Palaeohistoria* 8: 15-37.

Bonnichsen, R. 1977. *Models for deriving cultural information from stone tools.* Archaeological survey of Canada. Paper no. 60. Ottawa.

Bordes, F. 1961. *Typologie du Paleolithique Superieur et Moyen.* Bordeaux: Institut de Prehistoire, Universite de Bordeaux, France.

Bordes, F. 1969. 'Reflections on typology and techniques in the Paleolithic' *Arctic Anthropology* 6: 1-29.

Bordes, F. 1971. 'Physical evolution and technological evolution in man: a parallelism', *World Archaeology* 3/1.

Bordes, F. 1972. 'On old and new concepts of typology. Reply', *Current Anthropology* 13: 140-141.

Bordes, F. and D. de Sonneville-Bordes 1970. 'The significance of variability in Paleolithic assemblages', *World Archaeology* 2: 61-73.

Boriskovskii, P.I. (ed.) 1984. *Paleolit SSSR.* Moskva: Nauka.

Bosinski, G. 1967. *Die Mittelpaläolithische Funde im Westlichen Mitteleuropa.* Köln: Fundamenta Reihe A/4.

Bosinski, G. 1972. 'Comment on Chmielewski, W.: The continuity and discontinuity of the evolution of archaeological cultures in Central and Eastern Europe between the 55th and 25th millenaries BC.' in Bordes, F. (ed.) *Origine del'homme moderne.* Paris: Unesco.

Bosinski, G. 1982. 'The transition lower/middle Paleolithic in north-west Germany', in A. Ronen (ed.) *The Transition from Lower to Middle Paleolithic and the Origin of Modern Man.* Oxford: BAR International Series S151.

Bosinski, G. 1984. 'Die Interpretation der Gruben von Gönnersdorf', in Berke, H., J. Hahn and C.-J. Kind (eds.) *Jungpalaolitische Siedlung-strukturen in Europa.* Koloquium 1983 Reisenburg Gunzburg. Urgeschichtliche Materialhefte 6.

Bosinski, G. 1988. 'Upper and Final Paleolithic settlement patterns in the Rhineland', in H. Dibble and A. Montet-White (eds.) *Upper Pleistocene Prehistory of Western Eurasia.* Philadelphia: University of Pennsylvania.

Braun, D. P. 1983. 'Pots and tools', in J. Moore and A. Keene (eds.) *Archaeological Hammers and Theories.* New York: Academic Press.

Braun, D. P. and S. Plog 1982. 'Evolution of 'tribal' social networks: theory and prehistoric North American evidence', *American Antiquity* 47: 504-525.

Brew, J. 1946. 'The use and abuse of taxonomy', *The Archaeology of Alkali Ridge, Southern Utah.* Papers of the Peabody Museum 21. Cambridge, Mass.

Brill, R. H. (ed.) 1971. *Science and Archaeology.* Cambridge, Mass.: MIT Press.

Brink, J. 1978. *An Experimental Study of Microwear Formation on Endscrapers.* Mercury Series, Ottawa: National Museum of Man.

Broglio, A., and G. Laplace. 1966. 'Etudes de typologie analytique des complexes leptolithiques de l'Europe centrale. Les complexes aurignacoides de la Basse Autriche', *Rivista di Scienze Preistoriche.* 21/1: 61-121.

Brumfiel, E. M. 1992. 'Breaking and entering the ecosystem: gender, class, and faction steal the show', *American Anthropologist* 94: 551-567.

Burdukiewicz, J. M. 1986. *The Late Pleistocene Shouldered Point Assemblages in Western Europe.* Leiden: E.J. Brill.

Burdukiewicz, J. M. 1987. 'A review of the later Upper Paleolithic in the Oder and Vistula basins in the light of the recent research', *Oxford Journal of Archaeology* 6: 1-20.

Cahen, D., L. H. Keeley, and F. L. Van Noten. 1979. 'Stone tools, toolkits, and human behavior in prehistory', *Current Anthropology* 20: 661-683.

Campbell, J. M. 1968. 'Territoriality among ancient hunters: interpretations from ethnography and nature', in B. Meggers (ed.) *Anthropological Archaeology in the Americas.* Washington, D.C.: American Anthropological Society.

Champion, T., Gamble C., Shennan S., and A. Whittle (eds.) 1984. *Prehistoric Europe.* London: Academic Press.

Chang, K. C. 1962. 'A typology of settlement and community pattern in some circum-polar societies', *Arctic Anthropology* 1: 28-41.

Chang, K. C. 1967. *Rethinking Archaeology.* New York: Random House.

Chang, K. C. 1972. *Settlement Patterns in Archaeology.* Addison-Wesley Module in Anthropology No. 24.

Chatters, J. 1987. 'Hunter-gatherer adaptations and assemblage structure', *Journal of Anthropological Archaeology* 6: 336-375.

Chernysh, O. P. and V. P. Savich, 1987. *Arkheologiia Prikarpatia, Volyni i Zakarpatia; kamennyi vek.* Kiev: Nauk. dumka.

Childe, G. 1929. *The Danube in Prehistory.* Oxford: Clarendon Press.

Chippindale, C. 1988. 'The Invention of the words for the idea of 'prehistory'' *Proceedings of the Prehistoric Society* 54: 304-14.

Chmielewski, W. 1972. 'The continuity and discontinuity of the evolution of archaeological cultures in Central and Eastern Europe between the 55th and 25th millenaries BC.', in Bordes, F. (ed.) *Origine de l'homme moderne.* Paris: Unesco.

Clark, G. A. and J. M. Lindly 1989. 'The case of continuity: observations on the biocultural transition in Europe and Western Asia', in P. Mellars and C. Stringer (eds.) *The Human Revolution.* Princeton: Princeton University Press.

Clark, G. A. (ed.) 1991. *Perspectives on the Past.* Philadelphia: University of Pennsylvania Press.

Clark, J. D., and J. W. K. Harris. 1985. 'Fire and its roles in early hominid lifeways', *The African Archaeological Review.* 3: 3-27.

Clark, J. G. D. 1952. *Prehistoric Europe: The Economic Basis.* London: Methuen.

Clark, J. G. D. 1988. *Economic Prehistory: Papers in Archaeology.* Cambridge: Cambridge University Press.

Clark, J. G. D., and Thompson, M. 1953. 'The groove and splinter technique of working antler in Upper Paleolithic and Mesolithic Europe', *Proceedings of the Prehistoric Society* 20:148-160.

Clarke, D. L. 1968. *Analytical Archaeology.* London: Methuen & Co. Ltd.

Clarke, D. L. 1979. *Analytical Archaeologist.* New York: Academic Press.

Clarke, D. L., (ed.) 1972. *Models in Archaeology.* London: Methuen & Co.

Clay, R. B. 1976. 'Typological classification, attribute analysis, and lithic variability', *Journal of Field Archaeology* 3/3: 303-311.

Clermont, N. and Smith, P. 1990. 'Prehistoric, prehistory, prehistorian... who invented the terms?' *Antiquity* 64: 97-102.

Clifford, J. and Marcus, G. E. 1986. *Writing culture: The poetics and politics of ethnography.* Berkeley: University of California Press.

CLIMAP Project Members 1981. *Seasonal reconstructions of the earth's surface at the last glacial maximum.* Geological Society of America, Map and Chart Series MC 36.

Close, A. E. 1978. 'The identification of style in lithic artifacts', *World Archaeology* 10: 223-237.

Conkey, M. W. 1978. 'Style and information in cultural evolution: toward a predictive model for the Paleolithic', in C. L. Redman (cd.) *Social Archaeology.* New York: Academic Press.

Conkey, M. W. 1980. 'The identification of prehistoric hunter-gatherer aggregation sites: the case of Altamira', *Current Anthropology* 21: 609-630.

Conkey, M. W. 1982. 'Boundedness in art and society', in I. Hodder (ed.) *Symbolic and structural archaeology.* Cambridge: Cambridge University Press.

Conkey, M. W. 1983. 'On the origins of Paleolithic art: a review and some critical thoughts', in E. Trinkaus (ed.) *Mousterian Legacy: human biocultural change in the Upper Pleistocene.* BAR International Series.

Conkey, M. W. 1984. 'To Find Ourselves: Art and Social Geography of Prehistoric Hunter-Gatherers', in C. Schrire (ed.) *Past and Present in Hunter-Gatherer Studies.* New York: Academic Press.

Conkey, M. W. 1987. 'Interpretive problems in hunter-gatherer regional studies: The case of the European Upper Paleolithic', in O. Soffer (ed.) *The Pleistocene Old World: Regional Perspectives.* New York: Plenum Press.

Conkey, M. W. 1989a. 'The Place of Material Culture Studies in Contemporary Anthropology', in A. L. Hedlund (ed.) *Perspectives on Anthropological Collections from the American Southwest.* Arizona State University Anthropological Research papers No 40.

Conkey, M. W. 1989b. 'The structural analysis of Paleolithic art', in C. C. Lamberg-Karlovsky (ed.) *Archaeological Thought in America.* Cambridge: Cambridge University Press.

Conkey, M. W. 1991. 'Contexts of action, contexts for power: material culture and gender in Magdalenian times', in Conkey, M. W. and J. Gero (eds.) *Engendering Archaeology: Women and Prehistory.* Social Archaeology Series: Basil Blackwell.

Conkey, M. W. and Hastorf, C. (eds.) 1990. *The Uses of Style in Archaeology.* Cambridge: Cambridge University Press.

Conkey, M. W. and Spector, J. 1984. 'Archaeology and the study of gender', in M. B. Schiffer (ed.) *Advances in Archaeological Method and Theory,* v. 7. New York: Academic Press.

Conkey, M. W. and Williams, S. H. 1991. 'Original narratives: the political economy of gender in archaeology', in M. di Leonardo (ed.) *Gender at the Crossroads of Knowledge: Feminist Anthropology in the Postmodern Era.* Berkeley: University of California Press.

Cotterell, B., and Kamminga, J. 1987. 'The formation of flakes', *American Antiquity* 52: 675-708.

Cowgill, G. L. 1989. 'Formal approaches in archaeology', in *Archaeological Thought in America.* C. C. Lamberg-Karlovsky (ed.), Cambridge: Cambridge University Press.

Crabtree, D. 1972. *An Introduction to Flintworking.* Occasional Papers of the Idaho State University Museum, no.28. Pocatello: Idaho State University Museum.

Crawford, E. 1992. *Nationalism and Internationalism in Science, 1880-1939.* Cambridge: Cambridge University Press.

Cronk, L. 1991. 'Human behavioral ecology', *Annual Review of Anthropology* 20: 25-53.

Cross, J. 1983. 'Twigs, branches, trees and forests: problems of scale in lithic analysis', in J. A. Moore and A. S. Keene (eds.) *Archaeological Hammers and Theories.* London: Academic Press.

Cutliffe R. and S. Post (ed.) 1989. *Context, History and History of Technology.* Bethlehem: Lehigh University Press.

Damrosch, L. Jr. 1988. 'Myth and Fiction in Robinson Crusoe', in H. Bloom (ed.) *Daniel Defoe's Robinson Crusoe.* New York: Chelsea House Publishers.

Daniel, G. (ed.) 1981. *Towards a History of Archaeology.* London: Thames & Hudson.

Daniel, G. 1963. *Idea of Prehistory.* Cleveland: The World Publishing Co.

David, N. 1973. 'The Upper Paleolithic society, ecology and technological change: the Noaillan case', in C. Renfrew (ed.) *Explanation of culture change: models in prehistory.* London: Duckworth.

Davis, D. D. 1978. *Lithics and Subsistence.* Nashville, TN: Vanderbilt University Press.

De Atley, S. P. and F.J. Findlow (eds.) 1984. *Exploring the Limits: Frontiers and Boundaries in Prehistory.* BAR International Series 223.

Deetz, J. 1988. 'History and archaeological theory: W.W. Taylor revisited', *American Antiquity* 53: 13-22.

Deetz, J. 1989. 'Archaeography, archaeology, or archeology?' *American Journal of Archaeology* 93: 429-35.

Dennel, R. 1990. 'Progressive gradualism, imperialism and academic fashion: Lower Paleolithic archaeology in the 20th century', *Antiquity* 64: 549-58.

Dennell, R. 1983. *European Economic Prehistory: A New Approach.* New York: Academic Press.

Dennell, R. 1990. 'Progressive gradualism, imperialism and academic fashion: Lower Paleolithic archaeology in the 20th century', *Antiquity* 64: 549-58.

Desbrosses, R. and J. Kozlowski 1985. 'Le paléolithique final en Pologne', *L'Anthropologie* 89/4: 515-541.

Dibble, H. 1984. 'Interpreting typological variation of Middle Paleolithic scrapers: function, style or sequence of reduction?' *Journal of Field Archaeology* 11: 431-436.

Dibble, H. 1987. 'The interpretation of Middle Paleolithic scraper morphology', *American Antiquity.* 52: 109-117.

Dibble, H. 1988. 'Typological aspects of reduction and intensity of utilization of lithic resources in the French Mousterian', in H. Dibble and A. Montet-White (eds.) *Upper Pleistocene Prehistory of Western Eurasia.* Philadelphia: University Museum, University of Pennsylvania.

Dibble, H. 1990. 'A new synthesis of Middle Paleolithic variability', *American Antiquity* 55(3): 480-499.

Dibble, H. 1991. 'Local raw material exploitation and its effects on Lower and Middle Paleolithic assemblage variability', in A. Montet-White and S. Holen (eds.) *Raw Material Economies among Prehistoric Hunter-Gatherers*. Lawrence, Kansas: University of Kansas Press.

Dibble, H. and A. Montet-White (eds.) 1988. *Upper Pleistocene Prehistory of Western Eurasia.* Philadelphia: University Museum, University of Pennsylvania.

Dietler, M. 1994. 'Our ancestors the Gauls': archaeology, ethnic nationalism, and the manipulation of Celtic identity in modern Europe', *American Anthropologist* 96: 584-605.

Dincauze, D. F. 1987. 'Strategies for paleoenvironmental reconstruction in archaeology', in M. B. Schiffer (ed.) *.Advances in Archaeological Method and Theory* 11. New York: Academic Press.

Dobosi, V. T. 1991. 'Economy and raw material' in A. Montet-White and S. Holen (eds.) *Raw Material Economies among Prehistoric Hunter-Gatherers.* Lawrence, Kansas: University of Kansas Press.

Dolukhanov, P. M. 1979. *Geografiia kamennogo veka.* Moskva: Nauka.

Dolukhanov, P. M. 1986. 'Natural environment and the Holocene settlement pattern in the north-western part of the USSR', *Fennoscandia Archaeologica* 3: 3-16.

Dolukhanov, P. M. 1989. 'Cultural and ethnic processes in prehistory', in S. J. Shennan (ed.) *Archaeological Approaches to Cultural Identity.* London: Unwin Hyman.

Dolukhanov, P. M., J. K. Kozlowski, and S. K. Kozlowski 1980. *Multivariate Analysis of Upper Paleolithic and Mesolithic Stone Assemblages.* Warsaw: Panstwowe Wydawnictwo Naukowe.

Dumont, J. V. 1982. 'The quantification of microwear traces: a new use for interferometry', *World Archaeology* 14/2: 206-217.

Dumont, J. V. 1988. *A Microwear Analysis of Selected Artifact Types from the Mesolithic Sites of Star Carr and Mount Sandel.* BAR British Series 187.

Dunnell, R. C. 1971. *Systematics in Prehistory.* New York: Free Press.

Dunnell, R. C. 1973. 'Fire, air, earth, and water: a rational classification', *Mankind* 9: 127-131.

Dunnell, R. C. 1978. 'Style and function: a fundamental dichotomy', *American Antiquity* 43: 192-202.

Dunnell, R. C. 1982. 'Science, social science and common sense: the agonizing dilemma of modern archaeology', *Journal of Anthropological Research* 38/1: 1-25.

Dunnell, R. C. 1986. 'Methodological issues in Americanist artifact classification', in M. B. Schiffer (ed.) *Advances in Archaeological Method and Theory* 9. New York: Academic Press.

Edmonds, M. 1990. 'Description, understanding and the chaine operatoire', *Archaeological Review from Cambridge* 9/1: 55-70.

Efimenko, P.P. 1958. *Kostenki I.* Moskva: Nauka.

Eggers, H. J. 1986. *Einführung in die Vorgeschichte.* (3rd ed.) München: Piper.

Eggert, M. K. H. 1978. 'Zum Kulturkonzept in der prahistorischen Archaeologie', *Bonner Jahrbucher* 178: 1-20.

Ehrenberg, K. 1951. '30 Jahre paläobiologischer Forschung in Österreichischen Höhlen', *Quartär* 5.

Elias, N. 1978 (1939). *The History of Manners: The Civilizing Process v. 1.* New York: Pantheon Books.

Ellis, C. and J. Lothrop 1989. *Eastern Paleoindian Lithic Resource Use.* Boulder, Co.: Westview Press.

Ericson, J. E. and B. A. Purdy (eds.) 1984. *Prehistoric Quarries and Lithic Production.* Cambridge: Cambridge University Press.

Ermarth, M. 1978. *Wilhelm Dilthey: The Critique of Historical Reason.* Chicago: University of Chicago Press.

Fabian, J. 1983. *Time and the other: How anthropology makes its object.* New York: Columbia University Press.

Féblot-Augustins, J. 1993. 'Mobility strategies in the late middle Paleolithic of central Europe and western Europe: elements of stability and variability', *Journal of Anthropological Archaeology* 12: 211-265.

Felgenhauer, F. 1956-59. 'Willendorf in der Wachau', *Mitteilungen der Prahistorische Kommission der Osterreichische Akademie der Wissenschaft 4.*

Fink, J. 1962. 'Studien zur absoluten und relativen Chronologie der fossilen Boden in Osterreich', *Archeologia Austriaca* 31: 1-18.

Fish, P. 1981. 'Beyond tools: Middle Paleolithic debitage analysis and cultural inference', *Journal of Archaeological Research* 38: 374-386.

Fladmark, K. R. 1982. 'Microdebitage analysis: initial considerations', *Journal of Archaeological Science* 9: 205-220.

Flanagan, J. G. 1989. 'Hierarchy in simple egalitarian societies', *Annual Review of Anthropology* 18: 245-266.

Flenniken, J. J. 1984. 'The past, present and future of flintknapping: an anthropological perspective', *Annual Review of Anthropology* 13: 187-203.

Flenniken, J. J. and P. Wilke. 1989. 'Typology, technology, and chronology of Great Basin dart points', *American Anthropologist* 91: 149-158.

Ford, J. A. 1954. 'On the concept of types', *American Anthropologist,* 56/1: 42-54.

Ford, J. A. 1961. 'In favor of simple typology', *American Antiquity* 27: 113-114.

Fortescue, S. 1987. *The Communist Party and Soviet Science.* Baltimore: John Hopkins University Press.

Foucault, M. 1970. *The Order of Things: An Archaeology of Human Sciences.* New York: Vintage Books.

Foucault, M. 1980. *Power/Knowledge: Selected Interviews and Other Writings.* New York: Pantheon.

Fowler, D. D. 1987. 'Uses of the past: Archaeology in the service of the state', *American Antiquity* 52: 229-248.

Frayer, D. W. 1986. 'Cranial variation at Mladec and the relationship between Mousterian and Upper Paleolithic hominids', *Anthropos* 23: 243-256.

Freeman, L. G. 1991. 'What mean these stones? Remarks on raw material use in the Spanish Paleolithic', in A. Montet-White and S. Holen (eds.) *Raw Material Economies among Prehistoric Hunter-Gatherers.* Lawrence, Kansas: University of Kansas Press.

Frenzel, B. 1983. 'Die Umwelt des Menschen im Eiszeitalter', *Quartar* 35/36: 7-33.

Freund, G. 1944. 'Pschedmos*t*', unpublished PhD dissertation, German University at Prague.

Fridrich, J. 1982. *Stredopaleoliticke osidleni Cech.* Praha: Archeologicky ustav CSAV.

Gábori, M. 1976. *Les Civilisations du Paleolithique Moyen entre les Alpes et L'Oural.* Budapest: Akadémiai Kiadó.

Gambier, D. 1989. 'Fossil hominids from the early Upper Paleolithic (Aurignacian) of France', in P. Mellars and C. Stringer (eds.) *The Human Revolution.* Princeton: Princeton University Press.

Gamble, C. S. 1978. 'Resource exploitation and the spatial patterning of hunter-gatherers: a case study', in D. Green, C. Haselgrove and M. Spriggs (eds.) *Social Organization and Settlement.* Oxford: B.A.R.

Gamble, C. S. 1979. 'Hunting strategies in the central European Paleolithic', *Proceedings of the Prehistoric Society* 45: 35-52.

Gamble, C. S. 1982. 'Interaction and Alliance in Paleolithic Society', *Man* 17: 92-107.

Gamble, C. S. 1984. 'Regional variation in hunter-gatherer strategy in the Upper Pleistocene of Europe', in R. Foley (ed.) *Hominid Evolution and Community Ecology.* New York: Academic Press.

Gamble, C. S. 1986. *The Paleolithic Settlement of Europe.* Cambridge: Cambridge University Press.

Gamble, C. S. 1987. 'Archaeology, geography and time', *Progress in Human Geography* 11:227-246.

Gamble, C. S. 1991. 'An introduction to the living spaces of mobile peoples', in C. S. Gamble and W. A. Boismier (eds.) *Ethnoarchaeological Approaches to Mobile Campsites.* International Monographs in Prehistory, Ann Arbor, MI.

Gardin, J.-C. 1980. *Archaeological Constructs.* Cambridge: Cambridge University Press.

Garnsey, P. 1988. *Famine and Food Supply in the Graeco-Roman World.* Cambridge: Cambridge U. Press.

Garraty, J. and P. Gay, 1972. *The Columbia History of the World.* New York: Harper & Row.

Gathercole, P, and Lowenthal, D. (eds.) 1990. *Politics of the Past.* London: Unwin Hyman.

Gellner, E. 1983. *Nations and Nationalism.* Ithaca, N.Y.: Cornell University Press.

Geneste, J. M. 1990. 'Dévelopment des systémes de production lithique au cours du Paléolithique moyen en Aquitaine septentrionale', in C. Farizy (ed.) *Paleolithique supérieur ancien en Europe.* Nemours: Mémoires du Musée de Préhistoire d'Ille de France.

Gening, V. F. 1982. *Ocherki po istorii sovetskoi arkheologii : u istokov formirovaniia marksistskikh teoreticheskikh osnov sovetskoi.* Kiev: Nauk. dumka.

Gerasimov, I. P. and A. A. Velichko,(eds.)1982. *Palaeogegraphy of Europe during the Last 100,000 Years.* Moskva: Nauka.

Gero, J. M. 1983. 'Lithics and the representation of social complexity', paper presented at the 48th Annual Meeting of the SAA, Pitsburgh, Penn.

Gero, J. M. 1989. 'Assessing social information in material objects: how well do lithics measure up?' in R. Torrence (ed.) *Time, energy and stone tools.* Cambridge: Cambridge University Press.

Gero, J. M. and M. W. Conkey (eds.) 1991. *Engendering Archaeology.* Oxford: Basil Blackwell.

Gero, J. M., D. M. Lacy and M. L. Blakey (eds.) 1983. *The Socio-politics of archaeology.* Amherst: Dept. of Anthropology, University of Massachusetts, Research report no. 23.

Gilman, A. 1984. 'Explaining the upper Paleolithic revolution', in M. Spriggs (ed.) *Marxist Perspectives in Archaeology.* Cambridge: Cambridge University Press.

Gimbutas, M. 1956. 'Prehistory of East Europe. Part I.', *American School of Prehistoric Research Bulletin 20.*

Girtler, R. 1982. "Ethnos', 'Volk' und soziale Gruppe. Zum Problem eines zentralen Themas in den anthropologischen Wissenschaften', *Mitteilungen der Anthropologischen Geselschaft in Wien* 112: 42-57.

Gojda, M. 1991. *The Ancient Slavs: Settlement and Society.* Edinburgh: Edinburgh University Press.

González Morales, M. R. 1991. 'From hunter-gatherers to food producers in northern Spain: smooth adaptive shifts or revolutionary change in the Mesolithic', in G. Clark (ed.) *Perspectives on the Past: Theoretical Biases in Mediterranean Hunter-Gatherer Research.* Philadelphia: University of Pennsylvania Press.

Grahman, R. and Müller-Beck, H.-J. 1957. *Urgeschichte der Menschheit.* Stuttgart: W. Kohlhammer Verlag.

Gramley, R. 1980. 'Raw material source areas and curated tool assemblages', *American Antiquity* 45: 823-33.

Green, S. W. and S. M. Perlman (eds.) 1985. *The Archaeology of Frontiers and Boundaries.* Orlando: Academic Press.

Haesaerts, P. 1990. 'Nouvelles recherches au gisement de Willendorf (Basse Autriche)', *Bulletin van het Koninklijk Belgisch Insitut voor Naturr-wetenschappen, Aardwetenschappen,* 60: 203-218.

Haesaerts, P., F. Damblon, M. Bachner, and G. Trnka 1996. 'Revised stratigraphy and chronology of the Willendorf II sequence, Lower Austria', *Archaeologia Austriaca* 80: 25-42.

Hahn, J. 1972a. 'Das aurignacien in Mittel- und Osteuropa', *Acta Praehistorica et Archaeologica.* 3: 77-107.

Hahn, J. 1972b. 'Aurignacian signs, pendants and art objects in central and eastern Europe', *World Archaeology* 3: 252-66.

Hahn, J. 1977. *Aurignacien: das ältere Jungpaläolithikum in Mittel- und Osteuropa.* Köln: Fundamenta Reihe A/9.

Hahn, J. 1986. *Kraft und Aggression: Die Botschaft der Eiszeitkunst im Aurignacien Süddeutschlands?.* Tübingen: Verlag Archaeologica Venatoria.

Hahn, J. 1987. 'Aurignacian and Gravettian settlement patterns in Central Europe', in O. Soffer (ed.) *The Pleistocene Old World. Regional Perspectives.* New York: Plenum Press.

Halland, R. 1977. 'Archaeological classification and ethnic groups: a case study from Sudanese Nubia', *Norwegian Archaeological Review* 10: 1-17.

Haraway, D. 1989. *Primate Visions.* New York: Routledge.

Haraway, D. 1991. *Simians, Cyborgs, and Women.* New York: Routledge.

Harding, S. 1986. *The Science Question in Feminism.* Ithaca, N.Y.: Cornell University Press.

Hay, D. 1968. *Europe: The Emergence of an Idea.* Edinburgh: Edinburgh University Press.

Hayden, B. 1975. 'Curation: old and new', in J. Raymond (ed.) *Primitive Art and Technology.* Calgary: Department of Archaeology, University of Calgary.

Hayden, B. 1977. 'Stone tool function in the Western Desert, in R. V. S. Wright (ed.) *Stone Tools as Cultural Markers.* Canberra: Australian Institute of Aboriginal Studies.

Hayden, B. 1984. 'Are emic types relevant to archaeology?', *Ethnohistory* 31/2: 79-92.

Hayden, B. 1987. 'From chopper to celt: the evolution of resharpening techniques', *Lithic Technology* 16/2-3: 33-43.

Hayden, B. (ed.) 1979. *Lithic Use-Wear Analysis.* New York: Academic Press.

Hegmon, M. 1992. 'Archaeological research on style', *Annual Review of Anthropology* 21: 517-36.

Henry, D. O. and G. H. Odell (eds.) 1989. *Alternative Approaches to Lithic Analysis.* Archaeological Papers of the American Anthropological Association, no.1.

Hensel, W. 1977. 'A method of ethnic qualification of archaeological sources', *Archaeologia Polona* 18: 7-35.

Hensel, W. 1983. *Prahistoria ziem polskich 1975-1981.* Wroclaw: Polska Akademia Nauk.

Hill, J. N. and R. K. Evans 1972. 'A model for classification and typology', in D. Clarke (ed.) *Models in Archaeology*. London: Methuen.

Hobsbawm, E. and T. Ranger (eds.) 1988 (1983). *The Invention of Tradition.* Cambridge: Cambridge University Press.

Hobsbawm, E. J. 1992. 'Ethnicity and nationalism in Europe today', *Anthropology Today* 8/1: 3-8.

Hodder, I. 1978a. 'The maintenance of group identities in the Baringo District, Western Kenya', in D. Green, C. Hasselgrove, and M. Spriggs (eds.) *Social Organization and Settlement: Contributions from Anthropology, Archaeology and Geography.* Oxford: BAR International Series 47:

Hodder, I. 1978b. 'Simple correlations between material culture and society: a review', in I. Hodder (ed.) *The Spatial Organization of Culture.* London: Duckworth.

Hodder, I. 1978c. 'Some effects of distance on patterns of human interaction', in I. Hodder (ed.) *The Spatial Organization of Culture*. London: Duckworth.

Hodder, I. 1982. *Symbols in action. Ethnoarchaeological studies of material culture.* Cambridge, Cambridge University Press.

Hoffecker, J. F. and C. A. Wolf (eds.) 1988. *The Early Upper Paleolithic: Evidence from Europe and the Near East.* Oxford: BAR International Series, 437.

Horkheimer, M. and Adorno, T. W. 1987 (1944). *Dialectic of Enlightment.* New York: Continuum.

Hurcombe, L. 1992. 'Analyse des traces d'usure sur l'obsidienne', in *Anthropologie* 96/1: 179-185.

Ingold, T., D. Riches and J. Woodburn (eds.) 1988. *Hunters and Gatherers I: History, Evolution and Social Change.* Oxford: Berg.

Ivanova I. K. and N. D. Praslov (eds.) 1977. *Paleoekologiya Drevnevo Cheloveka.* Moskva: Nauka.

Jalevich, B. 1987. *Modern Austria: Empire and Republic, 1815-1986.* Cambridge: Cambridge University Press.

Jelinek, A. J. 1976. 'Form, function and style in lithic analysis', in C. E. Cleland (ed.) *Cultural Change and Continuity: Essays in Honor of James Bennet Griffin.* New York: Academic Press.

Jelinek, A. J. 1988. 'Technology, typology, and culture in the Middle Paleolithic', in H. L. Dibble and A. Montet-White (eds.) *Upper Pleistocene Western Eurasia.* Philadelphia: University Museum.

Jelinek, A. J. 1991. 'Observations on reduction patterns and raw materials in some Middle Paleolithic industries in the Perigord', in A. Montet-White and S. Holen (eds.) *Raw Material Economies among Prehistoric Hunter-Gatherers.* Lawrence, Kansas: University of Kansas.

Jelinek, J. 1991. 'Discovery of skeletal remains from the Moravian Gravettian population', *Anthropologie* 95/1: 137-153.

Jochim, M. A. 1976. *Hunter-gatherer subsistence and settlement: A predictive model.* New York, Academic Press.

Jochim, M. A. 1991. 'Archaeology as long-term ethnography', *American Anthropologist* 93/2: 308-321.

Jochim, M. A. 1998. *A Hunter-Gatherer Landscape.* New York, Plenum Press.

Johnson, J. 1989. 'The utility of production trajectory modeling as a framework for regional analysis', in D. O. Henry and G. H. Odell (eds.) *Alternative Approaches to Lithic Analysis.* Washington, D. C.: American Anthropological Association.

Johnson, J. K. and C.A. Morrow (eds.) 1987. *The Organization of Core Technology.* Westview Special Studies in Arch. Research, Boulder CO: Westview Press.

Jones, K. T. and D. B. Madsen 1989. 'Calculating the cost of resource transportation: a Great Basin example', *Current Anthropology* 30:529-534.

Jones, M. (ed.) 1983. *Integrating the Subsistence Economy.* BAR International Series 181.

Juel Jensen, H. 1988. 'Functional analysis of prehistoric flint tools by high-power microscopy: a review of West European research', *Journal of World Prehistory* 2: 53-88.

Kaminská, L. 1991. 'Význam surovinovej základne pre mladopaleolitickú spoločnost vo východo-karpatskej oblasti', *Slovenská Archeologia* 39:7-58.

Karp, I. and Lavine D. S. (eds.) 1991. *Exhibiting Cultures: The Poetics and Politics of Display.* Washington: Smithsonian Institution Press

Karp, I., Mullen, Kreamer, C. and D. S. Lavine (eds.) 1992. *Museums and Communities: The Politics of Public Culture.* Washington: Smithsonian Institution Press

Keeley, L. 1974. 'Technique and methodology in microwear studies: a critical review', *World Archaeology* 5/3: 323-336.

Keeley, L. 1980. *Experimental Determination of Stone Tool Uses.* Chicago: Chicago University Press.

Keeley, L. 1987. 'Hafting and retooling: effects on the archaeological record', *American Antiquity* 47/4: 798-80

Keene, A. S. 1986. 'Stories we tell: gatherer-hunters as ideology', paper presented at the IV International Congress on Hunters-Gatherers, London School of Economics.

Kehoe, A. B. 1990. 'Points and lines', in S. Nelson and A. Kehoe (eds.) *Powers of Observation.* Archaeological Papers of the American Anthropological Association No.2.

Keller, E. F. 1992. *Secrets of Life, Secrets of Death: Essays on Language, Gender and Science.* New York: Routledge.

Kelley, J. H. and Hanen, M. P. 1988. *Archaeology and the Methodology of Science.* Albuquerque: University of New Mexico Press.

Kelly, R. 1988 'The three sides of a biface', *American Antiquity* 53:717-734.

Kelly, R., and L. C. Todd 1988. 'Coming into the country: early Paleoindian hunting and mobility', *American Antiquity* 53:231-244.

Kirch, P. 1980. 'The archaeological study of adaptation: theoretical and methodological issues', in M. B. Schiffer (ed.) *Advances in Archaeological Method and Theory, Vol. 3.* New York: Academic Press.

Klejn, L. S. 1977. 'A panorama of theoretical archaeology', *Current Anthropology* 18: 1-42.

Klejn, L. S. 1982. *Archaeological Typology.* BAR International Series 153.

Klíma, B. 1951. 'Karel Jaroslav Maška', *Časopis Moravského Musea v Brne*, 36/2: 9-20.

Klíma, B. 1963. *Dolní Věstonice.* Praha: ČSAV.

Klíma, B. 1976. 'Die palaolithische station Pavlov II.', *Přírodovědné práce ústavu ČSAV v Brne* X/4. Praha: Academia.

Klíma, B. (ed.) 1976. *Périgordien et Gravettien en Europe.* Nice, U.I.S.P.P., IX Congrés, Colloque XV.

Kluckhohn, C. 1960 'The use of typology in anthropological theory', in A. F. C. Wallace (ed.) *Selected Papers from the Proceedings of the Vth International Congress of Anthropological and Ethnological Sciences* (1956). Philadelphia: University of Pennsylvania Press.

Kneblová, V. 1954. Fytopaleontologický rozbor uhlíků z paleolitického sídlište v Dolních Věstonicích', *Anthropozoikum* 3: 297-9.

Knecht, H. 1988. *Upper Paleolithic Burins: Type, Form and Function.* BAR International Series 434.

Knecht, H., Pike-Tay, A. and White, R. (eds.) 1993. *Before Lascaux: The Complex Record of the Early Upper Paleolithic.* Boca Raton: CRC Press.

Knor, A., Ložek, V., Pelíšek, J., Žebera, K., (eds.) 1953. *Dolní Věstonice.* Praha: CSAV.

Knutsson, K. 1988a. *Patterns of tool use.* Societas Archaeologica Upsaliensis. AUN 10. Uppsala.

Knutsson, K. 1988b. *Making and using stone tools.* Societas Archaeologica Upsaliensis. AUN 11. Uppsala.

Kohl, P. 1981. 'Materialist approaches in prehistory', *Annual Review of Anthropology* 10: 89-118.

Kozlowski, J. K. 1973. 'The origin of lithic raw materials used in the Paleolithic of the Carpathian countries', *Acta Archaeologica Carpathica* 13: 5-19.

Kozlowski, J. K. 1977. 'Prehistoric settlement in the northern part of central Europe in the light of palaeogeographic conditions prevailing during the earlier phase of the Würm Pleniglacial', *Folia Quaternaria* 49: 5-14.

Kozlowski, J. K. 1986. 'The Gravettian in Central and Eastern Europe', *Advances in World Archaeology* 5: 131-200.

Kozlowski, J. K. 1990a. 'A multiaspectual approach to the origins of the Upper Paleolithic in Europe', in P. Mellars *The Emergence of Modern Humans.* Ithaca, N.Y.: Cornell University Press.

Kozlowski, J. K. 1990b. 'Northern Central Europe c. 18,000 BP.', in O. Soffer and C. Gamble (eds.) *The World at 18 000 BP. Vol. I: High Altitudes.* London: Unwin Hyman.

Kozlowski, J. K. 1991a. 'Paleolithic quarry sites', in A. Montet-White and S. Holen (eds.) *Raw Material Economies among Prehistoric Hunter-Gatherers.* Lawrence, Kansas: University of Kansas Press.

Kozlowski, J. K. 1991b. 'Raw material procurement in the Upper Paleolithic of Central Europe', in A. Montet-White and S. Holen (eds.) *Raw Material Economies among Prehistoric Hunter-Gatherers.* Lawrence, Kansas: University of Kansas.

Kozlowski, J. K. (ed.) 1976. *L'Aurignacien en Europe.* Nice, U.I.S.P.P., IX Congrés, Colloque XVI.

Kozlowski, J. K., Drobniewicz, B., Sachse-Kozlowska, E., and B. van Vliet 1975. 'Upper Paleolithic site Kraków-Spadzista Street C', *Folia Quaternaria* 45.

Kozlowski, J. K. Kozlowski, S. K. 1979. *Upper Paleolithic and Mesolithic in Europe: Taxonomy and Palaeohistory.* Warszawa: Polska Akademia Nauk.

Königswald von, W. and Müller-Beck, H. 1975. 'Das pleistozän der Weinberghöhlen bei Mauern (Bayern)', *Quartär* 26: 107-118.

Kristiansen, K. 1990. 'National archaeology in the age of European integration', *Antiquity* 64: 825-8.

Kroeber, A. L. 1957. *Style and Civilizations.* Ithaca, N.Y.: Cornell University Press.

Kuhn, S. L. 1991. 'New problems, old glasses: methodological implications of an evolutionary paradigm for the study of paleolithic technologies', in G. Clark (ed.) *Perspectives on the Past: Theoretical Biases in Mediterranean Hunter-Gatherer Research.* Philadelphia: University of Pennsylvania Press.

Kuhn, S. L. 1992. 'Blank form and reduction as determinants of Mousterian scraper morphology', *American Antiquity* 57/1: 115-128.

Kuhn, S. L. 1994. 'Formal approach to the design and assembly of mobile toolkits', *American Antiquity* 59/3: 426-442.

Kukla, J. 1975. 'Loess stratigraphy of central Europe', in K. Butzer and G. Isaac (eds.) *After the Australopithecines.* The Hague: Mouton Press.

Kuper, A. 1988. *The Invention of Primitive Society.* London: Routledge.

Lamberg-Karlovsky, C. C. (ed.) 1990. *Archaeological Thought in America.* Cambridge: Cambridge University Press.

Lang, B. (ed.) 1979. *The Concept of Style.* Philadelphia: University of Pennsylvania Press.

Lanzinger, M. 1988. 'Flint exploitation and production at Monte Avena in the Dolomitic region of the Italian Alps', in H. Dibble and A. Montet-White (eds.) *Upper Pleistocene Prehistory of Western Eurasia.* Philadelphia: University Museum, University of Pennsylvania.

Laplace, G. 1970. 'L'industrie de Krems Hundssteig et le probleme de l'origine des complexes aurignaciens', *Frühe Menschheit und Umwelt.* Köln: Fundamenta Reihe A/2: 242-297.

Latour, B. 1987. *Science in Action: How to Follow Scientists and Engineers through Society.* Milton Keynes: Open University Press.

Layton, E. T. 1974. 'Technology as knowledge', *Technology and Culture* 15: 31-41.

Lechtman, H. 1977. 'Style in technology - some early thoughts', in H. Lechtman and R.S. Merrill (eds.) *Material Culture, Styles, Organization, and Dynamics of Technology.* St. Paul, MN: West Publishing.

Lemmonier, P. 1986. 'The study of material culture today: Towards an anthropology of technical systems', *Journal of Anthropological Archaeology* 5: 147-186.

Leone, M. 1982. 'Some opinions about recovering mind', *American Antiquity* 47: 742-760.

Levi-Sala, I. 1986. 'Use-wear and post-depositional surface modification: a note of caution', *Journal of Archaeological Science* 13: 203-218.

Longino, H. 1990. *Science as Social Knowledge.* Princeton, New Jersey: Princeton University Press.

Lowenthal, D. 1985. *The Past is a Foreign Country.* Cambridge: Cambridge University Press.

Luedtke, B. E. 1992. *An Archaeologists Guide to Chert and Flint.* Los Angeles: Institute of Archaeology.

Lurie, R. 1990. 'Lithic technology and mobility strategies: The Koster site Middle Archaic', in R. Torrence (ed.) *Time, Energy, and Stone Tool.* Cambridge University Press, Cambridge.

Lynch, M. and Woolgar, S. (eds.) 1990. *Representation in Scientific Practice.* Cambridge, Mass.: The MIT Press.

Magris, C. 1989. *Danube: a Journey through the Landscape, History, and Culture of Central Europe.* New York: Farrar Straus Giroux.

Malina, J. 1970. 'Die jungpaläolithische steinindustrie aus Mähren, ihre Rohstoffe und ihre Patina', *Acta Praehistorica et Archaeologica*, 1: 157-73.

Mania, D. and Toepfer, V. 1973. *Königsaue.* Berlin: Veröffentlichungen des Landesmuseums für Vorgeschichte in Halle, Band 26.

Mania, D., Toepfer, V. and Vlcek, E. 1980. *Bilzingsleben I.* Berlin: Veröffentlichungen des Landesmuseums für Vorgeschichte in Halle, Band 32.

Mansur, M. E. 1982. 'Microwear analysis of natural and use striations: new clues to the mechanism of striation formation', *Studia Praehistorica Belgica* 2: 213-233.

Marks, A. E., J. Shokler, and J. Zilhao. 1991. 'Raw material usage in the Paleolithic. The effects of local availability on selection and economy', in A. Montet-White and S. Holen (eds.) *Raw Material Economies among Prehistoric Hunter-Gatherers.* Lawrence, Kansas: University of Kansas Press.

Mason, S. L. R., Hather, J. G., and Hillman, G. C. 1994. 'Preliminary investigation of the plant macro-remains from Dolní Vestonice II, and its implications for the role of plant foods in Paleolithic and Mesolithic Europe', *Antiquity* 68: 48-57.

Meinander, C. F. 1981. 'The concept of culture in European archaeological literature', in G. Daniel (ed.) *Towards a history of archaeology*. London: Thames & Hudson.

Mellars, P. 1970. 'Some comments on the notion of functional variability in stone-tool assemblages', *World Archaeology* 2/1: 74-89

Mellars, P. (ed.) 1990. *The Emergence of Modern Humans*. Ithaca, N.Y.: Cornell University Press.

Mellars, P. 1996. *The Neanderthal Legacy*. Princeton: Princeton University Press.

Mellars, P., and Stringer, C. (eds.) 1989. *The Human Revolution*. Princeton: Princeton University Press.

Meltzer, D. 1981. 'A study of style and function in a class of tools', *Journal of Field Archaeology* 8: 313-326.

Metcalfe, D., and K. R. Barlow 1992. 'A model for exploring the optimal trade-off between field processing and transport', *American Anthropologist* 94:340-356.

Milisauskas, S. 1978. *European Prehistory*. New York: Academic Press.

Milisauskas, S. 1990. 'People's revolution of 1989 and archaeology in Eastern Europe', *Antiquity* 64: 283-5.

Miller. D. 1985. *Artifacts as Categories*. Cambridge: Cambridge University Press.

Mithen, S. J. 1989. 'Modelling hunter-gatherer decision making: complementing optimal foraging theory', *Human Ecology* 17/1: 59-83.

Monks, G. G. 1981. 'Seasonality studies', in M. B. Schiffer (ed.) *Advances in Archaeological Method and Theory* 4. New York: Academic Press.

Montet-White, A. 1988a. 'Recent excavations at Grubgraben, a Gravettian site in Lower Austria', *Archaologisches Korrespondenzblatt* 18: 213-218.

Montet-White, A. 1988b. 'Raw material economy among medium-sized late Paleolithic campsite of Central Europe', in H. Dibble and A. Montet-White (eds.) *Upper Pleistocene Prehistory of Western Eurasia*. Philadelphia: University Museum, University of Pennsylvania.

Montet-White, A. 1991. 'Lithic acquisition, settlements, and territory in the Epigravettian of Central Europe', in A. Montet-White and S. Holen (eds.) *Raw Material Economies among Prehistoric Hunter-Gatherers*. Lawrence, Kansas: University of Kansas Press.

Montet-White, A. 1994. 'Alternative interpretations of the Late Upper Paleolithic in Central Europe', *Annual Review of Anthropology* 23: 483-508.

Montet-White, A. and S. Holen (eds.) 1991. *Raw Material Economies among Prehistoric Hunter-Gatherers*. Lawrence, Kansas: University of Kansas Press.

Montet-White, A. and J. T. Williams 1994. 'Spatial organization at a winter campsite of the last glacial maximum: the case of Grubgraben', *Journal of Anthropological Archaeology* 13/2: 125-138.

Moore, J. A. 1983. 'The trouble with know-it-alls: information as a social and ecological resource', in J. A. Moore and A. S. Keene (eds.) *Archaeological Hammers and Theories*. London: Academic Press.

Moore, J. A. and A. S. Keene (eds.) 1983. *Archaeological Hammers and Theories*. London: Academic Press.

Moss, E. 1983. *The Functional Analysis of Flint Implements - Pincevent and Pont d'Ambon: Two Case Studies from the French Final Paleolithic*. BAR International Series 177.

Müller-Beck, H.-J. 1968. *Die Ältere und Mittlere Steinzeit*. Basel: Ur- und Frühgeschichtliche Archäologie der Schweiz, Band 1.

Müller-Beck, H.-J. (ed.) 1973. *Neue paläolithische und mesolithische Ausgrabungen in der Bundesrepublik Deutschland*. Tübingen, zum 9, INQUA-Kongress.

Muller-Wille, C. S. and D. B. Dickson 1991. 'An examination of some models of late Pleistocene Society in Southwestern Europe', in G. A. Clark (ed.) *Perspectives on the Past*. Philadelphia: University of Pennsylvania Press.

Musil, R. and Valoch, K. 1968. 'Stránská skála: its meaning for pleistocene studies', *Current Anthropology* 9: 534-9.

Nash, C. (ed.) 1990. *Narrative in Culture: The Uses of Storytelling in the Sciences, Philosophy, and Literature*. London: Routledge.

Neeley, M. P. and Barton, C. M. 1994. 'A new approach to interpreting late Pleistocene microlith industries in southwest Asia', *Antiquity* 68/259:275-289.

Nelson, M. 1991. 'The study of technological organization', in M. B. Schiffer (ed.) *Archaeological Method and Theory, vol. 3*. Tucson: University of Arizona Press.

Neugebauer-Maresch, Ch. (ed.) 1993. *Niederösterreich in Urgeschichte*. Wien.

Newcomer, M. 1974. 'Study and replication of bone tools from Ksar Akil (Lebanon)', *World Archaeology* 6:138-153.

O'Brien, P. 1989. 'Michel Foucault's History of Culture', in L. Hunt (ed.) *The New Cultural History*. Berkeley: University of California Press.

Odell, G. H. 1979. 'A new improved system for the retrieval of functional information from microscopic observation of chipped stone tools', in B. Hayden (ed.) *Lithic Use-Wear Analysis*. New York: Academic Press.

Odell, G. H. 1980. 'Toward a more behavioral approach to archaeological lithic concentrations', *American Antiquity* 45: 404-431.

Odell, G. H. 1982. 'Emerging directions in the analysis of prehistoric tool use', *Reviews in Anthropology* 9: 17-33.

Odell, G. H. 1988. 'Addressing prehistoric hunting practices through stone tool analysis', *American Anthropologist* 90: 335-356.

Odell, G. H. 1994a. 'The role of stone bladelets in Middle Woodland Society', *American Antiquity* 59: 102-120.

Odell, G. H. 1994b. 'Prehistoric hafting and mobility in the North American midcontinent: examples from Illinois', *Journal of Anthropological Archaeology* 13: 51-73.

Olausson, D. 1980. 'Starting from scratch: the history of edge-wear research from 1838 to 1978', *Lithic Technology* 9: 48-60.

Oliva, M. 1984. 'Technologie výroby a použité suroviny štípané industrie moravského Aurignacienu', *Archeologické Rozhledy*. 36: 601-628.

Oliva, M. 1988. 'Discovery of a Gravettian mammoth bone hut at Mílovice', *Journal of Human Evolution* 17/8: 787-790.

Oliva, M. 1989. 'Excavations in the Paleolithic site of Mílovice I (Southern Moravia)', *Anthropologie* 27: 265-271.

Oliva, M. 1991. 'Szeletian in Czechoslovakia', *Antiquity* 65/247: 318-325.

Olsen, D. J. 1986. *The City as a Work of Art.* New Haven: Yale University Press.

Otte, M. 1981. *Le Gravettien en Europe centrale.* Dissertationes Archaeologicae Gandenses 20.

Otte, M. 1983. 'Le paleolithique de Belgique: essai de synthese', *L'Anthropologie* 87: 291-321.

Otte, M. 1990. 'Révision de la séquence du Paléolithique Supérieur de Willendorf (Autriche)', *Bulletin van het Koninklijk Belgisch Insitut voor Naturrwetenschappen, Aardwetenschappen*, 60: 219-228.

Otte, M. 1991. 'Evolution in the relationship between raw materials and cultural tradition in the European Paleolithic', in A. Montet-White and S. Holen (eds.) *Raw Material Economies among Prehistoric Hunter-Gatherers*. Lawrence, Kansas: University of Kansas.

Otte, M. and Keeley L. H. 1990. 'The impact of regionalism on Paleolithic studies', *Current Anthropology* 31/5: 577-582.

Otte, M. (ed.) 1985. *La signification culturelle des industries lithiques : actes du Colloque de Liege du 3 au 7 octobre 1984.* Oxford: BAR International Series 239; Studia praehistorica Belgica 4.

Palladino, P. and Worboys, M. 1993. 'Science and Imperialism', *Isis* 84: 91-102.

Parkington, J. E. 1986. 'Stone-tool assemblages, raw material distributions and prehistoric subsistence activities: the Late Stone Age of South Africa', in G. N. Bailey and P. Callow (eds.) *Stone Age Prehistory*. Cambridge: Cambridge University Press.

Parry, W., and R. Kelly. 1987. 'Expedient core technology and sedentism' in J. Johnson and C. Morrow (eds.) *The Organization of Core Technology*. Boulder, Co.: Westview Press.

Patrick, L. 1985. 'Is there an archaeological record?', in M. B. Schiffer (ed.) *Advances in Archaeological Method and Theory vol.* 8. New York: Academic Press.

Pavuk, J. 1981. *Umenie a život doby kamennej. Dávnoveké umenie Slovenska:* Ars slovaca antiqua. Bratislava: SNP.

Pearce, S. M. 1993. *Museums, Objects, and Collections: a Cultural Study.* Washington: Smithsonian Institution Press

Pelegrin, J. 1990. 'Prehistoric lithic technology: some aspects of research', *Archaeological Review from Cambridge* 9/1: 116-125.

Pelegrin, J., C. Karlin, P. Bodu. 1988. 'Chaines Opératoires: un Outil pour le Prehistorien', in J. Tixier (ed.) *Technologie Préhistorique.* Paris: CNRS.

Perles, C. 1987. 'Les Industries lithiques taillées de Franchthi (Argolide, Grece)', in T. W. Jacobsen (ed.) *Excavations at Franchthi Cave, Greece, Fascicle 3, 5.* Bloomington: Indiana University Press.

Phillipson, D. W. 1990 (1985). *African Archaeology.* Cambridge: Cambridge University Press.

Pigeot, N. 1990. 'Technical and social actors: flintknapping specialists and apprentices at Magdalenian Etiolles', *Archaeological Review from Cambridge* 9/1: 126-141.

Pinsky, V. and A. Wylie (ed.) 1989. *Critical Traditions in Contemporary Archaeology*. New Directions in Archaeology. Cambridge: Cambridge University Press.

Plisson, H. 1985. *Etude Fonctionelle d'Outillages Lithiques Préhistorique par l'Analyse des Micro-usures: Recherche Méthodologique et Archaeologique*. Thése, Science Humaine, Université de Paris 1, Paris.

Pough, F. H. 1971. *A Field Guide to Rocks and Minerals*. London: Constable & Co. Ltd.

Pradel, L. 1971. 'A new classification of burins', *Current Anthropology* 12:562-563.

Praslov, N. D. 1977. *Problemy Paleolita Vostochnoi i Tsentralnoi Evropy*. Leningrad: Nauka.

Praslov, N. D. and A. N. Rogachev 1982. *Paleolit Kostenkovsko-Borschevskovo Raiona na Donu 1879-1979*. Leningrad: Nauka.

Price, T. D. and J. A. Brown (eds.) 1985. *Prehistoric Hunter-Gatherers*. Orlando: Academic Press.

Přichystal, A. 1989. 'Zdroje křišťálové suroviny v pravěku na Moravé', *Geologicky Pruzkum* 31.

Přichystal, A. 1991. 'Zdroje kamenných surovin', in J. Svoboda (ed.) *Paleolit Moravy a Slezska*. Archeologicky Ustav Brno: ČSAV.

Prüfer, O. H. 1958-59. 'The Upper Paleolithic cultures of the Lower Austrian Loess region', *Quartar* 10/11: 79-114.

Pyenson, L. 1985. *Cultural Imperialism and Exact Sciences: German Expansion Overseas, 1900-1930*. New York: Lang.

Pyenson, L. 1989. 'What is the good of history of science?' *History of Science* 27: 353-389.

Pyenson, L. 1993. 'Cultural imperialism and exact sciences revisited', *Isis* 84: 103-108.

Renfrcw, C. 1979 (1973). *Before Civilization: The Radiocarbon Revolution and Prehistoric Europe*. Cambridge: Cambridge University Press.

Renfrew, C. and Bahn, P. 1991. *Archaeology: Theories, Methods, and Practice*. New York: Thames and Hudson.

Rensink, E., J. Kolen, and A. Spieksma. 1991. 'Patterns of raw material distribution in the Upper Pleistocene of Northwestern and Central Europe', in A. Montet-White and S. Holen (eds.) *Raw Material Economies among Prehistoric Hunter-Gatherers*. Lawrence, Kansas: University of Kansas Press.

Riddington, R. 1982. 'Technology, world view and adaptive strategy in a northern hunting society', *Canadian Review of Sociology and Anthropology* 19: 469-481.

Riddington, R. 1988. 'Knowledge, power, and the individual in subarctic hunting societies', *American Anthropologist* 90: 98-110.

Riek, G. 1934. *Die Eiszeitjägerstation am Vogelherd, Band 1: Die Kulturen*. Tübingen: Heine.

Rigaud, A. 1972. 'La technologie du burin appliqué au matériél osseux de la Garrenne (Indre)', *Bulletin de la Société Préhistorique Française* 69:104-108.

Rolland, N. 1981. 'The interpratation of middle Paleolithic variability', *Man* 16: 15-42.

Rolland, N. 1988. 'Observations on some middle Paleolithic time series in southern France', in H. Dibble and A. Montet-White (eds.) *Upper Pleistocene Prehistory of Western Eurasia*. Philadelphia: University Museum, University of Pennsylvania.

Rolland, N. and H. Dibble 1990. 'A new synthesis of middle Paleolithic variability', *American Antiquity* 55/3: 480-499.

Rosen S. A. 1991. 'Paradigms and politics in the terminal Pleistocene Archaeology of the Levant', in G. A. Clark (ed.) *Perspectives on the Past*. Philadelphia: University of Pennsylvania Press.

Rowley-Conwy, P. 1983. 'Sedentary hunters: the Ertebolle example', in G. Bailey (ed.) *Hunter-Gatherer Economy in Prehistory*. Cambridge: Cambridge University Press.

Rudwick, M. 1985. *The Great Devonian Controversy: The Shaping of Scientific Knowledge among Gentlemanly Specialists*. Chicago: University of Chicago Press.

Rupke, N. A. 1983. 'The study of fossils in the romantic philosophy of history and nature', *History of Science* 21: 389-413.

Sachse-Kozlowska, E. 1976. 'The Aurignacian in Poland', in Kozlowski, J. K. (ed.) *L'Aurignacien en Europe*. Nice, U.I.S.P.P., IX Congrés, Colloque XVI.

Sachse-Kozlowska, E. 1978. 'Polish Aurignacian assemblages', *Folia Quaternaria* 50.

Sackett, J. R. 1977. 'The meaning of style in archaeology: a general model', *American Antiquity* 42:369-380.

Sackett, J. R. 1981. 'From de Mortillet to Bordes: a century of French Paleolithic research', in G. Daniel (ed.) *Towards a History of Archaeology*. London: Thames & Hudson.

Sackett, J. R. 1982. 'Approaches to style in lithic archaeology', *Journal of Anthropological Archaeology* 1: 59-112.

Sackett, J. R. 1985. 'Style, ethnicity and stone tools', in M. Thompson, M. T. Garcia and F.J. Kense (eds.) *Status, Structure, and Stratification: Current Archaeological Reconstructions.* Calgary: University of Calgary.

Sackett, J. R. 1986. 'Isochrestism and style: a clarification', *Journal of Anthropological Archaeology* 5: 266-277.

Sackett, J. R. 1989. 'Statistics, attributes, and the dynamics of burin typology', in D. O. Henry and G. H. Odell (eds.) *Alternative Approaches to Lithic Analysis.* Archaeological Papers of the American Anthropological Association, no.1.

Sackett, J. R. 1991. 'Straight archaeology French style: the phylogenetic paradigm in historic perspective', G. A. Clark (ed.) in *Perspectives on the Past.* Philadelphia: University of Pennsylvania Press.

Sahlins, M. 1968. 'Notes on the original affluent society', in R. B. Lee and I. DeVore *Man the Hunter.* Chicago: Aldine Publishing Company.

Sahlins, M. 1972. *Stone Age Economics.* Chicago: Aldine Publishing Company.

Schiffer, M. B. 1987. *Formation Processes of the Archeological Record.* Albuquerque: University of New Mexico Press.

Schild, R. 1976a. 'The final Paleolithic settlements of the European plain', *Scientific American* 243: 88-99.

Schild, R. 1976b. 'Flint mining and trade in Polish prehistory as seen from the perspective of the chocolate flint of central Poland. A second approach', *Acta Archaeologica Carpathica* 16:147-177.

Schild, R. 1984. 'Terminal Paleolithic of the North European plain: a review of lost chances, potential and hopes', *Advances in World Archaeology* 3: 193-274.

Schorske, C. E. 1980. *Fin-de-Siecle Vienna: Politics and Culture.* New York: Alfred A. Knopf.

Schrire, C. 1984. 'Wild surmises on savage thoughts', in C. Schrire (ed.) *Past and Present in Hunter Gartherer Studies.* New York: Academic Press.

Schrire, C. (ed.) 1984. *Past and Present in Hunter Gartherer Studies.* New York: Academic Press.

Schwabedissen, H. 1943. 'Stand und Aufgaben der Alt- und Mittelsteinzeitforschung im mährischen Raum', *Zeitschrift des Mährischen Landesmuseums,* Neue Folge, III. Band.

Semenov, S. A. 1964. *Prehistoric Technology.* Bath: Adams and Dart.

Shapin, S., and Schaffer, S. 1985. *Leviathan and the Air-Pump.* Princeton: Princeton University Press.

Shea, J. 1987. 'On accuracy and relevance in lithic use-wear analysis', *Lithic Technology* 16: 44-50.

Shea, J. 1993. 'Lithic use-wear evidence for hunting by Neandertals and early modern humans from the Levantine Mousterian', in G. L. Peterkin, H. Bricker, and P. Mellars (eds.) *Hunting and Animal Exploitation in the Later Paleolithic and Mesolithic of Eurasia.* Archeological Papers of the American Anthropological Association; no. 4.

Shea, J. 1998. Neanderthal and early modern human behavioral variability: a regional-scale approach to lithic evidence for hunting in the Levantine Mousterian', *Current Anthropology* 39S: 45-78.

Shennan, S. J. 1978. 'Archaeological cultures: an empirical investigation', in I. Hodder (ed.) *The Spatial Organization of Culture.* London: Duckworth

Shennan, S. J. 1989. 'Cultural transmission and cultural change', in S. Van der Leeuw and R. Torrence (eds.) *What's New? A Closer Look at the Process of Innovation.* London: Unwin Hyman.

Shennan, S. J. (ed.) 1989. *Archaeological Approaches to Cultural Identity.* London: Unwin Hyman.

Sherman, D. J. and Rogoff, I. (eds.) 1994. *Museum Culture: Histories, Discourses, Spectacles.* Media and Society Series. Minneapolis: University of Minnesota Press.

Shore, C., and Black, A. 1992. 'The European communities and the construction of Europe', *Anthropology Today* 8/3: 10-11.

Shott, M. 1986. 'Settlement mobility and technological organization: an ethnographic examination. *Journal of Anthropological Research* 42:15-51.

Sieveking, G. de G. and Hart M. B. 1986. *The Scientific Study of Flint and Chert.* Cambridge: Cambridge University Press.

Sieveking, G. de G. and Newcomer M. H. (eds.) 1987. *The Human Uses of Flint and Chert: Proceedings of the Fourth International Flint symposium Held at Brighton Polytechnic 10-15 April 1983,* New York: Cambridge University Press.

Sinclair, A. 1990. 'Technology as phenotype? An extended look at Time, Energy and Stone Tools', *Archaeological Review from Cambridge* 9/1:71-81.

Sklenář, K. 1981. 'The history of archaeology in Czechoslovakia', in G. Daniel (ed.) *Towards a history of archaeology.* London: Thames & Hudson.

Sklenář, K. 1985. *Archaeology in Central Europe.* Chicago: University of Chicago Press.

Sklenář, K. 1992. 'Kdo byl Lothar Zotz?' *Speleologie* 1992/3: 15-20.

Sliva R. J. and Keeley L. H. 1994. 'Frits and specialized hide preparation in the Belgian Early Neolithic', *Journal of Archaeological Science* 21: 91-99.

Slotta, R. 1980. 'Etymologisches zu den Begriffen Feuerstein, flint und Flinte' *Veröff. Deutsch. Bergbau-Museum Bochum* 22: 4-6.

Soffer, O. 1983. 'Politics of the Paleolithic in the USSR: a case of paradigms lost', in J.M. Gero, D. M. Lacy and M. L. Blakey (eds.) *The Socio-politics of archaeology.* Amherst: Dept. of Anthropology, University of Massachusetts, Research report no. 23.

Soffer, O. 1985a. *The Upper Paleolithic of Central Russian Plain.* New York: Academic Press.

Soffer, O. 1985b. 'Patterns of intensification as seen from the Upper Paleolithic of the Central Russian Plain. in D. Price and R. Brown (eds.) *Prehistoric Hunter-Gatherers.* New York: Academic Press.

Soffer, O. 1989a. 'Storage, sedentism and the Eurasian Paleolithic record', *Antiquity* 63: 719-732.

Soffer, O. 1989b. 'The Middle to Upper Paleolithic transition on the Russian Plain', in P. Mellars and C. Stringer (eds.) *The Human Revolution.* Princeton: Princeton University Press.

Soffer, O. 1991. 'Lithics and lifeways - the diversity in raw material procurement and settlement systems on the Upper Paleolithic East European Plain'. in A. Montet-White and S. Holen (eds.) *Raw Material Economies among Prehistoric Hunter-Gatherers.* Lawrence, Kansas: University of Kansas Press.

Soffer, O. (ed.) 1987. *The Pleistocene Old World: Regional Perspectives.* New York: Plenum Press.

Soffer, O., and Gamble, C. (eds.) 1990. *The World at 18 000 BP. Vol. I: High Altitudes.* London: Unwin Hyman.

Soffer, O., and Praslov, N. D. (eds.) 1993. *From Kostenki to Clovis: Upper Paleolithic - Paleo-Indian Adaptations.* New York: Plenum Press.

Soffer, O. and Vandiver, P. 1994. 'The Ceramics', in Svoboda, J. (ed.) *Pavlov 1952-53.* Monograph Series in Archaeology, University of Liege, Belgium.

Spaulding, A. C. 1953. 'Statistical techniques for the discovery of artifact types', *American Antiquity* 18: 305-314.

Spivak, G. C. 1988. *In Other Worlds: Essays in Cultural Politics.* New York: Routledge

Springer, E. 1979. *Geschichte und Kulturleben der Wiener Ringstrasse.* Wiesbaden: Franz Steiner Verlag Gmbh.

Stahl, A. B. 1993. 'Concepts of time and approaches to analogical reasoning in historical perspective', *American Antiquity* 58/2: 235.

Stiner, M. C. and S. L. Kuhn. 1992. 'Subsistence, technology, and adaptive variation in the Middle Paleolithic', *American Anthropologist* 94: 306-339.

Stoler, A. L. 1989. 'Rethinking colonial categories: European communities and the boundaries of rule', *Comparative Studies in Society and History* 13/1: 134-161.

Strathern, M. 1980. 'No nature, no culture', in MacCormack, C. and M. Strathern (eds.) *Nature, Culture, and Gender.* Cambridge: Cambridge University Press.

Straus, L. 1980. 'The role of raw materials in lithic assemblage variability', *Lithic Technology* 9: 68-72.

Sturdy, D. A. 1975. 'Some reindeer economies in prehistoric Europe', in E. S. Higgs (ed.) *Palaeoeconomy.* Cambridge: Cambridge University Press.

Sturdy, D. A. and D. P. Webley 1988. 'Paleolithic geography: or where are the deer?' *World Archaeolog* 19: 262-280.

Svoboda, J. 1983. 'Raw material sources in Early Upper Paleolithic Moravia', The concept of lithic exploitation areas', *Anthropologie* 21/2: 147-158.

Svoboda, J. 1989. 'Middle Pleistocene adaptations in Central Europe', *Journal of World Prehistory* 3: 33-70.

Svoboda, J. 1990. 'Moravia during the Upper Pleniglacial', in O. Soffer and C. Gamble (eds.) *The World at 18 000 BP. Vol. I: High Altitudes.* London: Unwin Hyman.

Svoboda, J. and Simán, K. 1989. 'Middle-Upper Paleolithic transition in southeastern central Europe', *Journal of World Prehistory* 3/3: 283-322.

Svoboda, J. (ed.) 1987. 'Stránská skála. Bohunický typ v brněnské kotline', *Studie AU ČSAV Brno* 14/1.

Svoboda, J. (ed.) 1991a. *Paleolit Moravy a Slezska.* Archeologický Ústav Brno: ČSAV.

Svoboda, J. (ed.) 1991b. *Dolní Věstonice II.* Monograph Series in Archaeology, University of Liege: Belgium.

Svoboda, J. (ed.) 1994. *Pavlov 1952-53.* Monograph Series in Archaeology, University of Liege, Belgium.

Svobodová H. 1991. 'Pollen analysis of the Upper Paleolithic triple burial at Dolní Věstonice', Archeologické Rozhledy 43/4: 505-510.

Svobodová H. and J. Svoboda 1988. 'La chronostratigraphie et paleoecologie du Paleolithique superieur morave d'apres les fouilles recentes', in A. Tuffreau (ed.) *Cultures et industries lithiques en millieu loessique, Revue archaeologique de Picardie.* Amiens: Direction des Antiquites.

Symens, N. 1986. 'A functional analysis of selected stone artifacts from the Magdalenian site at Verberie', *Journal of Field Archaeology* 13:213-222.

Taçon, P. S. C. 1991. 'The power of stone: symbolic aspects of stone use and tool development in Western Arnhem Land, Australia', *Antiquity* 65/3: 192-207.

Thomas, D. H. 1989. *Archaeology.* Holt, Reinhart and Winston, Inc., The Dryden Press.

Tixier, J. 1974. *Glossary for the Description of Stone Tools with Special Reference to the EpiPaleolithic of the Maghreb.* Special Publication No. 1. Newsletter of Lithic Technology, San Antonio.

Tainter, J. and B. Tainter (eds.) 1996. *Evolving Complexity and Environmental Risk in the Prehistoric Southwest.* New York: Addison-Wesley.

Tomášková, S. 1986. 'The low- and high-power approaches in lithic use-wear studies', B.A Honors Thesis on file Department of Anthropology, McGill University, Montreal Canada.

Tomášková, S. 1991. 'Use-wear analysis of the lithic material from Dolní Vestonice, Czechoslovakia', in J. Svoboda (ed.) *Dolní Vestonice II.* Monograph Series in Archaeology, University of Liege, Belgium.

Tomášková, S. 1994. 'Spatial analysis of the lithics from Pavlov I.' in J. Svoboda (ed.) *Pavlov 1952-53* Monograph Series in Archaeology, University of Liege, Belgium.

Tomášková, S. 1995. 'A site in history, Dolní Vestonice - Unterwisternitz', *Antiquity* 69/263:300-316.

Torrence, R. 1983. 'Time-budgeting and hunter-gatherer technology', in R. Torrence (ed.) *Time, Energy, and Stone Tools.* Cambridge: Cambridge University Press.

Torrence, R. 1986. *Production and Exchange of Stone Tools: Prehistoric Obsidian in the Aegean.* Cambridge: Cambridge University Press.

Torrence, R. (ed.) 1989. *Time, Energy and Stone Tools.* Cambridge: Cambridge University Press.

Trigger, B. 1978. *Time and Traditions.* New York: Columbia University Press.

Trigger, B. 1984. 'Alternative archaeologies. Nationalist, colonialist, imperialist', *Man* 19: 355-370.

Trigger, B. 1989a. *A History of Archaeological Thought.* Cambridge: Cambridge University Press.

Trigger, B. 1989b. 'History and contemporary American archaeology: a critical analysis', in C. C. Lamberg-Karlovsky (ed.) *Archaeological Thought in America.* Cambridge: Cambridge University Press.

Tringham, R. E. 1971. 'The function, technology, and typology of the chipped stone industry at Bylany, Czechoslovakia', *Alba Regia, Annales Musei Stephani Regis* 12: 143-148.

Tringham, R. E., G. Cooper, G. Odell, B. Voytek and A. Whitman 1974. 'Experimentation in the formation of edge damage: a new approach to lithic analysis', *Journal of Field Archaeology* 1/1-2: 171-196.

Tromnau, G. 1981. *Die Hamburger Kultur.* Archaeologia Interregionalis I.

Tromnau, G. 1984. 'Siedlungsmuster Hamburger Kultur; Beispiel: Teltwisch 1 (Ahrensburg Tunneltal)', in H. Berke, J. Hahn and C.-J. Kind (eds.) Jungpalaolitische Siedlungstrukturen in Europa. Koloquium 1983 Reisenburg Gunzburg. *Urgeschichtliche Materialhefte* 6.

Tyldesley, J. A. and Bahn, P. G. 1983. 'Use of plants in the European Paleolithic: a review of the evidence', *Quaternary Science Review* 2: 53-81.

Unrath, G., L. Owen, A. Van Gijn, A. Moss, H. Plisson, and P. Vaughan 1986. 'An evaluation of microwear studies: a multianalyst approach', *Early Man News.* 9-11: 117-176.

Valoch, K. 1968. 'Evolution of the Paleolithic in Central and Eastern Europe', *Current Anthropology* 9: 351-90.

Valoch, K. 1981. 'Beitrag zur Kenntnis des Pavlovien', *Archeologické Rozhledy.* 33: 279-298.

Valoch, K. 1984. 'The Early Paleolithic site Stránská skála I near Brno (Czechoslovakia)', *Antropologie* 25: 125-42.

Valoch, K. 1986a. 'Les questions du Pavlovien', *Antiquites Nationales* 18/19: 55-62.

Valoch, K. 1986b. 'Stone industries of the Middle/Upper Paleolithic transition', in M. Day, R. Foley and Wu Rukang (eds.) *The Pleistocene Perspective. vol.1.*

Valoch, K. 1987. 'Raw materials used in the Moravian Middle and Upper Paleolithic', *Proceedings of the International Conference on Prehistoric Flint Mining and Lithic Raw Material Identification in the Carpathian Basin.* Budapest: Sümeg.

Vandiver, P., O. Soffer, B. Klíma and J. Svoboda 1989. 'The Origins of Ceramic Technology at Dolní Věstonice, Czechoslovakia', *Science* 246: 1002-1008.

Vaughan, P. C. 1981. Lithic microwear experimentation and the functional analysis of a lower Magdalenian stone tool assemblage. Ph.D. thesis. University of Pennsylvania.

Vaughan, P. C. 1985a. *Use-wear Analysis of Flaked Stone Tools.* Tucson: University of Arizona.

Vaughan, P. C. 1985b. 'The burin-blow technique: Creator or eliminator?', *Journal of Field Archaeology* 12:488-496.

Veit, U. 1989. 'Ethnic concepts in German prehistory: a case study on the relationship between cultural identity and archaeological objectivity', in S. J. Shennan (ed.) *Archaeological Approaches to Cultural Identity.* London: Unwin Hyman.

Vencl, S. 1991. 'On the importance of spatio-temporal differences in the intensity of Paleolithic and Mesolithic settlement in Central Europe', *Antiquity* 65(3): 308-17.

Venclová, N. 1991. 'Introduction. Archaeology in Czechoslovakia: trends and approaches', *Antiquity* 65/3:306.

Vergo, P. (ed.) 1989. *The New Museology.* London: Routledge.

Wagner-Rieger, R. (ed.) 1969. *Die Wiener Ringstrasse, Bild einer Epoche.* Wien

Weniger, G. C. 1989. 'The Magadalenian in Western Central Europe. Settlement pattern and regionality', *Journal of World Prehistory.* 3/3: 323-372.

Weniger, G. C. 1991a. 'Notes on the mobility of groups of hunters during the Upper Paleolithic', *Saeculum* 42(1): 82-103.

Whallon, R. 1989. 'Elements of cultural change in the later Paleolithic', in P. Mellars Stringer, C. (ed.) *The Human Revolution. Behavioral and Biological Perspectives in the Origins of Modern Humans.* Princeton: Princeton University Press.

Whallon, R. and J. Brown (eds.) 1982. *Essays in Archaeological Typology.* Evanston: Center for American Archaeology Press.

White, J. P. and D. H. Thomas 1972. 'What mean these stones? Ethnotaxonomic models and archaeological interpretations in the New Guinea Highlands', in D. L. Clarke (ed.) *Models in Archaeology.* London: Methuen.

Wiessner, P. 1982. 'Beyond Willow Smoke and Dogs' Tails: A Comment on Binford's Analysis of Hunter-Gatherer Settlement Systems', *American Antiquity* 47/1: 171-178.

Wiessner, P. 1983. 'Style and social information in Kalahari San projectile points', *American Antiquity* 48: 253-276.

Williams O. and Nandris, J. G. 1977. 'The Hungarian and Slovak sources of archaeological obsidian: an interim report on further fieldwork, with a note on tektites', *Journal of Archaeological Science* 4: 207-219.

Williams-Thorpe O., Warren, S. E., and J. G. Nandris 1984. 'The distribution and provenance of archaeological obsidian in central and eastern Europe', *Journal of Archaeological Science* 11: 183-212.

Winkler, E.-M. 1983. 'Volk, Kultur, Ethnos, Population, Typus. Zur Methodik der 'ethnischen Deutung', *Mitteilungen der Anthropologischen Gesselschaft in Wien* 113: 5-14.

Winterhalder, B. and E. A. Smith 1981. *Hunter/Gatherer Foraging Strategies: Ethnographic and Archaeological Analyses.* Chicago: University of Chicago Press.

Wobst, H. M. 1974. 'Boundary conditions for Paleolithic social systems: a simulation approach', *American Antiquity* 39: 147-178.

Wobst, H. M. 1976. 'Locational relationships in Paleolithic society', *Journal of Human Evolution* 5: 49-58.

Wobst, H. M. 1978. 'The archaeo-ethnology of hunter-gatherers, or the tyranny of the ethnographic record in archaeology', *American Antiquity* 43/2: 303-309.

Wobst, H. M. 1989. 'Commentary: a socio-politics of socio-politics in archaeology', in V. Pinsky and A. Wylie (eds.) *Critical Traditions in Contemporary Archaeology.* New Directions in Archaeology. Cambridge: Cambridge University Press.

Wobst, H. M. 1990. 'Minitime and megaspace in the Paleolithic at 18K and otherwise', in O. Soffer and Gamble C. (eds.) *The World at 18,000 BP, vol.1, High Latitudes.* London: Unwin Hyman.

Wobst, H. M. and A. S. Keene 1983. 'Archaeological explanation as political economy', in J. Gero, D. M. Lacy and M. L. Blakey (eds.) *The Socio-politics of archaeology.* Amherst: Dept. of Anthropology, University of Massachusetts, Research report no. 23.

Wright, R. V. S. 1991. 'Technological styles: transferring a natural material into a cultural object', in W. Kingery and S. Luber (eds.) *Learning from Things: Working Papers on Material Culture.* Washington, D.C.: Smithsonian Institution Press.

Wright, R. V. S. (ed.) 1977. *Stone Tools as Cultural Markers.* Canberra: Australian Institute of Aboriginal Studies.

Yellen, J. 1989. 'The present and the future of hunter-gatherer studies', in C. C. Lamberg-Karlovsky (ed.) *Archaeological Thought in America.* Cambridge: Cambridge University Press.

Yoffee, N. and Sherratt A. (eds.) 1993. *Archaeological Theory: Who Sets the Agenda?* Cambridge: Cambridge University Press.

Young, D. and R. Bonnichsen 1985. *Understanding Stone Tools: a Cognitive Approach.* Maine, Ohio: Centre for the Study of Early Man.

Zotz, L. 1940. *Ist Böhmen-Mähren die Urheimat der Tschechen?*. Leipzig: Johann Ambrosius Barth.

Zotz, L. 1944. *Von den Mammutjägern zu den Wikingern: Ergebnisse und Aufgaben der böhmisch-mährischen Vorgeschichtskunde.* Leipzig: Johann Ambrosius Barth.

www.ingramcontent.com/pod-product-compliance
Lightning Source LLC
Chambersburg PA
CBHW041716180926
51615CB00022B/3078